Ian Wilson is a prolific, in̶t̶ ̶̶̶i̶r̶e̶lly pub̶l̶ ̶ ̶ ̶ ̶ ̶r̶ specia̶l̶̶i̶z̶i̶n̶g̶ in historical a̶n̶d̶ r̶e̶l̶i̶g̶ ̶ ̶ ̶ ̶ ̶ ̶ ̶ ̶n̶ sout̶h̶ ̶ ̶ ̶ ̶ ̶o̶n̶,̶ h̶e̶ graduated fr̶o̶ ̶ ̶ ̶ ̶ ̶g̶ ̶ ̶ ̶ ̶ ̶ College, Oxford University, in 1963 with honours in Modern History. The TV documentary that he co-scripted to accompany his 1978 Shroud book won a BAFTA award, and his later book *Jesus: The Evidence* was a bestseller on both sides of the Atlantic. Accompanied by his wife Judith, Wilson emigrated from England to Queensland, Australia, in 1995, where he enthusiastically continues wide-ranging research projects both at home and around the world.

Also by Ian Wilson

The Turin Shroud (1978)
Jesus: The Evidence (1984)
The Evidence of the Shroud (with Vernon Miller) (1986)
Holy Faces, Secret Places: the quest for Jesus's
true likeness (1991)
Jesus: the evidence, the latest research
and discoveries (1996)
The Blood and the Shroud (1998)
The Bible is History (1999)

THE SHROUD

Fresh light on the
2000-year-old mystery . . .

IAN WILSON

BANTAM BOOKS

LONDON • TORONTO • SYDNEY • AUCKLAND • JOHANNESBURG

TRANSWORLD PUBLISHERS
61–63 Uxbridge Road, London W5 5SA
A Random House Group Company
www.rbooks.co.uk

THE SHROUD
A BANTAM BOOK: 9780553824223

First published in Great Britain
in 2010 by Bantam Press
an imprint of Transworld Publishers
Corgi edition published 2011

Addresses for Random House Group Ltd companies outside the UK
can be found at: www.randomhouse.co.uk
The Random House Group Ltd Reg. No. 954009

The Random House Group Limited supports The Forest Stewardship Council
(FSC), the leading international forest certification organisation. All our titles
that are printed on Greenpeace approved FSC certified paper carry the FSC
logo. Our paper procurement policy can be found at
www.rbooks.co.uk/environment

Typeset in 11/13pt Sabon by Falcon Oast Graphic Art Ltd.
Printed in the UK by CPI Cox & Wyman, Reading, RG1 8EX.

2 4 6 8 10 9 7 5 3 1

Mixed Sources
Product group from well-managed
forests and other controlled sources
www.fsc.org Cert no. TT-COC-2139
© 1996 Forest Stewardship Council
FSC

Contents

Author's Preface

In a sense, this book has been fifty-five years in the making. Back in 1955, as a rabidly agnostic south London schoolboy, I gained top marks in 'Religious Instruction' – mainly because few others in the class were even interested. Emanuel School's Anglican chaplain, Revd Griffiths, for whom I must have been a very painful pupil, chose for my first-ever school prize American author Lloyd C. Douglas's *The Robe*. A novel about the seamless garment that Jesus had worn in life, and how after his death it went through many adventures that touched people's lives in very moving ways, in retrospect it can only be viewed as a highly prophetic choice.

During that same year, flicking through a popular weekly journal called *Picture Post* I came across an article by World War Two hero Group Captain Leonard Cheshire VC on the subject of the Turin Shroud. In my rampantly anti-Christian mindset of that time, any Roman Catholic relic automatically had to be a fake. But in this particular instance my strong interests in art and art history came into play. Priding myself that I could look at any section of an 'old master' and identify the artist from the style, the sight of the Shroud's 'photographic' face and body image

– to my eyes so self-evidently *not* the work of any artist – profoundly challenged my hitherto complacent agnosticism.

And so began what would become a life-long fascination with a time-scarred piece of cloth that, as I gradually learned, had had real-life adventures across two thousand years even more extraordinary and emotionally moving than anything Lloyd C. Douglas had conceived.

I began serious research on the Shroud in 1966. Most of those who helped and guided me at that time are sadly now deceased. Deeply influential were Guildford general practitioner Dr David Willis, Isle of Wight-based scouting pioneer Vera Barclay, and Ampleforth monk Dom Maurus Green, all of whom contributed enormously to my appreciation and understanding not only of the Shroud but also of Christianity itself. Similarly major influences were New York-based Roman Catholic priest Fr Peter Rinaldi, Anglican theologian Bishop John Robinson, and Swiss criminologist Dr Max Frei. From my Emanuel School classics teacher Bernard Slater and from Mrs Maria Jepps of Wells came pioneering translations from Byzantine Greek and modern Italian, which greatly helped me bring into being my first book on the subject. Most unexpectedly commissioned by Doubleday, New York, this was published in 1978, as *The Shroud of Turin* in the USA and *The Turin Shroud* in the UK and Commonwealth.

For this present book my aim has been a complete update of the 1978 one. Although much the same running order has been followed, and the occasional paragraph has been 'lifted' from earlier works, to all intents and purposes this is a brand-new production, presenting the subject as if started afresh at the present day. And for an arguably two-thousand-year-old cloth it is quite extraordinary how much there is that is fresh and new as a result of what has happened to it over the last three decades. The Shroud's proprietary status has changed: it is now actually owned by the Roman Catholic Church for the first time in its history. Its location has changed, the beautiful chapel in which it had been stored for three hundred years having been all but destroyed by fire in 1997. Even its physical appearance has been changed, because of conservators' recent removal of unsightly patches from a much earlier fire, in 1532, which so nearly destroyed the Shroud. The last three decades have also seen the Shroud subjected to a major international scientific examination which suggested it to be authentic, to be followed, in 1988, by the notorious radiocarbon dating which claimed to prove the opposite. The aim here has been to put into proper perspective the issues raised by such conflicting scientific findings – issues on which Connecticut chemistry professor the late Dr Alan Adler ever generously provided a now much-missed voice of scientific common sense for me as a historian.

As the prime focus of this present book, just as it was in 1978, is the Shroud's history, a most invaluable source of help has been the very recent publication of Mark Guscin's *The Image of Edessa*. The first comprehensive English-language scholarly study of the documentary background to the cloth that I identify as one and the same as the Shroud prior to 1204, Mark's book has made available many key texts for which a definitive English-language translation had until then been lacking. For my personal benefit, Mark most generously made his manuscript available to me several months before the book's publication, and it was a privilege to have him as a companion on the research trip to Cappadocia and eastern Turkey that was conducted for this book during May 2009. Lennox Manton, a veteran of many earlier explorations in Cappadocia, provided much valuable preliminary guidance for this trip.

Among other specialist help, I am indebted to British TV producer David Rolfe of Performance Films, and to television cameraman David Crute, for providing first-hand information on their high-definition filming of the Shroud in 2008; to textile conservator Dr Mechthild Flury-Lemberg of Bern, Switzerland, for much helpful correspondence on the latest understandings of the technical characteristics of the Shroud's weave; to Professor Stephen Mattingly of the University of Texas for advice on the issue of microbiological contamination of the Shroud; to Professor Christopher Ramsey of Oxford University's Oxford Radiocarbon Accelerator Unit for face-to-face discussions concerning ways in which the carbon dating result of 1988 may have been skewed; to Oxford University Syriac specialist Professor Sebastian Brock for his courtesy answering enquiries concerning the coinage of Edessa; to His Beatitude Mar Gewargis Sliwa, Metropolitan of Iraq and Russia for the Assyrian Church of the East, for many insights concerning early Christianity in Edessa; to Dr Mehmet Önal of Harran University, Şanliurfa, eastern Turkey, for allowing Mark Guscin and me to visit his excavations in Haleplibahce Park, Şanliurfa; to Nicosia-based lawyer George Apostolou and archaeologist Marc Fehlmann for gaining ground-breaking access on my behalf to hitherto inaccessible early Byzantine wall-paintings in a spring in Salamis, Cyprus, and providing photographs of early wall-paintings of the Image of Edessa in churches throughout Cyprus; to Dr Tamar Gegia, of the Institute of Manuscripts, Tbilisi, Georgia, for helpful information concerning recent researches in Georgia; to Serbian author Slavisa Djurdje Jevtic of Kragujevac, Serbia, for a wealth of correspondence and information concerning Serbian history during the Nemanjic dynasty, and the wall-paintings of the churches and monasteries that were founded at that time; to American numismatist Wayne G. Sayles of Gainesville, Missouri, for helpful information on the coinage of Edessa; to historian Professor Karlheinz Dietz of the University of Würzburg, Germany, for insights concerning the Jerusalem *sudarium* described by the pilgrim Arculf; to archivist Dr Barbara Frale of the Vatican Secret Archives, Rome, for making available her recent new findings concerning the Shroud's ownership by the Knights Templar; to Hugh Duncan, Marie-Dominique Germain and Asterid Le Gal for their kindness translating from Old French large sections of Shroud owner Geoffrey de Charny's *Livre Charny*; to art historian Dr Sarah Blick of Kenyon College, Gambier, Ohio, for advice concerning the Lirey pilgrims' badge; to Leonardo and Pietro Ferri of Rome for visiting the Capitoline Museum on my behalf and providing a close-up of the wound on the statue of the 'Dying Gaul'; and to Professor Gian Maria Zaccone of Turin and Antonio Cassanelli of Rome for information concerning aspects of the Shroud's more recent history.

Besides such specialist help, I have also benefited from many kindnesses by friends, relatives and acquaintances who have enabled me to overcome the difficulties presented by the widely scattered nature of some research materials – sometimes particularly daunting for an author living in Queensland, Australia. On my behalf, Melburnian Peter Slodowy has kindly visited the library of the University of Melbourne, and his wife Celina provided photos of the most recent appearance of the Shroud's repository following her visit to Turin. My daughter-in-law Leonie's mother Rosemary Hall has visited the British Library, London, for data from books unobtainable anywhere in Australia. Dr Stella Fletcher of the UK-based Ecclesiastical History Society kindly sent a similarly hard-to-obtain publication by her society. Bertram Foy of Malahide, Ireland, very kindly provided a copy of Alfred

O'Rahilly's now rare *The Crucified*. Pierre de Riedmattin of Paris helped track down a hitherto unknown medieval manuscript illustration. The national books interloan service operated by Libraries Australia has been magnificent, bringing to my local mobile library, at very modest cost, several highly specialist volumes from universities right across Australia. My thanks to the staff of the Brisbane City Council mobile library – particularly Robert and Wendy – for their unflagging patience and helpfulness with such requirements, also the staff of the library of the University of Queensland, St Lucia. During the trip to Turkey, particularly humbling was the helpfulness of some of the ordinary villagers and townspeople who went well out of their way to guide us to some particularly difficult-to-find locations, then refused any reward. And, as always, my thanks to my wife Judith who has most admirably suffered forty-two years of being married to the Shroud as well as to her husband. As chief photographer on the expedition to Turkey, Judith tackled many an awkward location, and checked every word of this book at manuscript stage.

My warmest thanks to my publisher Sally Gaminara of Transworld for taking on the subject of the Shroud despite a difficult book-buying market; to copy-editor Daniel Balado for his exemplary spotting of those grammatical and other slips an author all too easily fails to see, and his most eagle-eyed attention to every detail; to design manager Philip Lord for his great patience and conscientiousness in dealing with the complexities of the book's text illustrations and colour plates; to managing editor Katrina Whone for her ever-helpful guidance on editorial matters, and to Ann-Katrin Ziser and all others of the Transworld team who have helped bring this book to fruition.

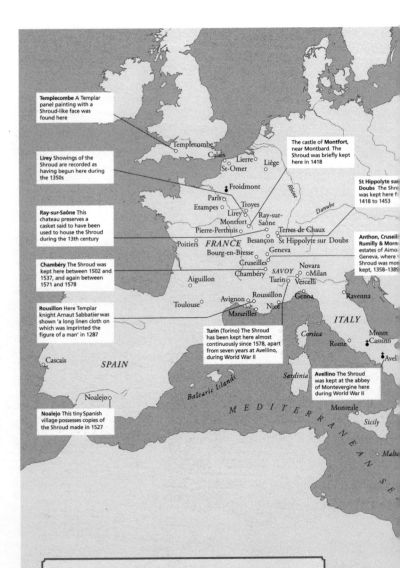

Templecombe A Templar panel painting with a Shroud-like face was found here

Lirey Showings of the Shroud are recorded as having begun here during the 1350s

Ray-sur-Saône This chateau preserves a casket said to have been used to house the Shroud during the 13th century

Chambéry The Shroud was kept here between 1502 and 1537, and again between 1571 and 1578

Roussillon Here Templar knight Arnaut Sabbatier was shown 'a long linen cloth on which was imprinted the figure of a man' in 1287

Noalejo This tiny Spanish village possesses copies of the Shroud made in 1527

The castle of **Montfort**, near Montbard. The Shroud was briefly kept here in 1418

St Hippolyte sur Doubs The Shroud was kept here from 1418 to 1453

Anthon, Cruseilles Rumilly & Morn estates of Aimo Geneva, where Shroud was mos kept, 1358–1389

Turin (Torino) The Shroud has been kept here almost continuously since 1578, apart from seven years at Avellino, during World War II

Avellino The Shroud was kept at the abbey of Montevergine here during World War II

Templecombe
Calais
Lierre
St-Omer
Liège
Froidmont
Paris
Lirey
Troyes
Etampes
Ray-sur-Saône
Montfort
Pierre-Perthuis
Terres de Chaux
Poitiers FRANCE Besançon St Hippolyte sur Doubs
Bourg-en-Bresse Geneva
Cruseilles
Aiguillon Chambéry SAVOY Novara
Turin Milan
Vercelli
Avignon Roussillon Nice Genoa Ravenna
Toulouse Marseilles
ITALY
Corsica
Monte Cassino
Rome
Avel
Cascais SPAIN
Sardinia
Noalejo
Balearic Islands
Monreale
MEDITERRANEAN SEA
Sicily
Malta

Rhône
Danube

THE WORLD OF THE SHROUD
AD30 TO THE PRESENT
based on the theory that it was one and the
same as the Image of Edessa of Eastern
Orthodox tradition

Monasteries and Abbeys

○ Other settlements

0   200km

0   200 miles

Constantinople If the Shroud was indeed the Image of Edessa of the Eastern Orthodox Church, it was brought here from Edessa in 944, and remained until French Crusaders captured and looted the city in 1204

Serbia 13th-century wall-paintings in certain royal-founded Serbian monasteries depict the Image of Edessa in a manner strikingly similar to the Shroud

Mount Athos –The monasteries here are a rich repository for early manuscripts on the Image of Edessa

Georgia Assyrian monks from Edessa brought copies of the face on the Image of Edessa to this region during the mid-6th century

Smyrna Shroud-owner Geoffrey I de Charney took part in the capture and defence of the harbour fortress here in 1344

Vostitza Greek barony which Shroud owner Geoffrey I de Charney's elder brother Dreux acquired by his marriage to Agnes de Charpigny

Cyprus. Copies of the face on the Image of Edessa were brought here in the 6th century

Edessa According to Eastern Orthodox tradition, the so-called Image of Edessa, a cloth imprinted with Jesus's likeness, was brought here from Jerusalem, c. AD30, and remained until 944

Jerusalem According to the Christain gospels, Jesus' burial shroud was left behind in his tomb here c. AD30

Masada In the 1960s israeli archeologists found here a first-century textile fragment with an unusual 'invisible' seam near identical to that on the Turin Shroud

St Catherine's Monastery, Sinai A rich repository for icons and manuscripts pertinent to the early history of the Image of Edessa/Shroud

Krokodilo In the 1990s French archaeologists found textile fragments here of similar herringbone weave to that of the Shroud

Kiev

CASPIAN SEA

Samtavisi *GEORGIA*   Martkopi

Rekha   Tbilisi

grade

ERBIA

Danube

BLACK SEA

Trabzon

Soumela

Studenica

Gradac

opocani

Thessaloniki

Constantinople (Istanbul)   *TURKEY*

Caesarea (Kayseri)

Arbela (Arbil)

Samosata (Samsat)

Edessa (Şanliurfa)

GREECE

Mount Athos

Lepanto

Vostitza (Aigion)

Athens

Smyrna (Izmir)

Hierapolis (Membij)

Antioch (Antakya)

Tigris

Euphrates

Lapithos   Salamis

Homs

Rhodes

Crete

Cyprus

Acre (Akko)

Jerusalem   *Dead Sea*

Masada

Alexandria

Sinai

St Catherine's monastery

Nile

EGYPT

RED SEA

Krokodilo

Göreme

A Note on Style

In a book spanning the histories of cultures that many English-language readers will find unfamiliar, spellings of personal names and similar issues have involved some occasionally difficult choices. In the case of Italy's Savoy family, should the form be the purist 'Duke Carlo-Emanuele' or the anglicized 'Duke Charles Emmanuel'? Mostly I have opted for the latter form, as the easier for English-language readers. Likewise, although the terms 'Melkite' and 'Chalcedonian' have become popular among historians to denote the denomination today represented by the Eastern Orthodox Church, here I have opted for 'Orthodox' because of its greater familiarity. Conversely, 'Chosroes' used to be the normal rendition for the name of certain Sassanian kings of Persia, yet 'Khosraw' (or 'Khusraw') has become the more favoured present-day rendering. So the latter has been followed here.

Among Shroud sceptics it has become fashionable to use a lower-case 's' when referring to the Turin Shroud, as if this somehow exhibits a more 'scientific' stance. The logic of this has always escaped me, because if the Shroud really is a fake then it was never a true 'shroud', as the lower case implies. The usage of the upper-case 'S' in this book is intended purely to identify the specific object preserved in Turin, just as it is normal to capitalize the 'Tower' of London, the Bayeux 'Tapestry' and the 'Great Wall' of

China without this denoting 'true belief' in any of these objects.

Ian Wilson
Kenmore Hills, Queensland, Australia
October 2009

Introduction

The Need for a Re-think

DR MEHMET ÖNAL sipped a glass of tea as we looked out over his excavation site. 'I have a surprise for you both,' he said. 'We have a mosaic of your "Image of Edessa" here in Şanliurfa.'

For me as a historian, and for my companion, Byzantine manuscripts specialist Mark Guscin, the Turkish archaeologist's news was indeed astonishing. We were in Şanliurfa because of a theory first advanced thirty years earlier, in my book *The Turin Shroud*, according to which Turin's 'Holy Shroud', with its ever-intriguing 'photographic' imprint of Jesus's face and crucified body, had spent much of its first nine hundred years here in this most eastern of Turkish cities.

Throughout most of the first millennium Şanliurfa, then called Edessa, had certainly possessed a closely guarded cloth bearing a mysterious imprint of Jesus. This was known to historians as the Image of Edessa. But while the Turin Shroud's identification with this Edessa cloth provided a logical explanation for where the Shroud had been for much of the period prior to its appearance in medieval France, the theory had some serious difficulties.

One of these was the fact that the Turks, on capturing Edessa in 1144, had systematically ransacked and razed every Christian church, mostly replacing each with a mosque. During this process every vestige of Christian imagery had been ruthlessly destroyed as an offence to the Koran, making any survival of pictorial evidence of the Image/Shroud's one-time presence in Şanliurfa highly unlikely.

A second difficulty was that in 1988 the Turin Shroud's credibility as a genuine two-thousand-year-old burial cloth had been very seriously undermined by highly publicized radiocarbon dating tests which had adjudged it to date to between 1260 and 1390. I was at the British Museum press conference when these results were announced, followed by the withering pronouncement from Oxford University physicist Professor Edward Hall: 'There was a multi-million-pound business in making forgeries during the fourteenth century. Someone just got a bit of linen, faked it up, and flogged it.'

With the odds against the Shroud dating to the time of Christ now declared 'astronomical', this seemingly hard scientific finding had led most people around the world, understandably, to assume that the Shroud simply must be a cynical medieval forgery. But so much of this was turned on its head by our Turkish archaeologist's mosaic. As Şanliurfa's museum director Erman Bediz explained to us, it was just a six-inch-by-eight-inch fragment some local citizen had found while making structural alterations to his house. He had hacked it out then sold it to the museum on a no-questions-asked basis. It was not even on public display, kept hidden away in one of the museum's store-rooms. Even so, as the very Islamic Dr Önal and his companions had already perceived, this was quite un-mistakably some early mosaicist's interpretation of the prophet Jesus's face as imprinted on this city's one-time

'Image of Edessa'. As such it was the sole survivor of the no doubt many hundreds of similar mosaics, frescoes and icons on this theme that had once proudly embellished Edessa's dozens of Christian churches, churches whose ghosts lay little more than a stone's throw from where we were calmly sipping tea.

The point also immediately apparent to Mark Guscin and me, from our familiarity with depictions of the Image of Edessa to be found elsewhere, was that stylistically this unique Şanliurfan example dated somewhere between the sixth and seventh centuries. It was therefore not only the earliest-known such depiction; it came from the very city from which the legend of this mysterious cloth had originated. It was like re-finding a vital missing piece from a giant jigsaw puzzle. And immediately from that flowed the need for a complete re-think of the links between this once historic lost Edessa cloth and Turin's Shroud with its otherwise 'lost' whereabouts prior to the fourteenth century.

That re-think is what this book is all about. For although it is now more than twenty years since scientists announced that the Shroud dated to between 1260 and 1390, not a single serious historian has come forward to buttress this with some sound, readily cogent elucidation of how someone in the Middle Ages could have come up with such an extraordinary object. In the English-language world at least, the only pro-fake 'explainers' of the radio-carbon date have been tabloid-type theorists of the 'Leonardo da Vinci did it' variety whom anyone with due critical sense would recognize as lacking serious substance.

In fairness, however, supporters of the Shroud's authenticity have scarcely offered anything more persuasive in the interim. In 2005 American scientist Dr Ray Rogers, dying from cancer, issued a peer-reviewed scientific paper claiming to prove that the Shroud sample used for the carbon dating was cut from an area rewoven during the Middle Ages, instead of having come from the true, original fabric. But as we will see later, even this argument lacks serious substance, and in general the 1988 radio-carbon dating finding has been able to rest unchallenged as science's overriding 'verdict' on the Shroud, snuffing out any evidence, from any discipline, that might suggest to the contrary.

By contrast, essentially unnoticed by the world at large because of their being cocooned in academic articles with obscure-sounding titles and terminology, there have been some significant advances in scholarly understanding of the Image of Edessa. The process began inauspiciously enough with London University academic Averil Cameron roundly attacking any identification of the Image with the Shroud in none other than her inaugural lecture as Professor of Ancient History in 1980. But Cameron chose to look only at texts, rather than texts and images in parallel. Subsequent scholars, mostly without any reference to the Shroud, have begun to adopt a rather broader approach towards the Image, taking due account of the extraordinary awe, mystique and secrecy with which it was associated in Byzantine thought. Important new insights have come from Georgia in the former Soviet Union, and elsewhere. In this process, the possibility of the Image's identification with the Shroud has become quite profoundly reinforced, not weakened. And within the last year this process has culminated in the publication of Mark Guscin's *The Image of Edessa*. The fruits of much scouring among the monastery libraries of Mount Athos, at last all the key texts on the Image have been made avail-able in English, with the original Greek alongside.

As if right on cue for such a re-think on the Shroud, on 2 June 2008 Pope Benedict XVI announced that fresh expositions would be held between 10 April and 23 May 2010. These could not be more timely, because they will give the world its first opportunity to see the Shroud since it was given a major 'make-over', in the form of extensive conservation work, during the summer of 2002.

When these expositions are held, no doubt some media commentators will characterize them as just for the superstitious and the gullible, clinging to blind belief in an object that science has proved to be a fake. That is certainly not the standpoint of this book. Let it be clearly understood that this book assumes no Christianity on the part of its readers. Wherever necessary the information that follows is fully referenced, and derived from reputable, reliable sources that the most sceptical reader is positively encouraged to check. Instead of belief, all that is required of the reader is the patience to listen to the evidence, followed by judgement, on whether or not the cloth historically known as the Image of Edessa, lost from Constantinople in 1204, was/is one and the same as the cloth that mysteriously appeared in northern France in the 1350s and is today known as the Turin Shroud.

Reader, should your verdict be that they were two separate objects, it is my hope that even so you will have found some entertainment in the colourful lives that each led throughout the centuries. But if, along with me, the only conclusion you can sensibly come to is that they were genuinely one and the same object, then the implications are quite extraordinary.

Chapter 1

The Shroud on View

*I'm not a religious person, but when we got into the room,
it made the hairs on the back of your neck stand up.*

David Crute, cameraman, 2008

LATE ON a Friday afternoon nearly two thousand years ago
a wealthy Jewish councillor purchased a large linen sheet.
His purpose was to bury the body of Jesus Christ, at that
time hanging dead on a cross just outside Jerusalem's
walls. Aided by a friend, he unfastened the body from the
cross and hurried it to the rock tomb he owned nearby.
They worked quickly because the Sabbath, when any
labour had to cease, would shortly begin, with the appearance of the first star. They wrapped the body in the sheet,
then rolled a heavy boulder across the tomb's entrance.
Thirty-six hours later the first visitors to the tomb found
the boulder rolled back and the body gone – except for the
sheet that had wrapped it, which had mysteriously been
left behind.

Any notion that the very same sheet exists today ought
to be extremely doubtful. If we add that the sheet bears a
photographic likeness, apparently imprinted from Jesus's

body and blood as he lay dead in the tomb, it seems even more unlikely. If we further strain any remaining credulity with the information that some well-respected radio-carbon dating laboratories have dated this sheet to the fourteenth century, then clearly the sceptic has every right to say, 'Enough. This cloth is quite obviously a medieval fake. Why are you giving it a moment's further thought, let alone writing a book about it?'

Such sentiments would seem entirely justified, were it not for the object itself. Italians know it as the Santa Sindone, the Holy Shroud. At most times it reposes unseen within the Cathedral of St John the Baptist in Turin, northern Italy, inside a three-ton air-conditioned container that is kept behind a bullet-proof glass screen in a compartmented area to the left of the nave (pl. 1b & 2b). Once the family heirloom of the dukes of Savoy, former rulers of Italy, since 1983 it has been owned by the reigning Pope, with the Archbishop of Turin, currently Cardinal Severino Poletto, as its official custodian.

But only the sight of the sheet itself, silent and inanimate though it is, can properly speak for the Shroud. Whenever it is publicly exhibited – which has happened only twice within the last quarter century, in 1998 and again in 2000 – security concerns necessitate that viewing happens at a 'safe' distance. And normally each individual's viewing cannot be prolonged, because the large numbers wanting to do the same need to be kept moving. I am therefore unusually privileged to have been able to study it in leisurely close-up on three separate occasions: in November 1973, when it was being filmed for colour television; in March 2000, when it was shown in daylight to specially invited attendees at a conference; and in September 2002, following some major conservation work.

Very soon, however, via the medium of high-resolution

digital photography and high-definition filming, anyone
anywhere in the world will be able, on their home computer, to study the sheet at similar leisure for themselves,
and at even closer range. And this is why this book has
been written – as a guide to the very latest insights.

The Shroud of Turin (fig. 1a & pl. 2a) is a sheet of linen
fourteen feet six inches long by three feet nine inches
wide,[1] these dimensions being a broad approximation
because of two missing corners. Most of those who have
had the opportunity to view it close up describe its general
background coloration as ivory. Even so, one of the first
surprises on any viewing is just how clean the fabric
appears for an object theoretically two thousand years old.
This was particularly notable during the 1973 showing,
when the strong lighting gave the Shroud's surface a
damask-like lustre or sheen. For the year 2000 showing,
when the conditions were typically subdued northern
European daylight, the impression was altogether duller.

Another surprise is the Shroud's general state of repair.
Any examination in close-up clearly reveals the cloth's
tight herringbone weave, and how fundamentally strong it
remains, with no sign of disintegration. Yet the texture is
not at all coarse in the manner of sailcloth or sacking.
Instead, as was possible to determine with a surreptitious
touch during the 1973 showing, it has the basic lightness
of a modern-day linen bed-sheet.

But what principally draws the eye during any direct
viewing is the Shroud's famous and all-important double
image. Like the subtlest of shadows, cast on the cloth can
be seen faint imprints of the back and front of the body of
a man with long hair and a beard. He seems to be quite
naked, bloodstained in places, and laid out in the attitude
of death.

To those unfamiliar with the Shroud, the head-to-head
arrangement of the two imprints (fig. 1a) can only appear

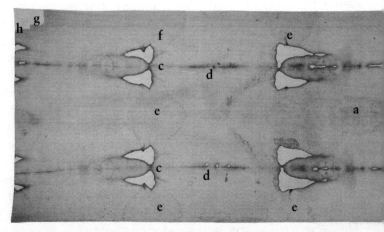

Fig. 1a PLAN OF THE SHROUD (above)

a Front-of-the-body imprint

b Back-of-the-body imprint

c Holes and scorches from fire damage sustained in 1532

d Triple holes from damage sustained in an apparently unrecorded incident prior to 1516

e Water stain damage, some associated with water used to douse the fire in 1532

f Near-invisible seam running the length of the cloth, used to join the strip above it to the main body of the cloth

g Top corner portions of the linen removed at some unknown period

h Area from which the sample was removed for radiocarbon dating purposes in 1988

Below: How the theoretical body was laid in the Shroud. Reconstruction after a painting by the seventeenth-century artist G. B. della Rovere.

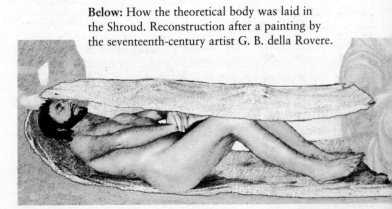

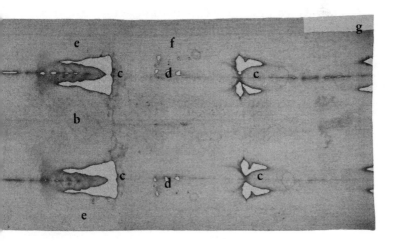

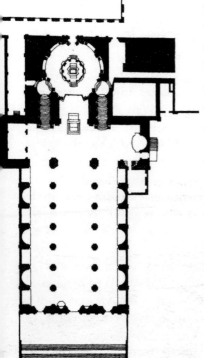

Fig. 1b PLAN OF TURIN CATHEDRAL (left)

a Cathedral steps

b Cathedral nave

c High altar

d Present repository of the Shroud

e Chapel of the Holy Shroud. The Shroud's home from 1694 to 1993. Ruined by fire in 1997 and currently being reconstructed

f Royal Palace of the Savoy Family, now the property of the Italian government

g Sacristy. The Shroud was brought in here for the taking of the sample for carbon dating in 1988. Nine years later the room was gutted during the fire of 1997 most curious without some explanation of the basic theory behind how they seem to have been formed. First the body the Shroud wrapped was laid on one half of the cloth, thereby creating the back-of-the-body imprint; the remaining half of the cloth was then drawn over the head and down to the feet, creating the front-of-the-body imprint. Given a corpse soaked in sweat and blood, each side of the body thereby acted like some kind of printing plate.[2]

Yet another of the surprises arising from viewing the Shroud directly rather than via a photograph is discovering just how pale and subtle the two body imprints appear. First-hand assessments of their colouring range from straw-yellow to sepia, much depending on the prevailing light conditions. Nevertheless there is universal agreement on their most enigmatic property: the closer one tries to examine them, the more they seem to melt like mist.

It is particularly important for anyone studying the Shroud from photographs to understand such subtlety. Any photography – whether colour stills, colour television or, more strikingly, black-and-white stills – seems to intensify the two imprints, making them show up several degrees more strongly than in real life. In the case of colour reproduction in books, this is because the usual reduction in scale brings otherwise diffuse elements into greater focus. In the case of black-and-white prints, a quirk of photographic chemistry causes the pale sepia/straw-yellow coloration to register particularly strongly when it is translated into tones of black. This property can be readily demonstrated by digitally scanning an old sepia-coloured family photo: it will almost always show up much more clearly when converted to black and white. The main point is that, except when viewed from a distance, the body imprints are extremely difficult to distinguish.

Similarly strange are their actual 'forms' or shapes. On

what may be labelled the 'frontal' (front-of-the-body) imprint, the face has a mask-like quality about it, with owl-like 'eyes' which some observers interpret to be open and staring, as if their owner were still alive. The region of the face and hair seems to be detached from the rest of the body because of the apparent absence of shoulders – a point some of the wilder 'fake' theorists have seized on as evidence that the body imprint and the face imprint were created from two different individuals. The crossing of the hands across the pelvic region is quite well defined, but the legs fade away below the knees. The upper feet are just a blur, and the toe region is absent because not enough length of cloth was allowed for these.

The back-of-the-body imprint has registered slightly more consistently, though in no way more strongly for its theoretically having received the weight of the body. In this case we can very faintly make out hair, shoulders, buttocks and calves, culminating in complete imprints for the soles of the feet and toes. A blank area of cloth extending beyond these indicates that an over-generous length was allowed for this half. If the Shroud is the work of an artist-forger this has to be a most paradoxical 'mistake' for him to have made, for we would surely expect him to have given priority to the more meaningful and more readily displayed front-of-the-body imprint.

Overall, the most everyday analogy for both imprints is their resemblance to the faint scorch marks typically found on a well-used linen cover for an ironing-board.

Anyone looking for evidence of artistic technique is therefore likely to find the Shroud imprints more than a little baffling. This was recognized even by atheist radio-carbon dating scientist the late Professor Edward Hall of Oxford. When in April 1988 he was invited to Turin to attend the taking of the samples for radiocarbon dating he very purposefully took out a hand-lens to check whether he could observe any solid particles of colouring matter, such as would be left by artists' pigments, in any of the Shroud's body imprint areas. By his own admission he found nothing.[3]

Part and parcel of the same curious phenomenon is the lack of any visible outline. Throughout the history of art, virtually until Turner and the Impressionists of the nineteenth century, artists relied to a greater or lesser degree on outlines to give shape to their work. The character of these, and the manner of modelling in any painting, has always provided reliable clues from which the art historian can usually make a confident judgement on date and origin.

But in the case of the Shroud imprints there really is no 'style' on which to base any art history judgement. Nor is there any artwork throughout the entirety of human history with which to sensibly compare them.

For this reason most art historians have tended to avoid any discussion of the Shroud, the most notable exception being American Gary Vikan, outspoken director of the Walters Art Gallery in Baltimore. In an article published in 1998,[4] Vikan claimed, without any substantiation of exactly what he meant, that 'three dozen' similar cloths competed with the Shroud during the Middle Ages, all of them bearing similar faint and elusive imprints.

A seriously loose statement to have come from a professional academic, it can only be presumed that he was referring to the fifty-odd artists' copies that were made, mostly during the sixteenth and seventeenth centuries.[5] And the irony is that these copies actually give the lie to Vikan's argument. Some make the Shroud face look grotesque and demon-like, others have the semblance of a gingerbread man (fig. 2 & pl. 3a). Some provide a loincloth for the figure, others do not. Although all are attempts at directly replicating what the eye sees on the

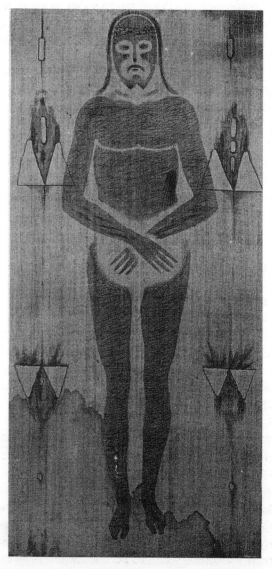

Fig. 2 TYPICAL ARTIST'S COPY OF THE SHROUD. Watercolour
on cloth, preserved in the Church of Notre-Dame de Chambéry.

Shroud, not one of them could even remotely be construed as anything but a work by the hand of an artist. They actually accentuate, not undermine, the Shroud image's uniqueness. Indeed, the more one considers the task of any artist trying to recreate the evanescent shadows that we see on the cloth, the more incomprehensible it becomes that any human being, in any century, could or would have done so, especially in eras during which all too few would have been able to appreciate the artist's subtlety.

Only a little less perplexing are the apparent bloodstains that accompany both the frontal and back-of-the-body imprints. On the upper forehead there are several distinctively shaped flows as if something had caused the scalp to bleed in several places. Similar flows can be seen coating the hair at the back of the head, following its curve. More flows of blood exude from the wrists and from the feet, some of the latter taking the form of globs that spilled directly on to the cloth. The chest has been pierced just below the chest-line in the form of a gaping wound, elliptic in shape and about two inches across, from which blood flows down near to the level where we would expect the umbilicus. A large spillage of blood that can be seen stretching right across the small of the back seems to have come from this same wound.

But both the colour and the character of these 'bloodstains' can only cause fresh scratching of the head. One might expect old blood to appear brown and encrusted, but this is not the case. In a subdued light the colour and consistency of the wounds can appear little different from the body stains. Only in clear daylight and under the strong television lighting that was used in 1973 did a quite separate colour become apparent, a clear, pale carmine, very slightly mauve, and surprisingly thin and 'clean'.

In addition to the stubbornly enigmatic body and blood imprints, the Shroud also bears a number of scars from moments when it itself sustained serious physical damage, scars that are actually rather more obtrusive to the eye.

Two whole pieces are missing from what we may refer to as the 'top' corners (fig. 1a). Whatever the occasion on which these pieces were removed, and whatever the purpose, it seems to have gone undocumented. There are also four sets of triple burn-holes (pl. 3b), some with ancillary damage, sustained during some other as yet undetermined incident. This must have taken place before 1516, because the holes can quite unmistakably be seen in one of the better artists' copies, created to one-third scale and dated to this year, which is preserved in the Church of St Gommaire in Lierre, Belgium (pl. 3a).

By far the most extensive scars, however, are scorches and actual holes from a historically well-recorded fire that occurred on the night of 4 December 1532, when the Shroud's home was the beautiful Sainte Chapelle in Alpine Chambéry. At that time the Shroud was very securely locked behind a metal grille in a niche behind the chapel's main altar. As flames engulfed the entire chapel, richly fuelled by all its furnishings and hangings, the holders of the grille's keys were too far away to be summoned. Saving the Shroud therefore fell to the efforts of a blacksmith, aided by two Franciscan friars. By the time they had forced the grille open, the folded-up Shroud's ornate silver casket had already become so hot that part of its lid dropped into its interior. This was quickly doused with water, which left its own set of stains on the cloth, in the form of repeating lozenge-shaped markings. Far worse, however, was the damage from the molten lid: an unsightly patchwork of holes and burns reminiscent of a paper-tearing act.

Yet by some extraordinary quirk of fate, the all-important imprint had scarcely been touched. From two life-size dated copies that were made only five years prior to the 1532 fire, and which are today preserved in Noalejo, Spain,[6] it can be deduced that the only area of imprint lost was from the lower back of the arms, together with two forked blood flows at the elbows. It could easily have been a lot worse.

Two years after the 1532 fire, four Poor Clare nuns from a local convent were asked to make the best repairs they could. Cutting a former altar-cloth into appropriate, mostly triangular shapes, they patched each hole, securing each patch to a strong backing cloth which they sewed on to the Shroud's underside. This arrangement, supplemented by a few ancillary patches that were added in 1694, remained in place up to the summer of 2002.

Between 1604 and 1998 the Shroud was kept rolled up inside a casket just four feet long by one foot wide. It was rolled and unrolled many times for exhibition purposes during the centuries, which caused visually distracting creases to form in ever-increasing numbers. Accordingly, at the end of 2001, by which time the Shroud was being stored flat, the decision was taken to implement remedial work.

Chosen to head this task was Dr Mechthild Flury-Lemberg (pl. 13a), a professional conservator from one of the world's finest collections of historic textiles, the Abegg Foundation, the former home of collectors Werner and Margaret Abegg, at Riggisberg, just south of Bern, Switzerland. Between 21 June and 23 July 2002, Dr Flury-Lemberg and Italian assistant Irene Tomedi secretly removed the sixteenth-century backing cloth and its accompanying patches.

This procedure duly revealed for the first time since the sixteenth century the Shroud's hitherto hidden underside (pl. 13b). It confirmed, as had long been anticipated, that the Shroud's double-image body imprints are effectively visible only on the side that is normally displayed – although to everyone's surprise some vestiges of a face turned out to be discernible on the underside.

This underside was photographed in its entirety using the latest digital photography. Flury-Lemberg and Tomedi then removed all loose carbonized debris from the surrounds of the burn-holes, carefully retaining such material for future scientific analysis. Sewing with only the very finest silk thread, which will break well before causing damage to the Shroud's linen threads, they finally stitched the Shroud on to a 'new' backing cloth,[7] leaving all the 1532 fire damage openly visible for whenever the Shroud is publicly exhibited on future occasions.

The results of Flury-Lemberg and Tomedi's handiwork were shown to the international press, and to other observers (including me), on 21 September 2002. Although some of those invited took great exception to the fact that such a radical 'make-over' had been carried out without sufficient consultation (some of the more vociferous critics even labelled Flury-Lemberg 'the Butcher of Bern'),[8] it is hard to fault the professionalism of the two conservators' work. The unsightly creases that once crisscrossed the cloth's surface have been almost completely eliminated. The exposed fire damage is no more displeasing and disruptive to the eye than the rather unsightly patches that formerly covered it. Although as yet the 'renovated' Shroud has never been seen at a public showing, this will happen during 2010, and the results are unlikely to disappoint.

In summary, the Shroud presents itself to us as an extraordinarily intriguing historical 'document', and its all-important imprints, even as seen by the naked eye, simply cannot be easily dismissed as the work of some artist-forger. When hardened professionals and those with absolutely no religious beliefs come face to face with the cloth, many find outright rejection impossible and their scepticism seriously challenged. These were the reactions of Hong Kong archaeologist Bill Meacham, who was one of those given a special close-up viewing in March 2000. 'In spite of all I had read about it,' he wrote, 'and all the photos I had seen of it, the real thing is haunting. The first glance tells you that this object is extraordinary . . .'[9] Likewise, seasoned television cameraman David Crute, who filmed the Shroud in high definition (HD) in January 2008, made these uncharacteristically emotional observations of the experience: 'I'm not a religious person, but when we got into the room, it made the hairs on the back of your neck stand up. It was a unique experience . . . and I was moved more than I had expected to be. The cloth itself was unbelievably beautiful.'[10]

So what are we to make of this cloth, with so many curious features that seem to defy any easy natural explanation? It bears no label, no trademark, nothing to tell us the age from which it comes, save for the imprinted figure, which in turn has no clothes to indicate its origin. Every sympathy must lie with the sceptic who regards it as too good to be true, because by all normal expectations it undoubtedly ought to be a fake.

But there is one more quite extraordinary visual property which has so far gone unmentioned. It is a property that up to the year 1898 lay hidden and totally unsuspected in the cloth – until it was most unexpectedly revealed by the eye of the camera.

Chapter 2

The World's Most Mysterious Image

I've been involved in the invention of many complicated visual processes and I can tell you that no one could have faked that image.

Leo Vala, photographer, 1967

IN TURIN'S MUSEUM of the Holy Shroud, just a short walk from the cathedral that houses the Shroud, is a late nineteenth-century plate camera with a Voigtländer precision lens. Back in 1898 it completely overturned previous perceptions of the Shroud's imprint.

That year marked the fiftieth anniversary of Italy's constitution. The Shroud had not been publicly exhibited for thirty years, and among Turin's scheduled celebrations was the staging of a fresh showing. For Don Nogier de Malijai, a twenty-seven-year-old Salesian priest who was also an enthusiastic amateur photographer, this seemed an ideal opportunity for the Shroud to be photographed for the very first time. The idea was not entirely new, for in 1842, when the Shroud had been shown from the Turin Royal Palace's balcony to celebrate the marriage of Savoy's Prince Victor Emmanuel, a local instrument-maker, Enrico Jest, had recently produced equipment that replicated France's new daguerreotype photographic process. Had the Shroud been shown on a brighter day and been held out for a little longer, Jest might possibly have had sufficient time to produce the first photograph.

As it happened, even half a century later the idea of allowing the Shroud to be photographed was regarded as too unseemly for something so holy. Certainly this was the opinion of the Shroud's then owner, Italy's ultra-conservative King Umberto I of Savoy, son of the Victor Emmanuel who had married back in 1842. It took a great deal of persuasion before King Umberto agreed to permit an official photograph, and when he did so the task fell at very short notice to Secondo Pia (pl. 4a), a forty-three-year-old lawyer who enjoyed a good reputation as an amateur photographer but who had never previously even set eyes on the Shroud.

As Pia quickly discovered, the assignment presented many technical difficulties. He was being allowed to photograph the Shroud only as it hung behind glass in a frame suspended above the cathedral altar. Because the cathedral's natural lighting was (and still is) rather dim, he needed electric lighting, a new and uncertain commodity in 1898. He also had to build a three-metre-high wooden platform for his camera in order to take the photograph from the right level.

It has often been said that he was unsuccessful during the first session, a trial one shortly before two p.m. on 25 May. Certainly he had problems with his electric lamps, the glass screens of which cracked under the heat. But he managed two exposures, and although they were less than perfect, already evident on these negatives was a rather strange effect.

Pia returned to the cathedral the night of 28 May, accompanied by Don Nogier and cathedral security guard Felice Fino, another amateur camera enthusiast. He began at around 9.30 p.m. with two trial exposures, and Don Nogier and Fino seem to have taken some unofficial photos of their own at this time. For the proper, definitive photos that he intended to take, Pia loaded his camera with the first of four 50 × 60 photographic plates, applied his most prized Voigtländer lens, then took four exposures, the maximum of fourteen minutes, the minimum of eight, only two of which he would officially record.[1]

In his darkroom later that night, as the best of the four plates revealed itself under the developer, Pia was able properly to verify the odd effect he had observed on his first trial negatives. There slowly appeared before him not the feeble ghosts of the shadowy imprints visible on the cloth, as anyone might expect, but something altogether more extraordinary.

In negative, the Shroud's head-to-head double figures could be seen to have undergone a dramatic change. Instead of their former difficult-to-interpret shadowiness, which so many of the copyists had 'seen' as grotesque, they had now acquired quite unmistakably natural light and dark shading, giving them meaningful relief and depth. The bloodstains, showing up in white, could be seen to flow very realistically from the hands and feet, from the right side, and from all around the crown of the head. Instead of a mask-like, almost gingerbread-man appearance, the man of the Shroud could be seen as a well-proportioned individual of an impressive build. Most striking of all was his face, so dignified in death, so incredibly lifelike against the black background – yet all on a photographic negative. With that eerie chill that can accompany such experiences, Pia found himself thinking that he was the first man for nearly 1,900 years to gaze on the actual appearance of the body of Christ as he had been laid in the tomb. He had discovered what could only be interpreted as a real photograph, hitherto hidden in the cloth, until it could be revealed by the eye of the camera.

Even over a century ago news could travel fast. The first media report, in *L'Italia Reale Corriere Nazionale*, appeared on 1 June, quickly followed by one of the unofficial photos. Not long after that the first doubters surfaced, the *Italia Corriere*, on 15 June, authoritatively claiming that the strange effect was due to Pia's use of a yellow filter. There followed further 'intelligent', scientific-sounding explanations for the phenomenon. It was just an accident of 'transparency', or of 'over-exposure', or of 'refraction'. More hurtfully for Pia, there were also sly insinuations that he must have 'retouched' his negative, the strange effect thereby being just a cynical hoax that he had perpetrated.

Within the next three years even some prominent Roman Catholic churchmen came out on the side of the doubters.[2] They heavily emphasized the Shroud's most suspicious lack of any historical provenance before its appearance in northern France in the mid-fourteenth century, when even back then it was beset by accusations that it was a fake. All this resulted in the reputation of both the Shroud and Secondo Pia falling under a dark cloud rather similar to that which followed the 1988 radiocarbon dating. That dark cloud would remain for thirty years, until a fresh occasion to exhibit the Shroud presented itself.

That occasion, in 1931, was the wedding of the then dashing young Prince Umberto of Piedmont (the later King Umberto II) to Princess Maria José of Belgium. It would attract some two million visitors to Turin, and to mark the event in what had become their traditional way, the Savoy family chose to put the Shroud on public display, from 3 to 14 May. Also, in order to check and hopefully improve upon the controversial Pia photograph, they appointed a fully professional Turin photographer, Giuseppe Enrie, to take some proper new official photographs.

Between 21 and 23 May Enrie duly took an impressive series of definitive black-and-white photographs, all of them with the Shroud unprotected by covering glass, and using photographic equipment that was technically a vast improvement on Secondo Pia's. For official purposes at least there were a dozen photographs in all: four of the entire Shroud, the Shroud in three sections, the complete back-of-the-body imprint, the face and chest, the face two-thirds the natural size of the original, the face the natural size of the original, and a direct sevenfold enlargement of the nail wound in the left wrist. They are all superb-quality, state-of-the-art examples of pre-digital-era black-and-white plate camera technology. As Enrie would later recall of the natural-size negative plate of the Shroud man's face (pl. 5), 'I will remember as one of the most beautiful moments of my life, certainly the most moving of my career, the instant in which I submitted my perfect plate to the avid look of the Archbishop and that select whole group of people.'[3] Among that 'select whole group of people' crowding around the plate was the crown prince Umberto in whose honour the Shroud had been displayed. According to Enrie's description, 'The young prince was almost beside himself with excitement and emotion.'[4]

That same glass plate, which I personally studied at Enrie's old studio in 1994, is now a historic object in its own right. Today it can be viewed in Turin's Museum of the Shroud. It speaks for itself.

In the light of that glass plate, and the literally

thousands of similar negative photographs of the Shroud face, both professional and amateur, that have since followed it, any suggestion that the phenomenon Pia first brought to light back in 1898 was some kind of hoax could now be dismissed out of hand. And, thankfully, Pia, although now seventy-six, was still alive to see his work thus vindicated. He had been invited to be present at the showing, along with a public notary and photographic experts to make absolutely sure that Enrie, in his turn, could not be accused of any trickery.

Four decades later, in 1973, Pope Paul VI, in a televised address accompanying the Shroud's first-ever filming for colour television, recalled his emotions upon first viewing Enrie's image when he was a young priest back in 1931. He said that the photograph seemed to him 'so true, so profound, so human and so divine, such as we have been unable to admire and venerate in any other image'.[5] A country and a language away, trendy London photographer Leo Vala, a complete agnostic, remarked of it in a photographic journal during the same decade, 'I've been involved in the invention of many complicated visual processes and I can tell you that no one could have faked that image. No one could do it today with the technology we have. It's a perfect negative. It has a photographic quality that is extremely precise.'[6]

Nor was this the only surprise the Shroud would spring for the world of photographic imagery during the course of the twentieth century. In 1976, by which time the Shroud had been photographed for the first time in colour,[7] the mystery of its photographic imprint had come to intrigue a group of young science teachers at the United States Air Force Academy in Colorado Springs, USA. One of the group, physicist Dr John Jackson, had been given access to a then state-of-the-art piece of equipment developed for the NASA space programme, an Interpretation Systems VP-8 Image

Analyzer, a device designed to enable shades of black and white to be translated into levels of vertical relief that can be viewed and adjusted via a television screen. A normal photograph records only variations in light and does not contain information about the camera's distance from the object being photographed. So when viewed via the VP-8, the result is almost invariably collapsed and distorted, the VP-8 not having been designed or intended to produce any 'true' 3D display, only a semblance of it. However, when the Shroud's negative image was placed beneath the machine, the result was nothing short of astonishing. A consistent 'true' 3D effect was produced (pl. 8a) around which it was possible, via the device's TV monitor, to move, viewing the contours of the body just as if viewing a range of mountains from a moving helicopter. The Shroud image's varying tones – the scientists called them 'intensity levels' – could thereby be seen not so much as true light and shade, but rather as encodings of the (still theoretical) body's relief in relation to its distance from the cloth at each related image point.

Even today it is difficult for laymen to appreciate just how astonishing this discovery was to the group of physicists and other technicians who first came across it. As John Jackson remarked, 'When I first saw this . . . I think I knew how Secondo Pia must have felt when in 1898 he saw his photographic image.'[8] The VP-8 Image Analyzer's inventor, Peter Schumacher, who personally delivered the machine to Jackson and his team, has recalled his similar emotions on seeing the Shroud's full-body image on his system's TV monitor for the very first time: 'A "true three-dimensional image" appeared on the monitor . . . The nose ramped in relief. The facial features were contoured properly. Body shapes of the arms, legs and chest and the basic human form . . . I had never heard of the Shroud of Turin before that moment. I had no idea what I was looking at. However, the results are unlike anything I have processed through the VP-8 Image Analyzer, before or since. Only the Shroud of Turin has [ever] produced these results from a VP-8 Image Analyzer.'[9] With regard to the idea of some unknown medieval artist-forger producing such an image, Schumacher had this to say:

> One must consider how and why an artist would embed three-dimensional information in the 'grey' shading of an image [when] no means of viewing this property of the image would be available for at least 650 years after this was done. One would have to ask why is this result not obtained in the analysis of other works? ... Why would the artist make only one such work requiring such special skills and talent, and not pass the technique along to others? How could the artist control the quality of the work when he or she could not 'see' grey scale as elevation? ... Would an artist produce this work before the device to show the results was [even] invented?[10]

Within just a few months of their discovery, John Jackson and some two dozen colleagues found themselves in Turin with permission to conduct the most extensive scientific examination of the Shroud ever. Their work included a variety of photographic techniques, from conventional colour photography to x-radiography, transmitted light photography, photography under ultraviolet light and photomicroscopy. One of the two professional photographers with this self-styled 'STURP'[11] team was Barrie Schwortz, a Los Angelean of Jewish parentage with positively no Christian affiliations. A self-confessed 'certified sceptic', by his own admission he fully expected to 'see the brushstrokes and go home'. In the event he became so fascinated he found himself working almost round the clock during the 120 hours the team had

had allotted for their work. And during the subsequent decades he became so convinced by the Shroud that he went on to found the world's first website on the subject, still active to this day.[12]

As a result of the STURP's efforts, a wealth of good-quality colour photographs of the Shroud, including detailed close-ups, were widely circulated, and some of the more scientific varieties of their photographic work will feature in later chapters of this book. But high-quality and exhaustive though the STURP photographic work was for its time, fresh photographic approaches to the Shroud have by no means stood still during more recent decades.

On 25 June 1997, only weeks after the Shroud had narrowly escaped destruction in a major fire in Turin Cathedral's Holy Shroud Chapel,[13] the cloth was brought unannounced to Turin's Church of the Confraternity of the Holy Shroud. There, once again a photographer found himself called upon at the shortest possible notice to take official photographs. On this occasion it was Turin-based professional Gian Carlo Durante who photographed the Shroud in its entirety and a close-up of the face in both colour and black and white on 13 × 18 and 10 × 12 format transparencies respectively.[14] Less than a year later Durante's superb-quality colour photo of the Shroud face was being viewed worldwide in more than four million homes via the front cover of *Time* magazine.[15]

Even as Durante carried out his unexpected assignment, the new age of digital photography was dawning, and he was not slow to adopt this new technology. His services were again called upon in 2002 when extensive digital photography and digital scanning were needed to document conservator Dr Mechthild Flury-Lemberg's removal of the Shroud's sixteenth-century patches and backing cloth. The latter, of course, revealed the Shroud's hitherto inaccessible underside for the first time in more than four centuries. A large-format colour publication was quickly produced from Durante's and other photographs to make these widely available.[16]

But further developments in digital image technology were in train. In 2007, in Novara, a city only fifty-five miles from Turin, a company titling themselves HAL9000 digitally photographed Leonardo da Vinci's vast twenty-nine-foot-long mural *The Last Supper* in 1,677 separate sections, following which they digitally stitched these together to create a giant multi-gigabyte image now accessible on the internet.[17] This enabled even as little as a square centimetre of Leonardo's sadly so ruined painting to be readily examined on a home computer.

Such an approach obviously lent itself to the Shroud. As a result, on 22 January 2008 the HAL9000 technicians were allowed a day of direct access to the cloth. During this time they captured 158 gigabytes' worth of photographic data enabling the viewing of details as tiny as one five-hundredth of a millimetre. A twenty-three-foot-long giant print of this has already been displayed in Novara, and internet access and/or a DVD similar to that for the da Vinci work is expected to be available by the time of this book's publication.

No less exciting have been recent developments in high-definition (HD) movie making, which has brought hitherto undreamed-of detail to moving images. Accordingly, in the footsteps of the HAL9000 technicians came a British television crew led by independent producer David Rolfe, veteran of a BAFTA-award-winning documentary on the Shroud from thirty years before. Until this time no English-speaking television documentary maker had ever been allowed to film the Shroud directly. On 24 January 2008 Rolfe's appointed cameraman, David Crute, together with a second back-up cameraman, took on the task, travelling to Turin with two state-of-the-art Sony HD

750 cameras (pl. 8b). Their filming was done in the same modern sacristy room in which I'd had the daylight viewing in 2000, except that for television purposes the sacristy's high windows were all blacked out, and the Shroud was lit with eight high-frequency Kino Flo fluorescent lights specially chosen for their cool running and their lack of flicker. Despite some unexpected difficulties with the power supply – perhaps shades of the same gremlin that had dogged Secondo Pia's work eleven decades earlier – Crute and his colleague were able to return to the UK with some superb-quality imagery of the Shroud. Some of this was used, as intended, for a BBC television documentary fronted by popular presenter Rageh Omaar which was screened on Easter Saturday in 2008.[18]

The altogether more comprehensive footage not yet seen by any television audience is arguably even more interesting.[19] Rolfe had asked Crute to film every detail of the Shroud's surface that he could, ranging from distance to high close-up. It is as the HD camera moves into close-up on areas of body image and blood image that this latest technology breaks intriguing new ground. On a computer monitor we are able to examine the lack of substance to the body image areas and the strange coloration of the bloodstains in such an easy and controlled way that it is better than if we had the cloth itself directly in front of us, examining it with a high-powered magnifying glass. As recognized by both Rolfe and Crute, the high degree of detail now available and the wealth of analysis techniques provided by the latest technological wizardry open up fresh areas of Shroud research that have not yet even begun to be tackled. In Crute's words, 'within the material we have there are other layers to the image which people have not seen'.[20]

Absolutely certain, however, is that the latest high-definition work verifies Secondo Pia and Giuseppe Enrie's hidden negative discoveries in the most powerful way possible. Even in the case of Enrie's superb life-size negative plate, endlessly replicated though it has been by innumerable subsequent photographs, there have been the occasional sceptics who have attributed some of the effect to Enrie's use of orthochromatic photographic plates, suggesting that this chemistry injected an impression of detail that might otherwise not be present.

With digital photography there neither is nor can be any intervening chemistry. Viewing on a computer monitor a Crute HD close-up of the Shroud face in its natural colour, with a single key stroke we are able to change the image to 'grey scale', or black and white. A second key stroke enables us to reverse these tones to negative. This then shows us the same mirror image of the theoretical original body that the Shroud itself appears to be. If we make one further key stroke to left-right reverse, or correct, this mirror-reversal, we find ourselves looking at exactly the same extraordinary face which decades ago so startled and enthralled Pia, Enrie and countless others. Near instantly and effortlessly we are able to verify for ourselves that the Shroud's remarkable 'hidden photograph' is there when-ever the Shroud's image is translated into black and white, and these shades reversed. On a high-tech editing suite as used in professional television production it is even possible to replicate the VP-8 Image Analyzer's viewing of the Shroud in three-dimensional relief. In the light of such data the onus therefore has to be very firmly on those who hold the Shroud to be a fake to explain away how some-one, at least as far back as the Middle Ages, some five centuries before photography's invention, could have produced such a 'hidden photograph effect'.

One of the earliest attempts came from Kentucky-based forensic anthropologist Dr Emily Craig, who with textile expert Randall Bresee proposed what may best be described as a 'dry transfer' technique. They suggested the hypothetical artist-forger made a preliminary sketch with a dry brush to indicate in shades of dark those that would normally have been light, and vice versa. He then applied a piece of cloth to this sketch and burnished in order to transfer it to the cloth in a manner suggesting no brush-strokes and no outlines. After Craig and Bresee had experimented with this method, they photographed the result, then published this and the 'negative' from it (pl. 4c) in a scientific journal.[21] Although they evidently found the effort satisfying as the explanation for the

Cloth coated with light-sensitive silver salts, acting as film. After the frontal image has developed the second half has to be set up to receive the back-of-the-body image

Quartz 'lens' set into the aperture of otherwise light-fast *camera obscura*

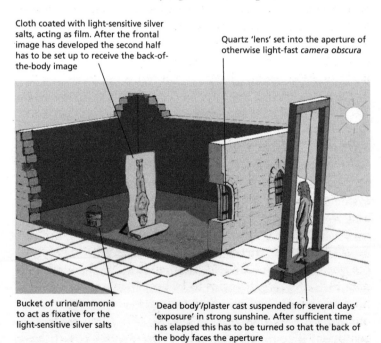

Bucket of urine/ammonia to act as fixative for the light-sensitive silver salts

'Dead body'/plaster cast suspended for several days' 'exposure' in strong sunshine. After sufficient time has elapsed this has to be turned so that the back of the body faces the aperture

Fig. 3 HOW A MEDIEVAL FORGER FAKED THE SHROUD? Professor Nicholas Allen's reconstructed 'camera' for producing the Shroud image, using basic 'photographic' materials and know-how theoretically available in the Middle Ages.

Shroud's imprint, it is hardly convincing and has failed to
attract any wide support.

Almost synchronous with Craig and Bresee came
London journalist Lynn Picknett and partner Clive Prince's
sensationalized theory that the Shroud is an early photograph created by Leonardo da Vinci.[22] Not only is there
not a shred of evidence that da Vinci's mostly mechanical
inventions included photography (and Picknett and
Prince's attempted replication of this was a positive dis-
grace to his memory), there is the further difficulty that the
Shroud was being reliably recorded, visually and verbally,
a hundred years before da Vinci was born. The theory may
satisfy some believers in tabloid journalism but it has
absolutely no serious credence.

Only a year later an altogether better-thought-out
version of the same theme – and thankfully without any
Leonardo da Vinci component – came from South African
art professor Nicholas Allen. Basing his method on a full-
size light-proof room, or 'camera', Allen suggested that the
hypothetical medieval forger used quartz for a lens, silver
salts for photographic emulsion and urine or ammonia for
a fixative. Outside the light-proof room he would have
suspended either an actual dead body or a plaster cast of
one in strong sunshine (fig. 3). Inside the room a shroud-
size length of cloth would have been mounted upright,
coated in silver salts and folded in two. With the room's
quartz 'lens' opened up, the image of the suspended body
would be projected on to the cloth inside. Several days'
'exposure' would be needed, and when this had elapsed
the forger would enter the room (necessarily during dark-
ness) and turn the cloth round so that its 'back' half faced
the lens. At the same time he would turn the body suspended outside 180 degrees to face in the opposite
direction. There would be several more days' wait for this
half of the cloth to be 'exposed'. Then a bucket of

ammonia or urine would be used to 'fix' the two images.

Whatever we may think of this complicated procedure, undeniably the method can work, as Allen's publication[23] of his replication of the negative readily demonstrates (pl. 4d). Exactly as in the case of the Shroud image, it looks convincingly like a positive photograph. However, whether six hundred years ago someone actually created the Shroud's image in such a way – a way that would surely have needed an understanding of some fundamental photographic principles that simply did not exist back then – is a very different matter. This is History Channel television producer Sean Heckman's account of how he tried to reproduce Allen's replication for a 2005 documentary[24] that specifically advanced this as the theory of how the Shroud was faked:

> I personally worked with artist Stephen Berkman to design, construct and test the theory that the Shroud was created by such a process. Stephen and I paid particular attention to building the camera and exposing the image with historical accuracy. We only used simple lenses that would have been available in the thirteenth century as well as exposed and fixed the image with chemicals that are known to have existed at the time. Put simply, the experiment failed. While it is theoretically possible to expose an image, there are a countless number of variables that make the process nearly impossible, a multitude of which caused our project to fail. Namely, in order to make a life-sized image, you would need to position the linen at least six or more feet away from the lens. Since light fades at an inverse square rate, a pinhole or simple lens only allows for an extremely faint amount of light to reach that distance, making it extremely difficult to expose the image. In our case it took 43 days to get a faint image, which completely disappeared once the image was fixed. Considering the

experiment was based on 200 years of *known* photographic technology, I find it difficult that such an image could have been created six hundred years ago, particularly an image that we'd still be able to see today.[25]

As recently as October 2009 came yet another claim to have 'reproduced' how the Shroud was faked. Luigi Garlaschelli, the bearded, pipe-smoking Professor of Organic Chemistry at the University of Pavia in Italy, has made something of a speciality of debunking claims of religious paranormal phenomena. Stigmata and bleeding statues have been among his earlier targets. In the case of the Shroud, Garlaschelli's method was to place a linen sheet flat over a volunteer model, then rub this with a pigment containing acid. The pigment was then artificially aged by heating the cloth in an oven, then the cloth was washed. This process removed the pigment from the surface but left an image reputedly similar to that of the Shroud.

Garlaschelli's claim, presented at a conference in northern Italy for atheists and agnostics, prompted a flurry of news headlines around the world. Yet even the most cursory comparison between his 'negative' (pl. 4e) and that on the Shroud reveals the former as hardly the 'definitive proof' of the Shroud's fraudulence that he has claimed for it. As remarked by one 'general public' commentator on the Reuters news story, 'Why isn't anyone saying the obvious? Compare these two images . . . the modern copy is garish, lacking any gradations of tone . . . it's completely inferior, especially when one contrasts the faces and the chest areas.'

Furthermore, both the Nicholas Allen and Luigi Garlaschelli theories require that after the hypothetical medieval artist-faker had so laboriously produced the 'body' image, he would then have had to daub on bloodstains for effect. If someone from the Middle Ages genuinely had worked in this way, we would surely expect modern-day forensic experts quickly to see through such a deception. But as we are about to see, this is very far from being the case.

Chapter 3

Home to a Real-Life Body

The markings on this image are so clear and so medically accurate that the pathological facts which they reflect concerning the suffering and death of the man depicted here are in my opinion beyond dispute.[1]
Dr Robert Bucklin, medical examiner, Los Angeles County, 1977

THE MOMENT that Secondo Pia's imperfect though nonetheless revelatory photograph became available, medical specialists were able to study critically the various injuries suffered by the purportedly real-life human body which the Shroud theoretically once wrapped. One of the earliest and in many ways the unlikeliest of these specialists was French anatomist Yves Delage, discoverer of the function of the semicircular canals in the inner ear.

An avowed agnostic, in 1902 Delage was forty-eight years old and at the zenith of his career as professor of comparative anatomy at the Sorbonne University in Paris. On 21 April that year he delivered a now historic lecture to the Paris Academy of Sciences in the very same theatre in which fifteen years earlier Louis Pasteur had announced his vaccine for rabies. The title Delage gave his lecture, 'The Image of Christ Visible on the Holy Shroud of Turin', ensured a particularly packed audience that day.

With the aid of lantern slides, Delage quickly demonstrated the anatomical flawlessness of the wounds and the other data that had been revealed by Pia's photographs. In the light of this medical exactitude, he explained that the only possible interpretation he could come to was that the injuries indicated were not the work of any artist. He pointed out how difficult it was for anyone, in any era, to try to convey the tones of a human body in negative. And for what possible reason, when back in the Middle Ages the technology simply had not been invented for anyone to appreciate such cleverness? Nor was there a sign of any pigments forming the image.

Delage further pointed out that for a genuine dead body to have been laid in the Shroud there had to have been a very fine line between the amount of time it would take for the image to form and that in which the body would have begun to decompose, ruining any image and ultimately the cloth itself. In which vein he reminded his listeners, 'Tradition – more or less apocryphal, I would say – tells us that this is precisely what happened to Christ, dead on Friday and disappeared on Sunday.'[2] He accordingly concluded that 'The man of the Shroud was the Christ'. Somehow the image had been formed by some unknown process while his body lay dead in the tomb.

Delage's lecture received some surprisingly favourable comments in the London *Times* and British medical journal *The Lancet*. However, in Paris the Academy of Sciences was dominated by rationalists and freethinkers, and they were not at all receptive to what they had heard. The Academy's secretary, Marcelin Berthelot, flatly refused to publish Delage's text among the Academy's proceedings, as would otherwise have been his normal procedure. And many of Delage's colleagues thought that he had severely jeopardized an otherwise distinguished scientific reputation.

Deeply wounded by the fuss, then frustrated by the Turin authorities' refusal to allow him any direct examination of the Shroud, not long after Delage dropped any further active work on the subject, instead devoting his energies mostly to marine biology studies at the Biology Station at Roscoff, Brittany, where he is remembered by an impressive relief sculpture. The next three decades were, in any case, those of that same 'dark cloud' period for Shroud studies that Secondo Pia was obliged to endure. Delage's only public expression of his sentiments was in a letter to his friend Charles Richet, editor of the *Revue Scientifique*:

> When I paid you a visit in your laboratory several months ago ... had you the presentiment of the impassioned quarrels which this question would arouse ...? I willingly recognize that none of these given arguments offer the features of an irrefutable demonstration, but it must be recognized that their whole constitutes a bundle of imposing probabilities, some of which are very near being proven ... A religious question has been needlessly injected into a problem which in itself is purely scientific, with the result that feelings have run high, and reason has been led astray. If, instead of Christ, there were a question of some person like a Sargon, an Achilles or one of the Pharaohs, no one would have thought of making any objection ... I have been faithful to the true spirit of science in treating this question, intent only on the truth, not concerned in the least whether it would affect the interests of any religious party ... I recognize Christ as a

historical personage and I see no reason why anyone should be scandalized that there still exist material traces of his earthly life.[3]

Exactly as in the case of Pia, it took until the 1930s before anyone of Delage's calibre reapproached the Shroud mystery from the medical angle. When they did so, it was as a result of the all-new Enrie photographs, and again it was a Frenchman, Dr Pierre Barbet, at the fore. Barbet, just like Delage, had taught anatomy for some years. However, his main career, extending over thirty-five years, was as chief surgeon at the Hospital of St Joseph in Paris, where he enjoyed the highest respect as a dedicated professional, sometimes performing as many as thirty operations per week.

By sheer chance, Barbet was asked by a friend to take a look at the body imprints on Enrie's recently released photographs, and he became hooked by what he saw. Two years later he visited Turin to study what he could of them on the actual cloth, while it was being displayed under glass in a huge gold frame on the cathedral's altar as part of celebrations marking the Holy Year of 1933. On 15 October, the last day of these showings, Barbet happened to be on the cathedral's outside steps when Turin's then archbishop, Cardinal Maurilio Fossati, made the impromptu decision to bring the Shroud outside for the large crowd to take one last look before it was returned to its casket (pl. 31c). As Barbet subsequently recalled:

> The sun had just gone down behind the houses on the other side of the square, and the bright but diffused light was ideal for studying it. I have thus seen the Shroud by the light of day, without any glass screening it, from a distance of less than a yard, and I suddenly experienced

one of the most powerful emotions of my life. For without expecting it, I saw that all the images of the wounds were of a colour quite different from the rest of the body; and this colour was that of dried blood which had sunk into the stuff. There was thus more than the brown stains on the Shroud reproducing the impression[4] of a corpse . . . It [the blood colour] was strongest at the side, at the head, the hands and the feet; it was paler, but nevertheless fully visible, in the innumerable marks of the scourging . . . A surgeon could understand, with no possibility of doubt, that it was blood which had sunk into the linen, and this blood was the blood of Christ.[5]

An interesting point about this passage, one to be returned to later in a different context, is how readily it conveys the difficulty of distinguishing between the body and blood imprints when viewing the Shroud in subdued interior lighting conditions, as inside the cathedral, compared to those in the open air during daytime, as on the cathedral steps. But the more fundamental point is that as a highly experienced surgeon Barbet, just like Delage the anatomist before him, found the appearance of the Shroud's bloodstains totally convincing.

Barbet would go on to complete several pioneering studies, including experiments on cadavers available to him from his teaching work, all aimed at interpreting every mark on the Shroud's imprint in relation to the injuries of Jesus as recorded in the Christian gospels. In the years before World War Two, during which he was legendary for treating casualties sometimes for eighteen hours at a stretch, he blazed a research trail which literally dozens of well-qualified medical practitioners right across the world would follow throughout the rest of the twentieth century.

One such practitioner was Dr David Willis, a family

doctor based mainly in Guildford, England, who hailed from an unbroken dynasty of physicians stretching back to the time of Charles II. Ancestor Dr Thomas Willis had discovered the Circle of Willis in the brain, and Dr Francis Willis had treated the madness of King George III. David Willis, like Delage, was an agnostic when he became interested in the Shroud, and it was my privilege to know him and to correspond with him intensively throughout a period of ten years before his untimely death in 1976. Another such practitioner, half a world away, was Dr Robert Bucklin, medical examiner with the Los Angeles coroner's office, and as such familiar with viewing more traumatic injuries in a single day than most of us will ever see in a lifetime. Bucklin memorably presented the main medical facts on the Shroud in David Rolfe's award-winning 1978 television documentary *The Silent Witness*. Both men independently followed on from Barbet in making the most exhaustive studies of the injuries visible on the Shroud.

Treating a life-size negative photograph of the imprint on the Shroud as if this was an actual body on the mortuary slab, Bucklin made this overall appraisal: 'The body is that of an adult male. The image as it appears here measures five feet ten inches in length, and the estimated body weight is 175 pounds. The body appears to be about thirty to thirty-five years of age.'[6] Of the visible injuries to this body, Bucklin remarked of those appearing on the head and face, 'On the right cheek over the malar area is a swelling which has resulted in partial closing of the right eye. There is also an area on the nose where there is a separation and possible fracture of the nasal cartilage.'[7]

David Willis also noted swellings to the left cheek and left side of the chin. With regard to the spillages of blood that are visible all around the top of the head, both on the front-of-the-body and the back-of-the-body imprints, Willis made these clinical observations:

On the back of the head, whose summit is not visible, there are dark irregular markings on the black-and-white photos. They could only be flows of blood and at least eight independent streams can be counted – some of which have divided. Only one has flowed almost vertically, at least seven have veered to the left and three to the right. They have been caused by independent puncture wounds of the scalp, which bleeds freely when injured, and they tend to expand as they descend. They have been halted on the nape of the neck along a line convex downwards which, assuming them to have been caused by something like a cap of thorns, would seem to be at the level where the thorn-branches had been secured to the back of the head. The different directions of the flows would suggest a tilting of the head at various times during the wearing of the spiky cap.

Turning to the front [pl. 9a], there are similar puncture wounds with their counter-drawings of blood flows but not so numerous as on the back. There are four or five that start from the top of the forehead moving down towards the eyes and the remainder are tangled in the masses of hair framing the face.

The most striking of these flows is one in the shape of a reversed three and repays detailed study, so true to life is it. It starts just below the hairline to the left of the midline from a single wound; the flow then moves down to the medial part of the arch above the left eye following a meandering course obliquely and outwards. As the stream descends it broadens and alters course twice, finally building up and spreading out horizontally to the medial line. Immediately below but separate is a 'tear' of blood close to the eyebrow, which is presumably part of the same flow, or possibly from an independent wound. The reason for the meandering course of this vivid mark indicates that it met some obstruction in its downward course, and most likely this was due to the reflex contraction of the muscles of the brow from the pain of the wounds, furrowing the surface.[8]

As noted by Dr Willis, it is actually quite impossible to talk sensibly about this particular set of bloodstains except in the context of a crown – or, as it seems most likely to have been, a cap of thorns.

Another group of visible injuries consists of around a hundred dumb-bell-shaped marks (pl. 9c) distributed all over the body, but particularly numerous on the back. As Dr Bucklin has described, 'These range from the top of the shoulders down to areas of the calf. They consist of double puncture-type wounds which appear to go in a direction from lateral to medial and downward. They have obviously been made by some implement with sharp edges, and the implement was applied to the skin in a flicking fashion in such a way as to pull out bits of skin.'[9]

From these markings' distribution, falling as they do on the front and back of the trunk and legs, even lay observation tells us that these were welts or contusions inflicted by a whip (pl. 9d). The tips of its lashes seem to have been fitted with double-pellet reinforcements, and we can see it to have been wielded first one way, then another, over the shoulders, then down to the calves, the movements of the hand behind it almost palpable.

Although any actual blood associated with these welts was not apparent to me during the 1973 viewing, Barbet noticed light flecks when he saw the Shroud on the cathedral steps. Los Angeles photographer Barrie Schwortz observed the same during his 1978 photography session with the STURP team, and during my daylight viewing in 2000 there was no mistaking their existence. They can be seen very clearly in some of the close-ups in

David Crute's high-definition filming of the Shroud in January 2008.

Somewhat unavoidably following the recorded treatment of Jesus, the next group of injuries consists of localized chafing to these whip-marks. As noted by David Willis,

> If we examine the back-of-the-body image and study the part between the large scorch marks and repairs just below the top of the shoulders, there is a quadrangular shading measuring four inches by three and a half inches over the right shoulder. Further down on the left side there is another area of excessive shading in the scapular region. This is rounded with a diameter of about five inches. These two areas represent broad excoriated wounds superimposed on the wounds from the whipping, which can be seen through them and have been widened and altered in form and perhaps in some cases obliterated compared with the marks alongside. These wounds could well have originated from the friction of some heavy object rubbing on an already damaged area of skin.[10]

The lay observer will need little prompting to associate these wounds with the carrying of a heavy beam across the shoulders.

The next group of injuries commanding attention is inevitably those indicative of crucifixion proper. The most illuminating of these is the set of blood flows coming from a wound in the left wrist. One of this injury's most important aspects is the angle of the two streams of blood closest to the hand, which flow on towards the inner border of the forearm. Other, interrupted streams run along the length of the arm as far as the elbow, dripping towards the edge of the arm at angles similar to the original flows. The first two flows are about ten degrees apart, the somewhat thinner one at an angle of about fifty-five degrees from the axis of the arm, and the broader one closer to the hand at about sixty-five degrees.

This again enables even the layman to make two key deductions. First, that at the time the blood flowed the arms must have been raised at positions varying between fifty-five and sixty-five degrees from the vertical – i.e. clearly a crucifixion position. Secondly, because of the ten-degree difference, the victim must have assumed slightly different positions on the cross. The position at sixty-five degrees represented full suspension of the body. The position at fifty-five degrees denoted a slightly more acute angle of the forearm produced by flexing the elbow to raise the body. The mode of suspension therefore self-evidently forced the victim into an up-and-down or seesaw motion on the cross (fig. 4). He may have had to do this in order to breathe, because the arms in such a position have to take a tension equal to nearly twice the weight of the body. Or he may have been trying to relieve one source of unbearable agony, the pain in the wrists, by pushing upwards on another source of pain, that in the feet.

Either way, the very thought is harrowing, and it is only intensified by the detailed studies that have been made by medical specialists of the actual wrist-wound area itself (pl. 10a). Although most lay people are usually surprised that the crucifixion nail should have been driven through the wrist rather than through the palm of the hand (as most artists have traditionally imagined), one very early student of the Shroud had observed it as far back as the late sixteenth century,[11] and Delage certainly noted it.

The first medical specialist to subject it to intensive study was Dr Pierre Barbet. Because of his work at a teaching hospital Barbet had cadavers and amputated limbs available to him for experimental purposes, and he did not hold back from using these for his research on the

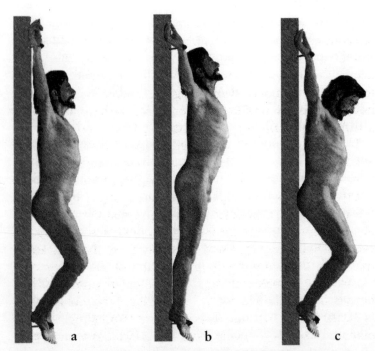

Fig. 4 Up and down motion of crucifixion victim. (a) With the arms at the crucifixion angle, each arm takes a weight equivalent to nearly twice that of the body, putting a massive strain on the chest muscles. (b) In order to breathe, the crucifixion victim has to press upwards on his feet, which, because they are nailed, causes excruciating pain. (c) In order to relieve this agony, the victim slumps again. Accompanied by profuse sweating, this torture continues until death intervenes.

Shroud. Using Enrie's photographs to calculate exactly where the nail had made its exit (for it is the backs of the hands that are seen on the Shroud), Barbet identified a space in the wrist bones that he called Destot's Space[12] through which a nail could easily pass (pl. 10b). When he suspended a cadaver from a nail driven through the wrist at this point, the body was easily held by the nail; but if he nailed the cadaver through the palms, the lack of any intervening bones to check its progress caused the flesh to tear through. Distressing in the extreme though Barbet's demonstration was, it graphically demonstrated how totally and uniquely convincing the Shroud is to show how crucifixion was actually practised rather than how artists traditionally imagined it during later centuries.

Inevitably related to the blood flows seen at the wrists are those at the feet. Because, as already observed, the front-of-the-body half of the Shroud only partially covered the feet, there is no complete front view of these and the full extent of the bloodstains in this region are therefore missing. In the case of the back-of-the-body imprint, however, Dr Bucklin made these observations of the foot area (pl. 10c): 'In examining the photograph of the right foot, we are able to make out an almost complete imprint. The border is slightly blurred in its middle part, but it still presents a very definite concavity corresponding to the plantar arch. More to the front the imprint is wider, and we can distinguish the imprint of five toes. The print is that which one might leave as he stepped on the flagstone with a wet foot.[13]

Unlike the hands, the actual point where the theoretical nail penetrated the feet is ill-defined. There is a darker patch between what would be the metatarsal bones which Barbet, Bucklin and Willis assumed to be the point of the nail, but this can only be conjectural rather than precise. A spillage of blood directly on to the cloth that can be seen in the region of the ankle might indicate ankle-nailing, for which, as we shall see, there is some archaeological evidence. But all this is a matter of fine detail.

Concluding the Shroud's catalogue of injuries is a significant flow of blood from an elliptic area measuring one and three-quarter inches long by seven-sixteenths of an inch high which can be seen on the right chest of the

Shroud body (pl. 11a), immediately next to one of the triangular-shaped holes from the 1532 fire. David Willis's clinical summary of this wound is again representative of a consensus of medical interpretation of this injury:

> It is generally agreed that this wound is in the space between the right fifth and sixth ribs ... The lower and inner extremity of the wound is at a level of about two-fifths of an inch below the tip of the sternum or breastbone, and just under two and a half inches from the midline. The blood flow from the wound spreads downward in an undulating and narrowing fashion for at least six inches and its inner edge is curiously cut about by some rounded indentations. It does not spread out in a homogeneous manner but is broken up by some clear areas, which are thought to indicate the mixture of a clear fluid with the blood. The indentations on the inner margin are probably due to the serrated muscular protrusion on the middle ribs, which correspond with these regular notchings. The body must have been erect when this chest wound was inflicted.[14]

Because laymen associate the heart with the left side of the body, medical specialists have been at pains to explain why this injury appears in the Shroud occupant's right chest. The plain fact is that the right side is the only side of the body from which blood could and would have spilled out if it had been penetrated by a bladed weapon after death. This is because the heart empties on that side. Had the piercing been in the left side, the left ventricles would already have been empty, resulting in little or no emission. The clear fluid that Dr Willis noted was identified by him and by most medical specialists as pericardial fluid which would have accumulated copiously in the chest from the percussive injuries sustained during the severe whipping.

What is seen on the front-of the-body image is not, however, the full extent of the blood that appears to have spilled from this same 'ultimate' wound. Dr Willis's summary continued:

> Turning to the back-of-the-body image – again at the level of the repair patches of 1534[15] – two meandering flows of blood extend horizontally across the loins but do not transgress the borders of the body image, except possibly to a very small extent on the left side, where it merges into the dark scorch marks. This blood has clearly come from the right chest, and, again, it appears to be mixed with a clear fluid. It would seem that this second flow came from the wound in the side after the body was taken from the cross and placed horizontally in the cloth or in the tomb.[16]

Even for the layman the most striking feature about this particular set of blood flows (pl. 11e) is its total veracity as a real-life spillage, as if from a quantity of fresh blood and pericardial fluid trickling out from the chest-wound when the body was tilted momentarily in the course of its being laid in the Shroud. Any comparison with how artists, both medieval and Renaissance, depicted wounds and blood flows only demonstrates the altogether greater naturalism evident in these stains on the Shroud. A 'Wounds Man' painted into a medical manuscript dating from circa 1500, and preserved in the Wellcome Institute for the History of Medicine, London, readily illustrates this.[17] Also worth noting is that people of the Middle Ages and later mostly failed to understand this particular set of bloodstains, interpreting them as injuries from a chain used to fasten Jesus.[18]

Meriting emphasis at this point is the fact that everything mentioned thus far concerns the convincing appearance of the wounds from a medical point of view, not their actual chemical composition, which will be considered in a later chapter. But even limiting ourselves to the wounds' appearance, the observations are far from exhausted.

As has long been recognized by medical specialists, when blood is spilled its cells thicken and concentrate, which on a linen surface causes them to shrink inwards within the bloodstain, leaving a rim of yellowish serum. Although some of the STURP scientists who worked on the Shroud in 1978 reported seeing such yellow-coloured rims, I was puzzled because nothing of this kind had been apparent during the 1973 showing. Then quite different daylight conditions at the March 2000 showing changed everything. A yellowish ring that had evidently been drowned out by the intense colour television lamps used in 1973 could clearly be seen around several of the major bloodstains. The same feature is also readily apparent in David Crute's high-definition close-ups of 2008. Dr Fred Zugibe,[19] a veteran medical examiner from New York, accompanied me to the 2000 exposition and was able to confirm with due authority that what we were seeing together was indeed blood serum.

Overall, then, some very well-qualified medical specialists find the Shroud totally convincing – at least to all outward appearances – as a cloth that wrapped the genuine, real-life body of a man who underwent crucifixion. To confound certain unqualified theorists who have ventured claims to the contrary,[20] the same specialists are equally confident that this man was dead. Barbet, Bucklin, Willis and Zugibe have all insisted on this, citing among other factors some sure signs of rigor mortis. To add to this list, Italian pathologist Professor Pierluigi Baima Bollone, head of forensic medicine at the University of Turin, who worked directly on the Shroud alongside the

STURP team during the 1978 examination, has stated that

> Forensic examination of the wounds and bloodstains on the Shroud allows us to affirm beyond all reasonable doubt that . . . we are dealing with the dead body of a man that was whipped, wounded on the head by a pointed instrument and nailed at the extremities before dying. While these injuries were indubitably inflicted on a living body, the extensive and complete muscular rigidity, the characteristics of the chest wound that are incompatible with survival, and the separated blood that issued forth, prove that the anatomical images are those of a dead person.[21]

This should be the point in the chapter at which to introduce any medical practitioners who have challenged the findings of those cited here in favour of the Shroud's authenticity. However, the fact is that with regard to medical findings, while one practitioner may disagree with another on some finer points of detail, for this particular branch of Shroud studies there is simply no opposition lobby. Undeniably, the radiocarbon dating findings in 1988 checked the flow of fresh medical specialists entering the subject. Also, very sadly, individuals of the stature of Willis and Bucklin have died, likewise Professor Taffy Cameron, head of forensic medicine at the London Hospital Medical School, who took a close interest in the subject for two decades.

Nonetheless, still active in the subject today are the aforementioned Dr Fred Zugibe and Professor Pierluigi Baima Bollone; World Health Organization consultant Dr Gilbert Lavoie of Boston, Massachusetts; obstetrician Dr August Accetta of Huntington Beach, southern California; leading plastic surgeon José Humberto Resende of Rio de Janeiro, Brazil; general physician Dr

Niels Svensson of Maribo, Denmark (pl. 9b);[22] and a number of others. It says a great deal for the strength of the medical evidence cited in this chapter that over a century since Yves Delage delivered his historic lecture at the Sorbonne not a single medical specialist has come forward with evidence that seriously refutes the findings stated. Confronted with such medical unanimity, even some of the theorists arguing that the Shroud is a fake have been obliged to accept that a genuine crucified human body must somehow have been involved in the forgery process.[23]

So if all the indications are there that the Shroud genuinely wrapped a crucified dead body, just how well does everything we see on it correspond to the last hours of Jesus's life as described in the Christian gospels?

Chapter 4

Window on the Passion

As to the identity of the body whose image is seen on the Shroud, no question is possible. The five wounds, the cruel flagellation, the punctures encircling the head, can still be clearly distinguished ... If this is not the impression of Christ, it was designed as the counterfeit of that impression. In no other person since the world began could these details be verified.[1]

Revd Herbert Thurston, Shroud sceptic, 1903

FOR THE HISTORIAN, the Christian gospels can often be very frustrating because of how sparing they are in terms of providing what we, with our twenty-first-century minds, would regard as essential background information. On matters such as Jesus's physical appearance, his family background, his crucifixion and his burial, there are just so many salient details that are downright lacking in the gospel texts. One of the great fascinations of the Shroud, therefore, is the potential it has to open up an unexpected window on at least some of these matters. But only, of course, if the data it reveals can actually check out satisfactorily with the gospels themselves, and with the many insights into New Testament times that have come from archaeology and other disciplines in recent decades.

As noted in the last chapter, the body that is visible on the Shroud exhibits some very realistic injuries which certainly suggest correspondence with those of the Jesus of the gospels. We have already established that the wounds themselves are medically convincing, indeed well beyond anything we might expect from the hand of a medieval forger. So our purpose now is to consider these same physical injuries in relation to the gospel texts and to any other available historical data.

The first set of injuries, we may recall, was the damage inflicted on the face, the most evident of them the heavily swollen right eye and its surrounds, though some damage was also observed to the nose, the left cheek and the chin. In this instance the injuries are rather too general to determine the implement or implements involved, which may well mostly have been fists. All that can be said is that they are in harmony with the gospel accounts of Jesus being struck repeatedly on the face. This battering was first sustained at the hands of the High Priest's men in the house of Jerusalem's High Priest: 'The men who guarded Jesus were mocking and beating him. They blindfolded him and questioned him, saying, "Prophesy! Who hit you then?" '[2] This was followed by further maltreatment at the hands of Roman governor Pontius Pilate's soldiers in an 'inner part' of the Roman military headquarters, or Praetorium: 'They struck him on the head with a reed, and spat on him.'[3]

The next group of injuries visible on the Shroud comprises the blood flows all around the head, back and front, as if something irregular and spiked had caused the scalp to bleed in many places. It is almost impossible to talk about these blood flows except in the context of a crown or cap of thorns of the kind reported during Jesus's maltreatment in the Roman Praetorium: 'They dressed him up in purple, twisted some thorns into a crown and put it on him. And they began saluting him, "Hail, king of the Jews!" '[4]

Worthy of emphasis here is that crowning or capping with thorns was never a normal part of Roman punishment procedure. Nor was it in any other culture throughout human history. On the occasion of which the Christian gospels write it seems to have arisen from some unknown Roman soldier's whim, perhaps after sighting some thorn-branches gathered for firewood,[5] which prompted him on the spur of the moment to add to his prisoner's pain and humiliation. And because Jesus is the only individual in known human history ever to have received such a punishment, to see this very injury on the Shroud – and to see it so graphically – renders highly unlikely the idea that the cloth might have belonged to some other unknown victim of crucifixion. The Shroud has either been deliberately faked as the shroud of Jesus or it is the genuine article. There is no viable option in between.

Next we turn to the Shroud's extensive group of injuries from a whipping with something hard and dumb-bell-shaped affixed to the tips of the whip's lashes. With the economy of detail that we have already noted, the Christian gospels say no more than that Jesus was scourged. Typical is the John gospel: 'Pilate then had Jesus taken away and scourged.'[6] Matthew has, simply, 'And having Jesus scourged, he handed him over to be crucified.'[7]

However, independently of the gospels, Roman-era historians and writers tell us something of just how horrifying scourging was as a punishment. Philo of Alexandria, a direct contemporary of Jesus, had this to say about how his local Roman governor scourged his fellow Jews: 'He ordered them all to be stripped and lacerated with scourges which are commonly used for the degradation of the vilest malefactors. In consequence of the flogging some had to be carried out on stretchers and died at once, while others lay sick for a long time despairing of recovery.'[8] As noted by Philo and other writers, victims were stripped naked for the scourging, and such nudity, particularly abhorrent to Jews, would certainly seem to be indicated on the Shroud.

But why could scourging be so lethal? Consultation with any dictionary of Greek and Roman antiquities gives us the answer. Look up the word *flagrum*, the Latin equivalent of our English word 'scourge', and you can see just what a fearsome implement the Roman scourge was (pl. 9d). It took various forms, but the common denominator was the fitting of the lashes with something extra to increase flesh damage. Common among these extras were *tali* (small sheep bones) and *plumbatae* (twin pellets of lead) – exactly as we see on the Shroud.

How many lashes were inflicted during a Roman scourging seems to have varied depending on the gravity of the crime and whether or not the victim was a Roman citizen. Something around a hundred individual dumb-bell percussions can be counted on the Shroud. And as the dumb-bell shapes tend to fall in pairs, the *flagrum* appears to have been fitted with at least two lashes, exactly as indicated in a Roman coin showing the *flagrum* being used in a gladiatorial contest (pl. 9e). St Paul, who was a Roman citizen, mentioned, 'Five times I have been given the thirty-nine lashes by the Jews.'[9] But as Roman law limited only the punishments inflicted by Jews, we have no firm guide to how many lashes Pilate may have ordered for Jesus – likewise, therefore, how many we might expect on the Shroud.

The next group of injuries noted on the Shroud consisted of the chafing on the shoulders that abraded some of the scourge wounds, as if from the carrying of a heavy beam. In this particular instance the gospel accounts are rather more informative than in the case of the scourging. According to the gospel of John, 'carrying his own cross, he [Jesus] went out to the Place of the Skull [where he would be crucified]'.[10] But we also know that Jesus did not carry this burden the whole distance, for according to Luke (and indeed the other two synoptics), 'As they were leading him away they seized on a man, Simon from Cyrene, who was coming in from the country, and made him shoulder the cross and carry it behind Jesus.'[11]

We should not be at all surprised at the Romans enlisting the help of passer-by Simon of Cyrene. Besides prescribing scourging as an automatic preliminary to crucifixion, Roman practice is known to have insisted that the crucifixion victim carry his *patibulum*, or crossbeam, through the streets to the execution site. This *patibulum* was not the full cross, because in normal Roman practice the upright, probably just a denuded tree-trunk, stayed permanently in position, ready for each fresh victim's crossbeam to be affixed to it. Even so, in the wake of the scourging the carrying of even a single large beam of wood would understandably have been a near impossible ordeal. The chafing visible on the Shroud is entirely consistent with what we may construe to have been Jesus's initial struggles to carry such a crossbeam on shoulders that would already have been horrendously raw and painful.

We now come to the prime evidence for the crucifixion itself as visible on the Shroud imprints – that is, the blood flows from the apparent nail-wounds in the wrists and feet. It may well be a surprise to many to learn just how scant are the physical details of Jesus's crucifixion in the gospels: 'when they had finished crucifying him' (Matthew) / 'Then they crucified him and shared out his clothing' (Mark) / 'they crucified him and the two criminals' (Luke) / 'they crucified him with two others, one on either side, Jesus being in the middle' (John). Note that there is not a single word to tell us whether nails were used in the procedure, let alone whether they went through the wrists or the palms. The only explicit information that nails were used at all is conveyed indirectly in disciple Doubting Thomas's famous remark, 'Unless I see the holes that the nails made in his hands,[12] and unless I can put my hand into his side, I refuse to believe.'[13]

Now for some independent historical guidance to what a Roman crucifixion involved. Roman historians and commentators have reported certain occasions of rebellion and revolt during which literally thousands were crucified at a time. For instance, as part of the putting down of the 'revolt of the slaves' in 73 BC, six thousand of Spartacus's fellow rebels were crucified at intervals along the Appian Way from Capua to Rome. In 4 BC two thousand Jewish rioters were crucified following disturbances that broke out after the death of Herod the Great. Up to five hundred rebels a day were crucified during the Jewish Revolt of AD 66–73. Yet even Roman authors were surprisingly sparing in their provision of practical details concerning how the procedure was actually carried out – except to convey how horrifying it was to watch. Cicero called it 'the most cruel and atrocious of punishments'. The Jewish historian Josephus, who lived just one generation after Jesus, told of an episode in the Jewish Revolt during which the Romans were besieging a Jewish-held citadel. Having captured a popular local hero, they prepared to crucify him in full view of the citadel's defenders. The defenders promptly surrendered, rather than have to see their hero die what Josephus called this 'most pitiable of deaths'.[14]

One of the very few available cross-references to what is seen on the Shroud is the only known skeleton of an individual who died from crucifixion, found in 1968 in an ossuary in an ancient cemetery just outside Jerusalem. Through the ankle bones was a large nail, still in situ – the key evidence that this was a victim of crucifixion. The ossuary's inscription bore the name Jehohanan, a young man who lived within a century of the time of Jesus. Unfortunately, Israeli anatomist Dr Nicu Haas's initial examination of Jehohanan's bones[15] was rather rushed, because Jerusalem's ultra-Jewish lobby always insists on the quick re-interment of any excavated bones. It also came in for much subsequent criticism and modification which Haas was never able to respond to because of his premature death in 1987.

But one of Haas's most interesting observations was an indentation on the radius or forearm bone very close to the point where it joins the wrist bones. This, he suggested, had been caused by the chafing of a crucifixion nail, hammered into the wrist slightly higher up than indicated on the Shroud, but none the less supporting the Shroud's otherwise unique indication that crucifixion victims were nailed by their wrists. The necessary qualification to this information, however, is that following Haas's death Israeli archaeologist Joseph Zias and medical specialist Eliezer Sekeles[16] noted similar indentations on other bones unconnected with any nailing. They therefore judged Haas's interpretation to be 'not convincing'.

But Zias and Sekeles have been rather more supportive of the Shroud's authenticity than they may have realized. Despite the nail remaining in situ through the ankle bones, Haas himself was far from confident about how Jehohanan had been nailed at the feet. After suggesting a couple of interpretations, ultimately he favoured a single nail driven through the two ankles positioned side by side as the body hung in some kind of side-saddle position. Here Zias and Sekeles, from their own fresh study, convincingly established that the nail-transfixed bones came from only one ankle, not two. As they demonstrated, the four-and-a-half-inch-long nail was simply not long enough to go through both ankles, then anchor them securely to an upright. According to their revised interpretation, initially Jehohanan's feet most likely dangled on either side of the upright while he was being lifted up by his executioners, with his hands already nailed to the cross-beam. His feet were then fastened to the upright by nails that were hammered through each ankle sideways on, effectively forcing him to straddle it.

Now, as we noted in the last chapter, exactly where the nails were driven in on the Shroud man's feet is not entirely clear. Most medical specialists favour the middle metatarsal area, and assume the left foot was laid over the right, with a single nail going through both insteps, much as can be seen on traditional crucifixes. What has rather too often been ignored, however, is the very distinctive rill of blood that can be seen to have spilled directly on to the foot region of the Shroud's back-of-the-body half. This rill presumably spilled on to the cloth at the time the body was laid in it, just like the spillage from the chest wound. From its position, it looks to have come from the soft area just above the heel and behind the ankle – i.e. the foot equivalent to the wrist. In which case it corresponds perfectly with Zias and Sekeles' revised findings in respect of Jehohanan's ankle nailing.

We now turn to the Shroud man's 'ultimate' injury, the bloody wound in his right chest (pl. 11a). As noted earlier, the exit point for this wound has an elliptic shape to its top edge, as if it was inflicted by a bladed weapon of this shape. Of the four gospels, it is only John that mentions this particular chest injury: 'When they [Jesus's executioners] came to Jesus, they saw he was already dead, and so instead of breaking his legs one of the soldiers pierced his side with a lance; and immediately there came out blood and water.'[17] It is important for us to remind ourselves that the gospel of John, like the rest of the New Testament, was composed in Greek. The Greek word the author used for 'lance' in this passage was *lonche*; the Latin equivalent was *lancea*. Archaeologically we know quite a lot about the *lancea*. With a long leaf-like blade thickening and rounding off towards the shaft, it was just the kind of general-purpose weapon that would have been standard issue for the small contingent of auxiliaries who took direct charge of Jesus's crucifixion (pl. 11b). In the Landesmuseum in Zurich there are several good examples with essentially exactly the same elliptic breadth to the blade that we can see on the Shroud.

The other feature of the chest wound, it may be remembered, was its location on the right side, where we might have expected the soldier to have aimed his lance at Jesus's left side in order to make sure the blow was a fatal one, straight to the heart. Yet again, however, we find that what we may not expect turns out to be correct after all. As part of their drill training Roman legionaries were specifically taught to aim their lances *sub alas*, 'below the armpit', just as indicated on the Shroud. They were also taught to aim at the right side of their opponent's body, because in any combat with a right-handed enemy carrying a shield it was the logical target. In Rome's Capitoline Museum there is a famous marble statue, the Dying Gaul, dated *circa* 220 BC (PL. 11C), in which the wound from which the Gaul is dying can be seen to be not only in the same right side as on the Shroud, but at the same point, immediately below the chest line (pl. 11d).

So, if the Shroud is a fake, the forger must have been an accomplished student of archaeology as well as of photography and medical science. Rather impressive for someone who, at least according to the carbon dating, lived between 1260 and 1390. Which leads us now to consider whether what we see on the Shroud is just as credible for its conformity to the gospel descriptions of Jesus's burial as it is in terms of what is known from archaeological and historical sources about how a Jew was buried in the New Testament era.

Here again we must confront how sparing, and sometimes downright confusing, the gospels can be with their information, quite irrespective of any Shroud considerations. The gospels were, remember, written in Greek, so variations in their translation into modern languages compound this confusion. If there were only the three synoptic writers, Matthew, Mark and Luke, all might be relatively straightforward. All three refer to the wealthy councillor Joseph of Arimathea obtaining Pontius Pilate's permission to take charge of Jesus's dead body, then procuring a clean *sindon* for the body's burial. Carrying the general meaning of a large linen cloth rather than the specifically sepulchral purpose conveyed by the English word 'shroud', *sindon* is readily equatable with Turin's Shroud, *sindone* notably being the Italian word that is used to describe it to this day. The big confusion arises with the John gospel. It is far more informative about Jesus's burial than the other three, but the word *sindon* is never used. Nor is Joseph's purchase of the cloth described. Instead John reports Joseph and companion Nicodemus 'binding' Jesus's body in *othonia* – a rather vague plural term seeming to convey a general assemblage of cloths.

John muddies the waters further with his apparently eyewitness reporting – for there is reasonable likelihood that he was the 'other disciple' present with Peter – of what he saw at Jesus's tomb on the first Easter Sunday morning. As the 'other disciple' he apparently arrived at the tomb first, peered in, saw the *othonia* but held back from going directly in. Peter was the first actually to enter the tomb. He too saw the *othonia*, but 'also the *soudarion* which had been over his head', which was 'not with the *othonia* but rolled up[18] in a place by itself'.[19] John/the 'other disciple' then followed Peter in, and 'saw and believed'.[20]

Now, whatever the pair 'saw and believed' in that particular moment, it must have been something hugely compelling. John very pointedly adds that until that moment 'they had still not understood the scripture, that he must rise from the dead'. This was, after all, the pivotal 'birth' moment of the entire religion we call Christianity. Peter became instantly transformed by it, losing all the fear he had exhibited thirty-six hours earlier, and going on to die a martyr's death for his new belief. And if our Turin Shroud is genuine, it has to have been one of those cloths right there in that rock-cut tomb in Jerusalem.

If so, which one was it? Nearly two thousand years on, and umpteen generations of scripture scholars later, there is no easy answer, again irrespective of any of the issues relating to the Turin Shroud. In the John gospel's earlier account of Jesus resurrecting Lazarus,[21] the author described Lazarus emerging from the tomb with his feet and hands bound with *keiriai* (which seem to be simple binding strips) and a *soudarion* over his face. In this context Lazarus's *soudarion* definitely seems not to have been a cloth of Turin Shroud-size proportions. It was more likely just a simple face cloth to screen the sight of death – which is how Jesus's *soudarion* has most commonly been interpreted.

But second thoughts are required. The root meaning of *soudarion* is 'sweat-cloth', from the Latin *sudor*, or sweat. Also, John's description of Jesus's *soudarion* having been 'rolled up' (or 'folded up') all by itself strongly smacks of something that was substantially larger than just a face cloth. And in the case of Jesus's as distinct from Lazarus's burial, John used a quite different word for how the cloth

was 'over' Jesus's head, *epi* rather than *peri*, the latter
having the alternative meaning of just 'around' the face.
All of which allows for (though certainly does not insist
upon) Jesus's *soudarion* having been a sweat-cloth that
was bound or wound over his head to cover his entire
body – precisely corresponding to the cloth we today
know as the Turin Shroud.

But why should Jesus have needed a sweat-cloth for his
entire body and not Lazarus? The answer lies in the fundamentally different circumstances of the two burials.
Lazarus died a natural death. In accordance with normal
Jewish practice he would have been washed, interred fully
dressed in his Sabbath best, tied up with a few binding
strips to keep his jaw and limbs suitably together, and provided with some kind of face cloth for screening purposes.
Jesus, in contrast, died a very bloody death, and stark
naked, his clothes having been removed from him at the
time of his crucifixion.[22] In his case Jewish law prescribed
something very different. As has been carefully explained
by Jewish-born Victor Tunkel[23] of the Faculty of Laws,
Queen Mary College, University of London, the belief
among the Pharisees of Jesus's time, shared by Jesus's own
followers, was that everyone's body would be physically
resurrected at the end of time. This meant that as far as
humanly possible everything that formed part of that
body, including particularly the life-blood, should be
buried with it.

As expressed in the Jewish Code of Laws,[24] 'One who
fell [e.g. in battle] and died instantly, if . . . blood flowed
from the wound, and there is apprehension that the blood
of the soul was absorbed in his clothes, he should not be
cleansed.'[25] In these circumstances, therefore, those
preparing the dead person for burial had to wrap a 'sheet
which is called a *sovev*' straight over any clothes, however
bloodstained. This *sovev* had to be an all-enveloping cloth,

that is a 'single sheet . . . used to go right round' the entire body. Such a *sovev* readily corresponds to the 'over the head' characteristics of Turin's Shroud. It also properly explains the John gospel's otherwise enigmatic description that Jesus's *soudarion* was a cloth that had been 'over his head' and then 'rolled up in a place by itself', as found at the time of Peter and John's arrival. We may thereby accept that what the synoptic writers called the *sindon* and what John called the *soudarion* was one and the same large linen cloth – a 'prime source' piece of confusion that, as we will discover, would echo down the centuries.

In summary, then, the Shroud makes very good sense in terms of its correspondence to the crucifixion and burial of Jesus as described in the gospels, actually explaining and illuminating aspects that would otherwise not be apparent. Perhaps one of the most unlikely of New Testament scholars to take an interest in the Shroud was the Anglican Bishop Dr John Robinson of Trinity College, Cambridge, who rattled many a vicarage teacup back in the 1960s with his controversial *Honest to God*. After making an exhaustive study of the Shroud, Robinson duly concluded, 'The more one went into it the more one realized there was so much about this thing that a forger would never really have thought of . . . In fact what we have fits extraordinarily well with the New Testament evidence. And from my point of view helps to make a great deal more sense than I saw before.'[26]

However, everything we have seen up to now has relied on the appearance of the Shroud's extraordinary images. It is now high time to take a look at how they fare when subjected to rather more direct and intrusive scientific scrutiny.

Chapter 5

Under the Microscope

*The chemical investigations are in complete agreement
with the image studies in concluding that the body images
are not composed of applied pigments, stains or dyes and
have been produced by a different process from that of the
blood marks.*[1]

Dr Alan Adler, Emeritus Professor of Chemistry,
Western Connecticut State University, USA, 2000

EVEN A HUNDRED years ago French anatomist Professor
Yves Delage had recognized that however convincing and
pigment-free the Shroud's body and blood imprints might
look to the naked eye, there could be no fully scientific
judgement without a proper 'hands-on' scientific examination involving microscopy and some direct analytical
approaches.

Delage, who died in 1920, would never see that ambition
realized. It was not until June 1969 that a small group of
Italian scientists made a first, very superficial direct examination, mainly as a preliminary check on the Shroud's
condition. Four years later, on 24 November 1973,
immediately after the Shroud's first-ever filming for colour

television, members of much the same group took some sample threads from which relatively few meaningful insights were gleaned.[2]

A far more exhaustive scientific examination – one for the first time involving scientists from across the Atlantic – followed five years later. This was the STURP project, in which Los Angeles photographer Barrie Schwortz so memorably participated.

It was a venture made possible almost entirely thanks to the patient diplomacy of Italian-born Fr Peter Rinaldi, parish priest of Corpus Christi, New York. In 1933 Rinaldi, already then resident in the United States, had acted as an interpreter during the showing of the Shroud – the occasion when Paris surgeon Dr Pierre Barbet had his revelatory close-up viewing on the cathedral steps. Rinaldi had similarly been allowed to view the Shroud in close-up just prior to its return to its casket, and the experience moved him so profoundly that it changed his life. Following the end of World War Two he increasingly took it upon himself to liaise between the Shroud's immediate clerical custodians in Turin, its legal owner, the exiled king of Italy Umberto II (living in Cascais, Portugal), and anyone else with a serious interest in the Shroud.

During the 1970s, prominent among Rinaldi's scientist protégés were Dr John Jackson and his close friend Dr Eric Jumper, the teachers at the US Air Force Academy in Colorado Springs who via the VP-8 Image Analyzer made the discovery of the Shroud image's three-dimensional properties. Having made good contacts with many specialists in image analysis technology around the USA, Jackson and Jumper formulated some impressive plans for a state-of-the-art scientific examination of the Shroud. They conveyed these to Rinaldi, and shortly before the public showings of the Shroud scheduled for 26 August to 8 October 1978 he was able to tell them that their programme had been approved. Their allotted time was the five days immediately following the end of the showings.

Calling themselves STURP, two dozen American scientists and technicians, together with eighty crates of equipment weighing eight tons, duly crossed the Atlantic to arrive at Turin's Royal Palace, where they had been allotted the magnificently frescoed Hall of the Visiting Princes for their work. Impressive as the hall might once have been for European princes, it had more than a few limitations for cutting-edge American scientists of the 1970s. Not least among its deficiencies was a completely inadequate power supply. None the less, by the time of the last Mass on the final day of the Shroud's public showings, all was at least broadly ready for the object of the exercise.

Turin's Cathedral of St John the Baptist and Royal Palace adjoin each other, linked by the then still intact Holy Shroud Chapel. At 10.45 p.m. that same Sunday night twelve young male Italians arrived at the hall carrying between them a sixteen-foot-long thick plywood board draped with a coverlet of plain red silk. When the silk was pulled back, the Shroud was revealed beneath, fastened to the plywood by some rather rusty-looking drawing-pins at two-foot intervals. Attendant Poor Clare nuns under the direction of the Shroud's then official 'carer' Monsignor José Cottino dextrously removed these. The Shroud, at that time surrounded on all sides by a nineteenth-century blue fabric frame, was then gently lifted off the plywood panel and transferred to the gleaming, purpose-built stainless-steel testing frame which the STURP team had brought over with them from the States. This had been specially designed to enable the Shroud to be studied and photographed both when lying flat and when tilted at a ninety-degree angle.

To the Americans' frustration, they could not immediately start their very full work programme. Instead they were obliged to watch while certain European scientists undertook their approved researches, some of them surprisingly invasive. At this time, remember, the 1532 fire patches were still in situ, and the Shroud was stitched to its sixteenth-century backing cloth. Early on in the proceedings Italian microscopy specialist Giovanni Riggi, a close friend of Turin's overall scientific coordinator Professor Luigi Gonella, directed the nuns to partly unstitch the Shroud from the backing cloth at one side so that he could slide between the two an endoscope camera of the kind normally used for medical examinations. The camera had a small light attached to it, the beam of which shone through the cloth whenever Riggi moved it to any point on the Shroud's underside. Positioned underneath an area of the Shroud's 'body' image, it showed up nothing particularly meaningful. It was a different matter when it was moved to directly beneath the distinctive forehead bloodstain in the shape of a reversed '3' – the one that had so entranced Dr David Willis. The stain immediately showed up very strongly, with a distinctively reddish colour.

This gave the watching Dr Eric Jumper the idea to have the entire Shroud lit from behind in a similar way – though necessarily shining the light through both the backing cloth and the Shroud proper. At Jumper's behest, photographer Barrie Schwortz quickly arranged some high-power photographic lamps as backlights. The experiment immediately showed up the shapes of the actual holes caused by the 1532 fire, otherwise still hidden by the triangular patches. Altogether more interesting, however, was what it revealed of the Shroud's body and blood imprints. In the case of the former, exactly as had been indicated by Riggi's endoscope light, these hardly showed up at all, which suggested they lacked any substance such as a pigment that might block the light. The bloodstains, in contrast, all now showed up dark and relatively solid against Schwortz's lamps, just as if the Shroud were some giant lantern slide. This indicated that they had some solid substance to them that blocked the light, which was therefore capable of being analysed.

But this experiment was merely an improvised add-on to the programme the STURP team had planned. They had three radiography technicians with them who very methodically X-rayed the Shroud in plate-size sections. Their most notable finding[3] was that neither the blood nor the body imprints showed up on their plates. This indicated that neither variety of imprint possessed sufficient atomic weight to interfere with their X-rays. As anyone familiar with X-rays of old master paintings will appreciate, solid lead-based pigments always show up very strongly. So if the Shroud is a medieval fake, the forger even anticipated his detection by X-ray.

Another technique commonly used in the examination of old masters is ultraviolet, or UV, photography. This is particularly useful for disclosing brushstrokes that may otherwise be invisible. In the case of the Shroud, however, nothing of this kind showed up.[4] Instead, its chief revelation was the serum rims surrounding the major blood flows. These serum rims, it may be remembered, I had failed to spot at the time of my examination in 1973. Under ultraviolet light, however, they fluoresce, and this is exactly what those on the bloodstains did when they were viewed under the STURP team's UV lighting.

A Wild portable photomicroscope was also part of the equipment the STURP team had brought with them, and with this they were able both to view and to photograph specially selected areas of the Shroud under magnifications up to 400×. Frustratingly, the movements of heavy vehicles in the outside street repeatedly disturbed this work. Although the Hall of the Visiting Princes is at first-floor level, the passing vehicles' vibrations would transmit through its wooden flooring, seriously impairing the microscope's ultra-fine focus. The hall was also subject to draughts wafting through the palace, so when the Shroud was in an upright mode on STURP's state-of-the-art viewing frame it acted like a sail, again seriously ruining focus.

Despite such on-site difficulties, the STURP team were able very clearly to see that the Shroud's blood imprints comprised a reddish-brown particulate material which had some definite semblance to crusty dried blood (pl. 12c). The body imprints, in contrast, lacked any evident substance, seeming to be no more than some inexplicable discoloration to the cloth's fibres (pl. 12a).

The best way to explore further exactly what constituted the body and blood imprints had to be for the STURP scientists to take away with them some samples for analysis in specialist laboratory facilities back in the States. Fortuitously, the European scientists who had made earlier studies in 1973 had set a useful precedent by taking several threads, and even two whole linen fragments. In the event, what the STURP team asked for, and were able to obtain, was a great deal more modest than the Europeans had gleaned. In expectation of their being allowed only the most minimal samples, they had brought with them a purpose-built, pressure-controlled torque applicator – in essence, a very high-tech variety of sticky-tape dispenser – for the extraction of whatever imprint materials they might be allowed. They had even persuaded the 3M Corporation to develop a specially formulated tape that would leave no glue residue when applied to the Shroud's surface. Thirty-two such sticky tapes were duly applied to the Shroud, several to blood imprint areas, several to body imprint areas, a few to water stain areas and some, as control, to areas perceived as lacking any kind of image.

With the benefit of hindsight, the STURP team were probably far too undemanding in their sample requirements. Their 'light touch' sticky tapes did little more than kiss the Shroud's surface. While they none the less picked up quite a quantity of microscopic materials from the Shroud's surface, this would cause a lot of confusion during the intense scientific scrutiny to which they were about to be subjected.

For back in the States the first port of call for the tapes, unfortunately, was Chicago, home of the laboratory of microanalyst Dr Walter McCrone, who in the 1970s had attracted much publicity for his work on Yale University's Vinland Map. By what seemed, at that time, some brilliant microanalysis, McCrone had shown the map to be a fake.[5] The expectation (not least on my part) was therefore that by some equally brilliant feat he might be able to show the Shroud to be authentic. It was not to be. On the basis of what he found on the sticky tapes McCrone forcefully and very publicly declared the Shroud's image to be just a conventionally created painting. According to him, its body imprint had been painted with a fine iron-oxide pigment in a gelatin protein binding medium. The bloodstains had been created in much the same way, simply with the addition of the medieval and Renaissance artist's pigment cinnabar or vermilion.[6]

To the members of the STURP team, who unlike McCrone had examined the Shroud directly on-site in Turin, McCrone's findings simply did not make sense. They sought a second opinion and were very relieved when Yale University chemist Dr John Heller and Jewish-born Dr Alan Adler, the latter a world authority on blood chemistry, produced some altogether different interpretations from exactly the same tapes.[7]

As Heller and Adler showed, McCrone had been correct in stating that the Shroud has iron oxide particles scattered across its surface, and they too found particles of vermilion and other artists' pigments. However, they were emphatic that neither of these sets of materials was responsible for the Shroud's body and blood images. The artists' pigments, for instance, they found to be random and quantitatively distributed no more strongly in the Shroud's image than in its non-image areas. Their presence was readily explained as mere strays left on the Shroud's surface from episodes during which it was in the vicinity of artists' materials. The scientists who worked in the Hall of the Visiting Princes recalled how paint particles would quite regularly drop on them from the ceiling frescoes. Probably far more pertinently, the sixteenth- and seventeenth-century artists who made the earlier-mentioned life-size copies of the Shroud were in the habit of pressing these against the original to give them something of its special holiness. One example preserved in Toledo, and by no means the only one, is specifically worded (my italics): 'This picture was made as closely as possible to the precious relic ... at Chambéry [i.e. the Turin Shroud] *and was laid upon it* in June 1568.'

After studying the tapes taken from the Shroud's body imprint areas, Heller and Adler concluded that there was no identifiable substance added to them that could be considered responsible for what the eye sees as the reverse shadow of the Shroud man's body. It was as if the fibres had simply degraded, or 'aged', at those places where this type of imprint appears. The process involved resembled that of the rapid yellowing of a newspaper when exposed to strong sunlight, except that in the Shroud's case the yellowing occurred selectively, relative to the theoretical body's distance from the cloth at any one point.

During the 1990s Adler was given the opportunity to

examine the Shroud at first hand. Reinforced by this experience, up to his sudden death in 2000 he consistently maintained that the body image areas are superficial in the extreme. They lie only on the very top of the Shroud threads. They do not penetrate the cloth, nor do they exhibit any capillarity or absorptive properties. They are more brittle than their non-image counterparts, as if whatever formed them lightly corroded them. They are uniform in coloration. They are not cemented together, neither are they 'diffused' as they would be if they derived from some dye or stain. They do not 'fluoresce' or reflect back any light. Most emphatically, they are not made by any pigment.

With regard to the Shroud's bloodstains, Heller and Adler's equally authoritative finding was that they derived from genuine wounds that had clotted some time before their transfer to the Shroud's surface.[8] They passed eleven different diagnostic tests that would allow them to be pronounced as true blood in any court of law. Their constituents included proteins, albumen, haem products, and the bile pigment bilirubin (on which Adler was an acknowledged world expert) – all components of true blood. One remarkable feature noted by Adler was that where blood occurred in the same region as body image, the cloth fibres lacked body image characteristics below the bloodstain. This suggested that the blood was on the cloth before the body image-making process began – hardly the way any artist-forger might be expected to work.

As for the blood's clear, thin, overly reddish colour, Adler explained that whenever someone is severely beaten, or otherwise suffers severe traumatic shock, the haemoglobin from the broken blood cells goes through the liver. The liver converts the haemoglobin into bile pigments such as bilirubin, which is yellow-orange in colour. When this becomes mixed with other blood products that have oxidized brown, the result is very credibly the unusually reddish colour that is still visible on the Shroud.

Although McCrone continued vigorously to dispute Heller and Adler's findings right up to his death in 2002, during the 1980s his ostensibly brilliant Vinland Map findings were seriously discredited by the newly developed technique of proton-induced X-ray analysis.[9] As a result Yale University reinstated their Vinland Map as very likely genuine. Even the scientists who carried out the Shroud radiocarbon dating, to their considerable credit, declined to take any interest in McCrone's claims about the image's formation.

Although the incidental debris on the Shroud's surface that was picked up by the sticky tapes caused such confusion, in the event it turned out to have its own independent diagnostic value. Rather ill-advisedly, the Turin microanalyst Giovanni Riggi had used a mini vacuum cleaner to obtain rather more of this dust than he should have done – a procedure that was undoubtedly far too indiscriminate. However, when this debris was subjected to microanalysis it most usefully revealed an intriguing time capsule of what was blowing around in the atmosphere at different times of the Shroud's history, one that even now remains far from fully evaluated.

Some of the analysed debris was found to be power-station fly ash, readily attributable to the general atmospheric pollution due to Turin being the main manufacturing centre for the Fiat motor car. There were also tiny particles of iron, bronze, silver and even gold. Some of these particles would have come from the caskets in which the Shroud is known to have been kept over the centuries. One example is the highly expensive and relatively new silver casket ruined beyond repair by the 1532 fire, another an iron chest in which it was temporarily housed after the fire. But there would have been several more whose composition we do not know. Among other metal debris should be mentioned a number of wire fragments which showed up in STURP's X-rays and have yet to be properly investigated and explained.

Fabric threads, fibres and particles represent another substantial constituent. Even on their kiss-the-surface sticky tapes Heller and Adler noted the presence of red silk, blue linen, plain cotton, plain wool and even pink nylon. The historical fact that the Shroud has had at least two red silk coverlets, one in the fifteenth century and another still in situ at the time of the 1978 examination, readily explains the silk. The blue material could have come from the now-removed nineteenth-century blue fabric frame that surrounded the Shroud at the time of the 1978 examination. With regard to the cotton, in 1978 alone no fewer than four different varieties of cotton gloves were worn by those handling the Shroud. Who knows where the pink nylon came from, but it is simple common sense that innumerable ecclesiastical vestments and other garments would have been brushed against the Shroud over the decades, some of these inevitably leaving behind small micro-traces of themselves.

Invisible in the air around us there are always pollen grains, and in the 1970s the discovery of many dozens of these on sticky tapes (other than STURP's) that had been applied to the Shroud was regarded as one of the most promising new areas of research. This discovery was single-handedly due to Swiss criminologist Dr Max Frei, a botanist by training, who had taken his own sticky samples from the Shroud's surface during the European preliminary examination of 1973, and more in 1978, much to the disdain of the Americans, who took serious exception to his 'dime store' plastic dispenser.

The principle behind Frei's method lay in the fact that pollen grains, which are extremely durable – capable of surviving for millions of years, in fact – differ one from another according to the type of plant from which they have come. By identifying the different pollen grains present on a piece of fabric one can build a picture of the vegetation of the areas through which that fabric has moved over time, thereby catching a glimpse of the countries in which it has been kept.

Frei's major breakthrough, of which the first hints began in 1976, was to identify the pollens of no fewer than fifty-eight different species of plant (fig. 5). Some of these plants were European, as was only to be expected from the Shroud's known movements in France and Italy since the mid-fourteenth century, but there were others that simply had to be Middle Eastern. Some of these were from Turkey, others from the environs of Israel, notably halo-phytes, desert plants that are specifically adapted to live in soils with the high salt content found almost exclusively around the Dead Sea. As Frei provisionally summarized his findings concerning the latter, 'These plants are of great diagnostic value for our geographical studies as identical desert plants are missing in all the other countries where the Shroud is believed to have been exposed to the open air. Consequently a forgery, produced somewhere in France during the Middle Ages in a country lacking these typical halophytes, could not contain such characteristic pollen grains from the desert regions of Palestine.'[10]

Sadly, Frei died suddenly in 1983 having neither com-pleted nor published his research. The chart of his pollen finds reproduced here is simply a posthumous assemblage by his friend the German scholar Dr Werner Bulst. And although Frei's collection of the sticky tapes that bear these pollen grains was purchased from his widow by an American Shroud enthusiast, psychiatrist Dr Alan Whanger of Durham, North Carolina, its subsequent history is not an edifying one.

Plant pollens corresponding to the Shroud's historically definite travels around western Europe and the western Mediterranean

1 **Varieties best known under their everyday names**
Anemone (Anemone coronaria L.)
Beech (Fagus silvatica)
Castor-oil plant (Ricinus communis L.)
Cedar of Lebanon (Cedrus libanotica Lk.)
Cypress (Cupressus sempervirens L.)
Hazel (Corylus avellana L.)
Juniper (Juniperus oxycedrus L.)
Laurel (Laurus nobilis L.)
Love-lies-bleeding (Amaranthus lividus L.)
Oriental plane tree (Platanus orientalis L.)
Pine (Pinus halepensis L.)
Pistacia (Pistacia halepensis L. & Pistacia lentiscus L.)
Rice (Oryza sativa L.)
Rock rose (Cistus creticus L.)
Rush (Scirpus triquetrus L.)
Rye (Secale spec.)
Spina Christi (Paliurus Spina-Christi Mill.)
Thistle (Carduus personata Jacq.)
Yew (Taxus baccata L.)

2 **Other varieties**
Alnus glutinosa Vill.
Capparis spec.
Carpinus betulus L.
Lythrum salicaria L.
Phyllirea angustifolia L.
Poterium spinosum L.
Ridolfia segetum moris

Plant pollens indicative of the Shroud at some time in its history having been in Near Eastern environs, particularly Anatolia and Israel

1 **Varieties found in desert and semi-desert terrain**
Acacia (Acacia albida Del.)
Artemisia herba-alba A.
Atraphaxis spinosa L.
Haplophyllum tuberculatum J.
Helianthemum versicarium B.
Oligomejus subulata Boiss.
Peganum harmala L.
Pteranthus dichotomus Forsk.
Scabiosa prolifera L.

2 **Varieties found in steppe type terrain, as in eastern Turkey and southern Israel**
Glaucium grandiflorum
Hyoscyamus reticulatus L.
Ixiolirion montanum Herb.
Linum mucronatum Bert.
Roemeria hybrida (L.) DC
Silene conoida L.

3 **Varieties particularly typical of Anatolian (Turkish) environs**
Epimedium pubigerum DC
Prunus spartiodes Spach.

4 **Varieties found in salty environs, particularly the Dead Sea**
Althea officinalis L.
Anabasis aphylla L.
Gundelia tournefortii L.
Haloxylon persicum Bg.
Prosopis farcta Macbr.
Reaumuria hirtella J. & Sp.
Suaeda aegyptiaca Zoh.
Tamarix nilotica Bunge

5 **Varieties particularly typical of the environs of Jerusalem**
Bassia muricata Asch.
Echinops glaberrimus DC
Fagonia mollis Del.
Hyoscyamus aureus L.
Onosma syriacum Labil.
Zygophyllum dumosum B.

Note: *All regionizations are approximations*

Fig. 5 PLANTS WHOSE POLLEN GRAINS HAVE BEEN FOUND AMONG THE SHROUD'S SURFACE DUST, as identified by Dr Max Frei. Dr Frei did not live to complete his research, and this list has been adapted from one published posthumously on Dr Frei's behalf by Professor Heinrich Pfeiffer.

To his great credit, Whanger attracted the interest of Dr Avinoam Danin, Professor of Botany at the Hebrew University, Jerusalem, a veteran of the Six Day War and foremost authority on the flora of Israel. But Whanger and Danin's newly shared enthusiasms became concentrated on identifying images of plants that they perceived on the Shroud, neglecting the Frei pollen collection. As Danin is not a specialist in pollen identification, he passed this collection to Dr Uri Baruch, a palynologist with the Israel Antiquities Authority whose work on it initially seemed very impressive. In 1999 the highly respected Missouri Botanical Garden Press published a combined study by Danin, Baruch and Whanger that included charts showing the precise quantities of each variety of pollen grain that Baruch found to be present on Max Frei's tapes.[11] According to this study a surprisingly high percentage of these grains were from *Gundelia tournefortii*, a species of thorn-bush with fearsome spikes which does not grow in Europe and is distinctly Middle Eastern in its distribution, extending from Turkey to Israel, Syria, northern Iraq and Iran. Baruch identified no fewer than ninety-one specimens of *Gundelia* on the Frei tapes.

But then some of the Frei tapes were passed to Dr Thomas Litt, a specialist in pollen analysis at the University of Bonn. Litt quickly noted that Baruch, unlike Frei, had not removed any of the pollen grains from the sticky tapes, instead making his identifications with the grains still in situ. Yet as Litt was aware, the only way to make a sure identification to the species level of, say, a *Gundelia tournefortii* was very painstakingly to extract it from the tape (which requires the most delicate manipulation), clean it, then make a direct comparison with an equivalent field-gathered specimen from a present-day plant. That Frei had done all this was quite evident from tiny incisions he had made in some of his tapes to extract the specimens he identified. He had also gone on special field trips to this end.[12] But Baruch had never had access to these specimens of Frei's – they had disappeared in the wake of his death twenty years earlier – and he had neglected to carry out the extremely time-consuming procedure himself. Bluntly, Baruch's work was scientifically unsafe, in which light Danin saw no option but to disown the publication of 1999 that bore his name alongside Baruch's. Today, Baruch is no longer employed by the Israel Antiquities Authority and has since left the field of pollen analysis to work in computer science.

The upshot of this is that Frei's work, although untrammelled by the problems affecting Baruch's, remains to this day in the uncompleted state in which Frei left it on his death in 1983. An offer by me to arrange a properly qualified specialist to make a fresh approach to Frei's collection was flatly refused by Whanger. While there remains good reason to believe the fundamental soundness of Frei's identifications, the argument for the presence of Middle Eastern pollen grains on the Shroud can no longer be advanced with quite the assurance it could thirty years ago.

However, even this by no means exhausts the varieties of extraneous debris that have been detected on the Shroud's surface. Included among the twenty-four-strong STURP expedition were Roger and Marty Gilbert, a husband-and-wife team whose speciality was reflectance spectroscopy, used to help identify materials, whether solids, liquids or even gases, from the wavelengths, or spectra, that they emit. As the duo ran their equipment up and down the Shroud's body and blood images the wavelengths they obtained proved mostly regular, until they reached the sole of the foot imprint on the back-of-the-body half of the cloth. Suddenly the spectra changed dramatically. Something in the foot area, and particularly around the heel, was giving a markedly stronger signal than elsewhere.

Clearly it was a task for the Wild portable microscope, and when optical physicist Sam Pellicori was called on to deploy this, his answer was as brief as it was revelatory: 'It's dirt!' As could only be expected of an individual who had had even his sandals taken away from him, apparently the man of the Shroud had dirty feet. At the time of my 1973 viewing of the Shroud, the upright mode of display meant that this top end part of the cloth was a full body length above my head, too far away to be studied in sufficient detail. But in 2000 not only was the Shroud displayed in landscape mode, the lighting conditions were clear daylight. And there, at the sole of the foot, even with the unaided eye, could be seen significantly more dirt than anywhere else on the cloth. Furthermore, the dirt convincingly underlay the serum-haloed bloodstains that otherwise coat the same soles. This can readily be seen in David Crute's high-definition film of 2008 (pl. 10c).

Are we looking at the very dirt from Jerusalem's streets that once adhered to the feet of Jesus two thousand years ago? Another related study, too often overlooked, certainly supports such a suggestion. In 1982 another member of the STURP team, Dr Ray Rogers, took some of the expedition's sticky-tape samples to his old friend Dr Joseph Kohlbeck, an optical crystallographer working for Hercules Aerospace in Utah. All Rogers wanted was for Kohlbeck to make some photomicrographs from the sticky tapes, Kohlbeck's facilities being particularly well equipped for such work.

Kohlbeck became keenly interested in some of the

calcium carbonate or limestone particles that he noticed among the materials on the tapes. Crystals of limestone have a 'signature' associated with where they come from, much like a spent bullet carries a signature of the gun barrel from which it has been fired. The rocks around Jerusalem are limestone, and Kohlbeck persuaded an archaeologist who had made a special study of its rock tombs to send him some samples. When he studied these, he quickly found they had some of the distinctive characteristics he had been hoping for. As he explained, 'This particular limestone was primarily travertine aragonite deposited from springs rather than the more common calcite . . . Aragonite is formed under a much narrower range of conditions than calcite. In addition to the aragonite, our Jerusalem samples also contained small quantities of iron and strontium, but no lead.'[13]

For the best possible specimen from the Shroud, Kohlbeck chose a sample from the very same area where Roger and Marty Gilbert's foot dirt had been found. This was selected because it included the largest available concentration of limestone. When Kohlbeck made his comparison, he found it to be of exactly the same rare aragonite variety as that sent from Jerusalem, and with the same trace amounts of iron and strontium.

To narrow the field further in the hope of that 'signature' match, Kohlbeck took both sets of samples to Dr Ricardo Levi-Setti of the University of Chicago's Enrico Fermi Institute. Among Levi-Setti's equipment was a then state-of-the-art scanning ion microprobe with which the wavelengths emitted by the two different samples could be readily compared. Studying together the spectra emitted by the two different samples, the two scientists saw there was an unusually close match (fig. 6). The only disparity was a slight organic variation explicable as minute pieces of flax embedded in the Shroud's limestone. While it could not be claimed as 100 per cent proof that the Shroud aragonite came from Jerusalem, when combined with the other scientific evidence assembled thus far it was tantalizingly close.

There remains one very crucial aspect of the Shroud to be addressed. That is, the fabric itself, its weave and its technical composition. Can this truly be compatible with what we should expect of a cloth fabricated some-where around first-century Jerusalem?

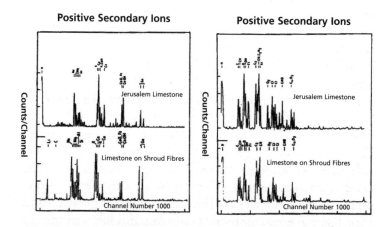

Fig. 6 SHROUD LIMESTONE AND JERUSALEM LIMESTONE COMPARED. The two graphs show the relative amounts of the chemical com-pounds found in thin sections of aragonite, the particular crystalline form of calcium carbonate present both on the Shroud fibres and in Jerusalem tombs dating from the first century. The peaks, which indicate the ions from a particular chemical compound, reveal such a strikingly similar pattern that crystallographer Joseph Kohlbeck has argued that there's a strong possibility the Shroud limestone is of Jerusalem provenance.

Chapter 6

The Cloth's Own Tale

It can be said that the linen cloth of the Shroud of Turin does not display any weaving or sewing techniques which speak against its origin as a high-quality product of the textile workers of the first century AD.[1]

Mechthild Flury-Lemberg, retired curator,
Abegg Foundation Textile Museum, Switzerland, 2001

ANY PIECE OF CLOTH, because manufactured by humans, carries certain intrinsic clues to how and when it was made, and the manner in which it has been kept over time. For fabrics from our own age of mechanized textile production, such clues are usually a matter of reading the label. 'Made in China' may well tell us all we need to know. But the Shroud carries no such label. Unquestionably it was woven by hand on a loom of a kind that existed before the invention of machines such as the Spinning Jenny. But because such weaving methods stayed much the same over thousands of years, extracting all the intrinsic information we might ideally want in order to determine where and when the Shroud cloth was woven involves some considerable difficulties.

Experts in the history of textiles are clearly required, and the Shroud has at least had no shortage of them. At the time of the 1933 showing, Italy's Virginio Timossi was able to examine it closely and wrote a report that was awarded a gold medal at the National Meeting of Textile Experts in Rome in 1938.[2] Forty years later, Gilbert Raes of the Ghent Institute of Textile Technology in Belgium was invited to join the panel of otherwise mainly Italian specialists who examined the Shroud between eight a.m. and four p.m. on 24 November 1973. Because Raes's flight from Paris was delayed he did not arrive until noon. He was nevertheless allowed to take back to his Ghent laboratory two samples, postage-stamp-size, cut from the Shroud's top left-hand corner (see fig. 13, p. 127), on which he subsequently wrote a detailed report.[3]

Five years later came the five-day, much more intensive STURP examination described in the last chapter, which from the viewpoint of textile analysis was definitely a missed opportunity. Among the two dozen in the American contingent, representing every conceivable variety of image analysis technique, not a single textile specialist was included.

In 1988, Turin's then archbishop, Cardinal Anastasio Ballestrero, invited Gabriel Vial, curator of the International Centre for the Study of Ancient Textiles (CIETA) at the Textile Museum in Lyon, France, to attend as an expert witness when the sample for carbon dating was taken on 21 April of that year. Vial, who used the opportunity to make a very careful examination of the Shroud's technical features, also wrote a report which remains one of the best available.[4]

At much the same time as Vial's involvement, the name of Dr Mechthild Flury-Lemberg of the Abegg Foundation, Switzerland's counterpart to the Lyon Museum, was being put forward as another specialist whose expertise could prove valuable, particularly from the viewpoint of the Shroud's conservation, a word that was becoming increasingly fashionable at that time. Accordingly, in the early 1990s the Turin authorities co-opted Dr Flury-Lemberg on to their conservation committee to consider how best to preserve the Shroud for future generations. This ultimately led to her being given overall responsibility for the extensive 'make-over' that was carried out in 2002, a task that undoubtedly gave her and her helper Irene Tomedi a much greater direct acquaintance with the Shroud than anyone else of equivalent expertise throughout the last four centuries.

From Vial's, Flury-Lemberg's and to a lesser extent the other specialists' published findings, certain reliable facts can be determined about the Shroud. For instance, even to the untrained eye the fabric can be seen to be linen. This means that somewhere, at some point of history, bunches of flax from the plant *Linum usitatissimum* were harvested and spun into thread to make the linen sheet we now see. As a plant, flax is native to a region from the eastern Mediterranean to India, and was used particularly extensively for linen clothing in ancient Egypt, but also throughout the ancient classical world, including Roman Palestine.

The specialists define the spin that was used as a 'Z' twist, meaning simply that whoever held the original spindle rotated it clockwise. This establishes the Shroud as definitely not of ancient Egyptian manufacture, since their cloth-makers spun their linen yarn 'S' twist, i.e. anti-clockwise. Vial estimated the Shroud's thread count at approximately eighty threads for the warp and forty for the weft, the fabric being not at all thick or coarse in texture but soft, supple and showing no signs of splitting.

Measurements of the Shroud's overall dimensions have tended to vary, even during relatively recent decades, because of its traditional storage mode, rolled up around a velvet-covered rod. Every time it was unrolled and rolled up again there was a slightly different calculation of its size. However, following the flat storage method and the stabilization achieved by Dr Flury-Lemberg's conservation measures of 2002, the dimensions should now stay constant at the official fourteen feet six inches long by three feet nine inches wide.[5] By any standards the Shroud is a cloth of substantial size, and the loom on which it was woven must have been relatively large – a requirement that certainly presents no problem for the Shroud dating from antiquity. For instance, in Turin's Egyptian Museum there is a bed-sheet from the 12th Dynasty – that is, from around the twentieth century BC – that is seven metres long and of much the same width as the Shroud. Twice as old as the Shroud (if it genuinely dates from the time of Christ), the bed-sheet's condition is notably still perfect.[6] It may seem surprising that linen can last so long, but moth grubs ignore the material because it lacks the keratin their diet needs, and most other insects find it too hard to masticate.

One genuinely big surprise from Dr Flury-Lemberg's findings within the last decade is that the original cloth from which the Shroud derived was very likely substantially larger. The clue to this lies in the beautifully crafted seam that runs the Shroud's full length, just under three and a half inches below its top edge. At the Shroud's top and bottom edges there is selvedge,[7] indicating that these edges formed the cloth's original, weaver-finished 'sides' as it originally grew lengthwise on the loom. When the weaving was finished someone seems to have very accurately cut the Shroud along its length, then most expertly joined up the two raw edges left by this process via the beautifully crafted seam. As argued by Dr Flury-Lemberg, the only possible purpose behind

such highly professional tailoring would have been that
when the original piece of fabric was on the loom, it was
woven substantially wider than its present width. In her
opinion, it could well have been up to three times the
present width, which would have made it eleven feet
wide by nearly fifteen feet long. The width was then
narrowed to that required for its usage as a shroud by
cutting twice along its length, using the two sections with
selvedge for the Shroud, then joining up the raw edges
(fig. 7). 'The ancient Egyptians used looms up to eleven
and a half feet wide,' Dr Flury-Lemberg explained. 'They
needed the wide looms particularly for the production of
the *tunica inconsutilis*, the seamless tunics . . . also worn
by Jesus Christ, according to the gospel writers[8] . . .
Naturally these wide looms could be used to produce
fabrics of a smaller width. But it made more sense to use
the full width of the loom, as only a little more work
input resulted in the production of a broad length of
fabric which could then be cut into two or three
narrower pieces as required.'[9]

Just as interesting are Dr Flury-Lemberg's observations
concerning the beautifully crafted seam by which the raw
edges were joined up. On the Shroud's exposed or 'face'
side – that is, the side that received the imprint – the seam's
sewing is so neat and so professional that analysis of its
technical details was never possible while the Shroud
remained fastened to its sixteenth-century backing cloth.
When the conservation work freed it from its backing
cloth and opened up its underside, Dr Flury-Lemberg
became the first person in modern times to be able to study
the seam's underside. And this proved quite a revelation.

She found the seam to have some highly unusual technical characteristics that in four decades of working on
historic textiles she had come across only once before, on
first-century textiles found at Masada, the historic Dead

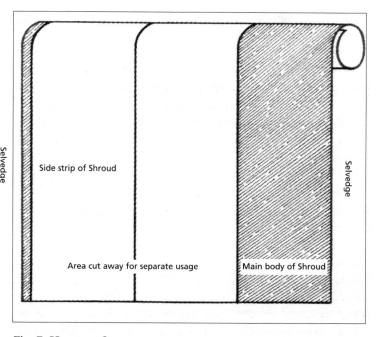

Selvedge

Side strip of Shroud

Selvedge

Area cut away for separate usage

Main body of Shroud

Selvedge

Fig. 7 HOW THE SHROUD WAS ORIGINALLY WOVEN MUCH WIDER THAN ITS PRESENT WIDTH. Reconstruction of the likely size of the bolt of cloth of which the two lengths of the Shroud (shaded) formed part. This wider cloth was very expertly cut lengthwise, then the raw (i.e. non-selvedge) edges of the shaded segments joined together by a very professional seam to form the Shroud we know today.

Sea fortress where over nine hundred Jewish rebels who lived just one generation later than Jesus made a famous last stand against the Romans at the end of the Jewish Revolt, in AD 72–3. When the site was excavated by the famous Israeli commander Yigael Yadin back in the 1950s, substantial scraps of the defenders' clothing came to light, the dry air of the surrounding Judaean desert having preserved them well. In 1994 Yadin's successors at last published the technical report on these clothing scraps, and right there in that report is a technical drawing of

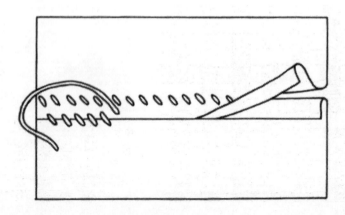

Fig. 8 VERY RARE TYPE OF 'INVISIBLE SEAM' as found on cloth fragment excavated at the first-century Jewish fortress of Masada, overlooking the Dead Sea. According to Dr Mechthild Flury-Lemberg, exactly this type of seam was used to join up the segments of the Shroud.

what the excavators adjudged to be a very unusual seam – one which in Dr Flury-Lemberg's opinion is essentially identical to the one visible on the Turin Shroud (fig. 8).[10] Also found at Masada were examples of exactly the same two double thread selvedge as seen on the Shroud, a mode of construction which Gabriel Vial back in the 1980s had described as 'tout à fait inhabituelle' – most unusual.[11]

A further highly unusual feature of the Shroud's linen is the weave itself. One of the big sources of frustration for historians of textiles is that relatively few examples have survived from antiquity. Even with regard to those that have survived, another problem is the quite disproportionate number that are mummy wrappings from ancient Egypt – far from representative of the many fabrics from many other ancient cultures that have been irretrievably lost. Egyptian mummy wrappings are almost invariably

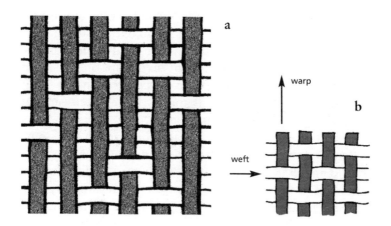

Fig. 9 HERRINGBONE WEAVE VERSUS PLAIN WEAVE. (a) The Shroud's three-to-one herringbone weave, showing its greater complexity compared to the plain weave (b) at right.

woven plain weave, that is, in a simple 'one over, one under' style (fig. 9b). The Shroud's weave, by contrast, is an altogether more complex three-to-one herringbone twill (fig. 9a & pl. 12a).[12] To make it, the weaver would have had to pass each weft (or transverse) thread alternately under three warp (or vertical) threads, then over one, creating diagonal lines. At regular intervals he or she would then have had to reverse direction to create the distinctive zigzags. In any age prior to the Industrial Revolution this would have been something rare and expensive, the work of a highly skilled professional.

Even among textile experts, therefore, the search for parallels to the Shroud, whether from the Middle Ages or from further back in antiquity, has not been easy. This difficulty was made very evident when the British Museum's Dr Michael Tite, the official invigilator for the 1988 carbon dating work, was looking for some historical samples of linen resembling the Shroud's weave to use for controls. His plan was that the carbon dating laboratories should not know which of the samples had come from the actual Shroud. He even sought my help on this. But the plan failed. In order to provide controls that were at least all of linen he had to abandon the requirement that their weave should be herringbone.

French specialist Gabriel Vial found much the same difficulty following his hands-on examination of the Shroud in 1988. There was literally no parallel that he could cite from the Middle Ages. Between antiquity and the modern age the closest example he could find, at least one that had been reliably analysed and published, was an artist's canvas from the second half of the sixteenth century.[13] And even this was technically 'beaucoup plus simple' (a lot more simple) than that of the Turin Shroud.

Vial found the era of antiquity itself – that is, around the time of Christ – significantly more productive, albeit not quite as much as anyone favouring the Shroud's antiquity might wish. Examples of herringbone weave can definitely be found from the Roman era. One specimen, from Palmyra, Syria, is precisely dateable, from its context, to before AD 276. There is another example from a child's coffin excavated at Holborough in Kent, England. Yet other examples have been found at Trier, Conthey, Riveauville and Cologne. The downside is these were all woven in silk, not linen, but at least they show that this type of weave was used at this early period.

As Dr Flury-Lemberg has further pointed out, there are also some good specimens of three-to-one twill dating even closer to Jesus's time which have come to light subsequent to Vial's researches. In the late 1990s French archaeologists were sifting for papyri among ancient rubbish dumps at the Roman fort of Krokodilo, a staging post in Egypt's Eastern Desert between the Nile and the Red Sea, when they came across substantial quantities of discarded textiles. Among these were fragments in wool that had been woven to specifications very similar to those of the Turin Shroud. And from the context in which these fragments were found they have been dated with absolute certainty to between AD 100 and 120.[14]

So, although any absolutely exact first-century parallel to the Shroud's linen remains elusive, in the light of the most recent textile findings the Shroud actually has rather more going for a manufacture date some time around the time of Jesus than for its being a product of the Middle Ages. The very high degree of professionalism that is evident from the mega-wide loom that appears to have been used, the very expert lengthwise cutting and seaming (to all appearances done while still 'in-factory'), the unusual selvedge, and the very unusual, expensive style of weaving – all of these point to production in a major, sophisticated cloth-making 'factory' of the kind that certainly existed during the Roman era for making the large sheet-like garments that were then fashionable. The Middle Ages, in contrast, were not noted for such operations. As Dr Flury-Lemberg remarked,

During the Middle Ages I do not know of any reason for the use of looms of that kind of width. Tapestries during the fourteenth and fifteenth centuries were very small – only between three and six feet high – compared to their sixteenth- and seventeenth-century counterparts, precisely because of the looms, the tapestry's height indicating its loom's width. And the *tunica inconsutilis* was only produced in ancient times, never in the Middle Ages. I doubt that there would have been a linen factory on that kind of scale in the Middle Ages. If you find linen bed sheets from that time (which is rare), you will find a seam in the middle of the sheet. Two loom pieces will have been sewn together at their selvedges to make them wide enough for the bed.[15]

While still on the subject of the Shroud's fabric, worthy of at least some consideration are certain interesting clues, still present within the cloth, to some of the different ways in which it has been folded over the centuries.

Today, as mentioned earlier, the Shroud lies stretched out flat in its high-tech container, but it has only been stored this way since 1998. From that year all the way back to 1604 it is known to have been kept rolled up 'Swiss roll'-style around a velvet-covered staff. Broadly this was an excellent mode of conservation for times prior to the age of advanced technology. It minimized the image's exposure to the air, and although surface irregularities from the presence of the 1532 fire's patches caused light creases to form (now rapidly fading), it avoided any of the unsightly fold marks that might otherwise have formed from long-term folding.

Looking back to the sixteenth century, the period of the chapel fire of 1532, it is the marks of the fire itself that tell us how the Shroud was being stored at that time. And back then it certainly was being kept folded. Because of the way the distinctively shaped burn-holes and scorches repeat across the cloth, in the manner of a 'paper-tearing act', Dr Flury-Lemberg and Turin photographer Aldo Guerreschi have independently reconstructed the actual number and arrangement of these folds. With the image on the inside it comprised a parcel of sixteen layers, each measuring about eleven by thirty inches, partly folded over itself to make thirty-two layers of cloth (fig. 10).[16] Interestingly, with the exception of one strong fold-mark line running centrally the length of the cloth – which importantly again shows that the image was protectively kept on the inside of the folding arrangement – it is only

The fire damage sustained on 4 December 1532

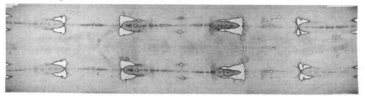

First folding (length-wise)

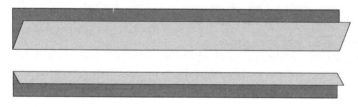

Second folding (length-wise)

Third folding (width-wise)   Fourth folding (width-wise)

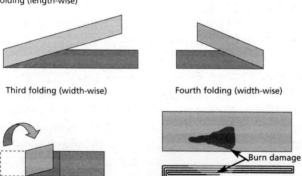

Burn damage

Final folding

Fig. 10 FOLDING ARRANGEMENTS AT THE TIME OF THE 1532 FIRE.
From the pattern of burn marks it is possible to reconstruct the exact
mode in which the Shroud was folded at the time of the chapel fire at
Chambéry in 1532. It was first folded twice lengthwise, then twice
across the width, with one end partially folded over one more time to
a length of fourteen inches. The folded arrangement would have
measured approximately thirty inches by eleven inches, thirty-two
layers thick at the thicker end seen on the left.

the fire marks and not any surviving crease lines that disclose to us this one-time storage arrangement for the fabric. This may be accounted a good indication of the Shroud linen's overall suppleness and resilience. No readily apparent surface fold or crease line indicates how the Shroud was stored long-term before the sixteenth century.

The very last item on STURP's 1978 research programme was to light and photograph the Shroud via raking light. This technique involves lighting the object being studied from one side, and at a very low angle to the surface, specifically to show up any surface anomalies such as crease or fold lines. Although the method is a searching one, neither STURP's rather rushed attempt of 1978, nor a technically better version by Giovanni Pisano in 1992,[17] nor even the excellent high-definition version by cameraman David Crute in 2008 (which very clearly showed up Flury-Lemberg and Tomedi's 2002 needlework), revealed anything that could be called obvious and significant ancient fold lines.

Otherwise, only two other pre-sixteenth-century folding arrangements are definitely indicated on the Shroud.

The first of these has as its basis the mysterious 'triple burn-hole' fire damage referred to in chapter 1 (pl. 3b). Whatever caused this damage is unknown. Superficially it looks as though something like a red-hot poker, coated with some substance causing it to sputter, was driven through the cloth three times – though three large droplets of an extremely caustic liquid might have had the same effect. When it happened is also unknown. All we know is that an artist's copy of the Shroud with the holes dated 1516 (pl. 3a) means the damage must have been sustained before that year.

The one certainty is how the Shroud was folded at the time. This is because the Shroud bears no fewer than four sets of the same holes, mirroring each other and backing

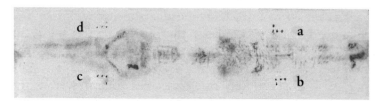

The burn-hole damage (a–d) sustained in an unknown incident pre-1516

First folding (length-wise)

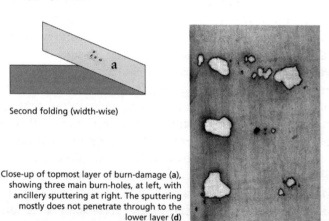

Second folding (width-wise)

Close-up of topmost layer of burn-damage (a), showing three main burn-holes, at left, with ancillary sputtering at right. The sputtering mostly does not penetrate through to the lower layer (d)

Fig. 11 FOLDING ARRANGEMENT AT THE TIME OF THE TRIPLE BURN-HOLE DAMAGE. From the descending order of penetration of the four poker holes (a–d), it can be determined that at the time of this incident the Shroud was folded once lengthwise and once width-wise. Something burning-hot and sputtering dropped three times forming a line of holes in the dead centre of this folding arrangement. The impression is of the cloth having been run through three times with a poker covered in pitch, though another alternative, favoured by Dr Flury-Lemberg, is the dropping of a highly corrosive liquid. For a possible explanation of when this damage occurred, see chapter 11.

each other up in a descending order of penetration. At the time the damage was inflicted the Shroud has to have been folded in four, forming a package roughly seven feet by two feet (fig. 11). This is not exactly a practical storage size, so it is very likely this particular folding was for a limited time only, possibly even momentary. We will be returning to this damage later in the book.

The second pre-sixteenth-century folding arrangement has as its basis the water stains also referred to in chapter 1. Just like the 1532 fire damage, these stains repeat and mirror-image each other paper-tearing-act style. Until quite recently the widespread assumption (one shared by me) was that they were all sustained when water was thrown on to the Shroud to douse it when it was rescued from the fire of 1532. However, independent studies by Italian photographer Aldo Guerreschi and Dr Flury-Lemberg[18] identified a second, distinctively different – and earlier – group of water stains. These cannot be associated with the 1532 fire incident because they do not match its folding pattern. Nor can they be related to the 'triple burn-hole' incident, because the latter exhibits no apparent accompanying attempt to extinguish it. When the quite separate folding arrangement of these 'second incident' water stains is reconstructed (fig. 12, overleaf), a most intriguing accordion-type folding arrangement becomes evident. This comprised fifty-two segments, making a bundle around one foot by one foot one inch square – in this shape nearly impossible to be kept together without something to hold it. Unlike the other folding arrangements this one was quite loose, the edges being very rounded. Some time while the Shroud was folded in this way these edges became soaked in water that had pooled at the bottom of whatever container was holding the Shroud in the concertina shape.

So when might such an accordion style of folding have

been applied to the Shroud? As noted by Flury-Lemberg, there is an example of this folding mode from antiquity. The Etruscan equivalent to ancient Egypt's Book of the Dead, the *liber linteus*, which was specifically made of linen cloth, was folded in just this way. The National Museum of Zagreb in Croatia[19] has an example that has survived solely due to its re-usage for the wrappings of an Egyptian mummy, and the accordion style in which it was originally folded is very clearly depicted in relief on an Etruscan burial urn in the Berlin Museum dating to the fourth century BC.

What of the container that held the Shroud accordion style? On this topic Turin photographer Aldo Guerreschi has come up with the most interesting suggestion – accompanied by a particularly compelling demonstration. Because the water stains indicated a loose mode of storage, the basic suggested shape strongly reminded Guerreschi of the tall, round earthenware storage jars in which the famous Dead Sea Scrolls were found. Such jars, we should be aware, date from around the time of Christ. After obtaining an exact replica of one of these jars, Guerreschi tried fitting into this a cloth of exactly the same dimensions as the Shroud, folding it in the accordion-style arrangement the 'second incident' water stains indicated. It was a perfect fit, and held the arrangement perfectly. Guerreschi then repeated the experiment, this time with a puddle of water in the bottom of the jar. When he removed the cloth and opened it out, there was an identical pattern of water stains.

This is no anecdote. Guerreschi repeated it in April 2004, with me acting as his assistant, for a British-made television documentary produced by Pioneer Productions.[20] As the production team can confirm, the filming occurred at the very end of the day, with no possible opportunity for a second 'take'. Again, an identical pattern was produced.

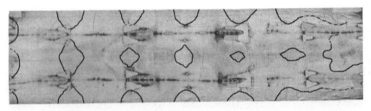

The water stain damage sustained at some time prior to the 1532 fire

First folding (length-wise)

Second folding (length-wise)

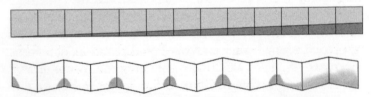

Further foldings, accordion-style, showing ascending water penetration

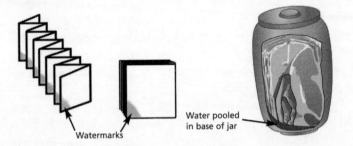

Watermarks

Water pooled in base of jar

Fig. 12 HOW THE SHROUD WAS FOLDED AT THE TIME OF THE CREATION OF THE 'SECOND INCIDENT' WATER STAINS. From the pattern of this second group of water stains, unconnected with the 1532 fire, it can be deduced that when this damage occurred the Shroud was folded accordion-style. The folding was also loose, suggesting the cloth's storage at the time in something resembling a large jar.

It would be quite wrong to suggest this as any kind of proof that the Shroud dates to the first century, and no such suggestion is intended. Even if Guerreschi's reconstruction is correct we have as yet no idea when or why the Shroud should have been kept in such a storage jar, nor how long it may have remained there, nor why there should have been water in its bottom. It remains a suggestion only.

But after all we have seen thus far – the compelling photographic evidence, the chilling medical evidence, the close correspondence with archaeology and the New Testament, the uncompromisingly scientific insights into the body and blood stains, the new understandings in the light of recent textile findings – let us recall those contemptuous words of radiocarbon dating scientist Professor Edward Hall: 'Someone just got a bit of linen, faked it up, and flogged it.'

Really? Can radiocarbon dating truly be so good, so utterly infallible, so overwhelmingly superior to any other approach that all other evidential considerations should be swept aside as valueless? Just how 'astronomically' accurate was the radiocarbon dating test that was carried out on the Shroud in 1988?

Chapter 7

What's in a Date?

Rogue dates are common in archaeology and geology ...
Such has been my experience as an archaeologist who has
excavated, submitted and interpreted more than one
hundred carbon 14 samples from Neolithic, Bronze Age
and Early Historical sites. Of these dates obtained, 78
were considered credible, 26 were rejected as unreliable
and 11 were problematic. I mention this merely to inform
the non-specialist ...[1]

William Meacham, archaeologist, Centre of Asian Studies,
University of Hong Kong, 2000

WHEN ON 13 OCTOBER 1988 a trio of radiocarbon dating
scientists, sitting on a podium in a dingy basement room
in London's British Museum, announced that the Shroud
dated to between 1260 and 1390, suddenly, for the general
populace, it had been proved a medieval fake. The London
Times duly ran the headline 'Turin Shroud Shown to be a
Fake', and according to the two journalists who wrote the
accompanying news story the Shroud could now
confidently be lumped in with 'a feather from the
Archangel Gabriel ... vials containing the last breath of

Saint Joseph [and] several heads of Saint John the Baptist' as just another of the 'thousands of relics' which could be ascribed to 'mediaeval tricksters'.[2] Ever since that time the world at large has blandly accepted this 'verdict of science' as the last word on the subject.

Which gives rise to the question, can any radiocarbon dating result, even one carrying the authority of three internationally well-respected laboratories, truly justify such unquestioning rejection of all other data as worthless? And if the answer is no, how, back in 1988, did the science of radiocarbon dating manage to achieve such an all-powerful spin?

Immediately to be acknowledged is that radiocarbon dating has long proved its worth to archaeology in innumerable instances during the decades since its invention by Chicago physicist Willard F. Libby shortly after World War Two. The principle behind it is that all living things, animal or vegetable, take in the very mildly radioactive isotope carbon 14, but only while they are alive. When they die this isotope 'decays', or loses its radioactivity, at a steady rate relative to the stable carbon 12. Libby's brilliant achievement, for which he was awarded a Nobel Prize, was to develop a form of Geiger counter to measure this decay in organic materials such as bone from an ancient skeleton, or wood from some historic boat, or linen from what had once been a flax plant. As if from an atomic clock, Libby's counter could 'read' the date on which the once living organism died.

The early years of radiocarbon dating indicated the need for certain adjustments, which are now routinely included in all measurements. Once such glitches were attended to, very quickly archaeology was revolutionized by the new method. Where all had previously been supposition based on pottery sequencing and other imprecise data, now it became possible to date to within a few centuries the building of Stonehenge, or the era of a whole prehistoric culture.

Throughout the first quarter century following Libby's invention, its usage for the Shroud would not have been advisable. Indeed in the early 1960s scientists at Britain's then premier radiocarbon dating laboratory, part of the Atomic Energy Research Establishment, Harwell, specifically advised against it.[3] One major reason they gave was that at least one pocket-handkerchief-size portion of the Shroud would need to be irretrievably destroyed for their sampling purposes. As they anticipated, the loss of such a substantial portion would most likely, and understandably, be considered unacceptable by those responsible for the Shroud's care.

This major obstacle prevailed for nearly two decades. Then in June 1977 an article in *Time* magazine announced the development in the USA of a new method of carbon dating, accelerator mass spectroscopy (AMS), whereby very much smaller samples could be radiocarbon dated, and with much the same precision as the old Libby method. In the case of the Shroud, the new method reduced the sample size required to that of a postage stamp.

From London, American clergyman the Revd David Sox, secretary of the then newly formed British Society for the Turin Shroud, immediately wrote to Rochester University's Professor Harry Gove, inventor of the AMS method, enquiring about its applicability to the Shroud. Gove, who by his own admission had never heard of the Shroud, responded encouragingly. None the less he warned Sox that it was 'probably a bit too soon to apply so recently developed a technique to such a renowned object'.[4]

The next eleven years saw some extremely complex politics in two quite different arenas. In the radiocarbon dating field, the inception of the new method, which needed some heavy investment and some lengthy installation time by laboratories opting for it (such as Oxford in England), was hotly followed by the traditional Libby laboratories, led by Harwell, developing a rival version which they claimed to be nearly as capable of accurate radiocarbon dating from very tiny samples. And Harwell and its counterparts had decades of experience behind them, which the AMS laboratories lacked. Behind the ostensibly aloof façades of their very high-tech science facilities, the two methods became engaged in a very bloody war to the death during which the opportunity to radiocarbon-date the Shroud, with all its attendant publicity, escalated from an incidental scientific exercise to a highly sought-after prize.

Meanwhile in the Shroud arena, the death in 1983 of its traditional owner ex-King Umberto of Italy saw its ownership pass to the then young and vibrant new Polish pope John Paul II. Up until that point Turin, as immediate custodians, and with Umberto near powerless in exile, had become used to calling all the main shots on Shroud matters. With the Vatican having become 'owners', suddenly the Vatican's scientific advisers, in the form of the Pontifical Academy of Sciences, wanted rather more say in matters than was welcomed by their opposite number in Turin, Cardinal Anastasio Ballestrero's chief scientific adviser Professor Luigi Gonella.

When the Vatican's Brazilian-born Professor Carlos Chagas put forward a well-researched protocol involving seven named laboratories in radiocarbon-dating the Shroud, some of them using the old Libby method, others the new AMS method, Gonella went feral. On all sides stances were taken and angry messages exchanged. Even I found myself an intermediary amid the crossfire. When the dust began to settle, it was Turin who had won. On 10 October 1987 Cardinal Ballestrero calmly faxed all interested parties that the seven laboratories had been reduced to three, chosen on the basis of their 'experience in the field of archaeological radiocarbon dating'.[5] The three were Oxford, Zurich, and Tucson (Arizona) – all of them AMS laboratories. The stipulated 'experience' criterion was the very reverse of the truth, the rejected Harwell alone having vastly more experience than all three of the chosen ones put together.

Although whatever went on behind the scenes remains far from clear, Gonella had got his way, and at daybreak on 21 April 1988 representatives of the three chosen laboratories duly assembled at Turin Cathedral's old sacristy, a centuries-old room that would literally go up in flames nine years later (pl. 1a). While the scientists watched from pews, the Shroud was solemnly brought into the room and unrolled before them.

Incredibly, it was only at this point that Gonella and his close friend Giovanni Riggi proceeded at length to deliberate on the best location from which to take the sample that was to be apportioned between the three laboratories. Some have claimed that the debate between the two scientists took two hours. However long it took, the eventual choice was the top left corner. Riggi, immaculate in a white coat, and with his own appointed cameraman filming his every move, snipped off a three-inch-by-half-an-inch sliver (pl. 13c) from which he trimmed a section of seam,[6] then divided the rest so that each of the three laboratories received a portion weighing roughly fifty milligrams (fig. 13, portions a–d). Approved 'referee' of the exercise Dr Michael Tite of the British Museum, together with Cardinal Ballestrero, then took the three portions to a side room where, with no camera present, they placed them, together with the pre-arranged 'controls', into sealed canisters carefully labelled for the laboratory representatives to take away with them.

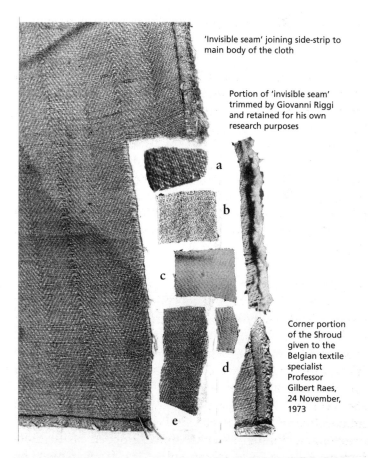

'Invisible seam' joining side-strip to main body of the cloth

Portion of 'invisible seam' trimmed by Giovanni Riggi and retained for his own research purposes

Corner portion of the Shroud given to the Belgian textile specialist Professor Gilbert Raes, 24 November, 1973

Fig. 13 THE SHROUD'S SAMPLES CORNER, SHOWING WHERE SAMPLES WERE TAKEN. Viewed from the underside, the area of the Shroud's top left-hand corner from which all samples have been taken for scientific purposes, first in 1973 (at bottom right), then for radiocarbon dating purposes in 1988. (a) 40mg sample provided for Arizona laboratory; (b) 50mg sample provided for Zurich laboratory; (c) 50mg sample provided for Oxford laboratory; (d) 10mg sample provided for Arizona laboratory to supplement sample (a); (e) 135mg portion retained in the care of the Turin Archdiocese. Because the individual photos a–d were taken by the laboratories, not all were photographed from the underside view, hampering an exact match.

A 135-milligram portion of the sliver that was surplus to the laboratories' requirements was left over.

The first of the laboratories to achieve a result was Arizona at the beginning of June, followed by Zurich in July, then Oxford on 8 August. Their findings were supposed to be kept secret, but before the end of August London's *Evening Standard* was on the streets with the banner headline 'Turin Shroud is a Fake', the carbon dating having reportedly found the Shroud to date from 1350. The *Standard*'s source was a Cambridge librarian with no known connection either to the Shroud or to the carbon dating laboratories who simply declared scientific laboratories to be 'leaky institutions'.

On Thursday 13 October, press conferences for the official results were held near simultaneously in London and in Turin. Dr Michael Tite, together with Professor Edward Hall and Dr Robert Hedges of the Oxford radiocarbon dating laboratory, fronted the London gathering, which was held in the British Museum's basement press room. On a blackboard behind the three scientists was scrawled in very large numerals '1260–1390!' – a gesture that, though apparently anonymous, was readily attributable to the ebullient Professor Hall. Almost nonchalantly the three scientists, with Hall as leading spokesman, informed the gathered journalists that this set of dates carried a 95 per cent likelihood of accuracy, and represented an average of the findings between the three laboratories, all of which were close to each other. They declared the odds 'astronomical' against the Shroud dating from the first century. The following February the highly respected scientific journal *Nature* carried a detailed paper under the joint authorship of all twenty-one scientists who had participated in these results, declaring them 'conclusive evidence that the linen of the Shroud of Turin is mediaeval'.[7]

Not long after this, Hall's Oxford AMS laboratory received funding of a million pounds, bringing into being a permanent professorship post that was immediately filled by the British Museum's Dr Michael Tite, while Harwell's Libby method laboratory quietly went out of business. For the wider world, however, the main story was that the Shroud's credibility for dating from the time of Christ had been shot dead in the water – a situation that essentially remains to this day.

Sad to relate, some supporters of the Shroud's authenticity, particularly the French, quickly began to allege conspiracy. French priest Brother Bonnet-Eymard of the very right-wing Catholic group the 'Catholic Counter Reformation in the Twentieth Century' claimed that Dr Tite, when out of sight of cameras in the sacristy's side-room, had slyly replaced the Shroud with some samples that had been brought as medieval 'controls'. In Germany, sensationalist author Holger Kersten suggested much the same,[8] surmising that Dr Tite had been secretly in league with the Vatican to make sure the Shroud was proved a fake. Purportedly the Vatican wanted to keep from the world the Shroud's 'proof' that Jesus was still alive when he was brought down from the cross, a finding that would destroy Christians' belief in the Resurrection.

Such ideas may quickly be dismissed as pure moonshine. Anyone who knows Dr Tite personally can only find entirely risible any suggestion that he was part of some cloak-and-dagger plot with the Vatican. And all three radiocarbon laboratories had astutely photographed the samples they were given before submitting them to the procedures that would destroy them. In these photos, the samples that were radiocarbon-dated clearly show the Shroud's rare and unmistakable herringbone weave.

More recently it has been claimed, with some considerable attendant publicity, that the area of the Shroud from which the radiocarbon dating sample was taken was one which had been extensively rewoven in the Middle Ages.[9] Purportedly the 'new' area had been invisibly matched to the rest of the cloth, so the radiocarbon dating laboratories had unwittingly dated from a piece that was not part of the original cloth, but added thirteen centuries later.

Despite this idea having been put forward, even in peer-reviewed scientific journals, it has little more going for it than the 'clandestine switch' theory. During my eight-hour examination of the Shroud back in 1973, high among the priorities was a very careful search for any anomalies that might disclose an early historical display method, any ancient fold marks, signs of fastenings, etc. Every area of the linen's surface that was accessible on that occasion[10] was carefully surveyed from a glancing angle for this purpose, and the area from which the radiocarbon dating sample was later taken was well within my viewing range.[11] Not the slightest sign of any compositional difference was apparent.

A medieval 'reweaver' would therefore have needed to be incredibly clever to match his new patch so perfectly with the original cloth. Which raises the historical question, if invisible mending was so good in medieval Europe, why the crude patches that were sewn on after the 1532 fire? The Savoy family was, after all, acutely embarrassed about the damage the Shroud sustained while in their care. If 'invisible' repairs had been at all possible, they would undoubtedly have had them carried out, whatever the expense. As it was, the patches were not even of the same weave as the Shroud proper.

Doubts concerning the reweaving hypothesis have been strongly endorsed by those with the appropriate professional expertise – textile specialists with first-hand knowledge of the Shroud. The main purpose of Gabriel Vial's attendance at the 1988 sample-taking was to make sure that the sample given to the laboratories came from an indisputably representative part of the Shroud. He had therefore checked the area chosen specifically from this very point of view. For the same reason Riggi had trimmed off any seam from the samples just in case the seam might have been added later. Mechthild Flury-Lemberg, with her unrivalled hands-on knowledge of the Shroud's every square centimetre, similarly rejects the idea of reweaving in the area from which the carbon dating sample was taken. As she has pointed out, it would have been technically impossible not to leave some traces of such workmanship, particularly on the Shroud's underside, and she found nothing of the kind.[12]

All this said, the reweave hypothesis does have one virtue: it draws attention to the possibility that an error in the carbon dating findings might have had nothing to do with either the integrity or competence of the laboratories, or the efficacy of the radiocarbon method, rather something to do with the sample that was provided – a choice that was notably made by Turin, not by the laboratories.

Worth introducing to the debate at this point is a second remark made in the 1960s by the Harwell scientists concerning why they thought it inadvisable to submit the Shroud to carbon dating. These are the words of senior Harwell scientist P. J. Anderson:

> The history of the Shroud does not encourage one to put a great deal of reliance upon the validity of my carbon 14 dating. The whole principle of the method depends upon the specimen not undergoing any exchange of carbon between its own molecules and atmospheric dioxide, etc. The cellulose of the linen itself would be good from this point of view, but the effect of the fires and subsequent drenching with water . . . and the possibility of contamination during early times, would, I think, make the results

doubtful. Any microbiological action upon the Shroud (fungi, moulds, etc., which might arise from damp conditions) might have important effects upon the carbon 14 content. This possibility could not be ruled out.[13]

Where in all the hype put out by the AMS scientists in 1988 about 'astronomical' odds was there even the slightest word of caution that contamination might have affected the dating result?

It is not as if such a possible source of interference had somehow been eliminated in the years between the early 1960s and 1988. In 1990, thirty years on from Anderson's words of caution, British Museum scientist Dr Sheridan Bowman included this reiteration in the information booklet *Radiocarbon Dating*: 'One of the fundamental assumptions of radiocarbon dating is that no process other than radioactive decay has altered the level of carbon 14 in a sample since its removal from the biosphere.'[14] Dr Bowman further noted (my italics), 'It is important not to introduce any contamination when collecting and packing the sample ... Many materials used for preserving or conserving samples contain carbon that it may be impossible to remove subsequently. Many ordinary packing materials such as paper, cardboard, cotton wool and string contain carbon and are potential contaminants. *Cigarette ash is also taboo.*'[15]

Now, in the case of the Shroud we are talking of a cloth that has been stored in any number of containers, mostly unknown, and wrapped in a variety of other fabrics over just the known centuries. It has also been exposed to two fires, the more serious of these, in 1532, accompanied by the combustion of many rich furnishings, hangings, etc. As recently pointed out to me by a correspondent, David Panagos,

When a Fire Department is called to a blaze they fight it with water. I presume people did the same when they fought the fire which the Shroud was exposed to in 1532. When water is used to douse the flames it forms steam which carries soot, i.e. carbon, *to all parts of the structure*. There was a house fire in my neighbourhood in 1969 which the Fire Department quickly restricted to the lounge – only the sofas and stuffed chairs were burnt. However all the unburnt furnishings, i.e. curtains, drapes and carpets, even in the other parts of the house, were soiled by the soot which was carried all over by the steam . . . the fire had not only reduced the burnt furniture to carbon, but had also uniformly contaminated all the other unburnt furnishings by the action of the steam.[16]

Manchester-based textile specialist the late John Tyrer emphasized that during such conditions contaminants would be forced into the molecular structure of the flax's fibres,[17] making them therefore part of the Shroud's permanent structure, impossible to remove by any of the routine 'pre-treatment' procedures used in carbon dating. Quite aside from the fires, on virtually every occasion when the Shroud was shown publicly during the six centuries of its known history in Europe, clouds of incense were wafted in its presence. So are we expected to take seriously Sheridan Bowman's cautions regarding even smoking in the vicinity of an object to be carbon dated, yet dismiss as of no account the serious smoke to which the Shroud has undoubtedly been subjected over the centuries?

All this said, smoke from fires, incense, etc. may be only a partial contributor to the Shroud's contamination rather than the main culprit. Harwell's P. J. Anderson, it may be recalled, mentioned microbiological action. As is readily demonstrable by any microbiologist, even the cleanest hands leave a microbiological residue on whatever they

Corner from which the sample was taken for radiocarbon dating in 1988

Fig. 14 LATE SEVENTEENTH-CENTURY SHOWING OF THE SHROUD (*detail*), illustrating how the cloth was held to crowds by the very top corner area from which the sample was cut for radiocarbon dating purposes in 1988. From the 1350s through to the mid nineteenth century the Shroud was always held up in this way for public showings.

have handled. The constituent micro-organisms in this residue can multiply astonishingly under the right conditions yet be invisible, even under the microscope, except to experienced microbiologists. A case in point is the 'shine' that can develop on rock surfaces that are subject to repeated handling. The popular supposition is that this is just 'polished' rock. As a microbiologist will assure you, it is actually a microbiological accretion.[18]

This prompts us to reconsider the area on the Shroud from which the single sliver was cut amid so much argument, then divided between the three radiocarbon laboratories. The area was in the top left-hand corner, the very spot, together with the opposite right-hand corner, by which generations of bishops and archbishops held up the Shroud to the crowds during the hundreds of occasions on which the cloth was exhibited over the centuries (fig. 14). Such expositions quite often took place on hot days (4 May was a traditional date), and they could be prolonged. The areas of the Shroud most likely to be contaminated from repeated handling are these two top corners. From the viewpoint of ensuring the best reliability for radiocarbon dating, the outcome of that heated argument between the two Turin scientists Gonella and Riggi was the worst decision possible.

But surely the radiocarbon dating laboratories would have made some preliminary check of their samples to make sure that no contamination of any such kind was present? The straight answer is no. The Libby method laboratories, with their much greater experience back in 1988, would almost certainly have been more prudent, but the three AMS laboratories dated blind. They went through certain routine pre-treatment procedures that were theoretically supposed to eliminate contaminants, but in all other respects they went ahead simply assuming that they were testing pure Shroud, without actually knowing the chemical constituents of the object they were destroying. As remarked by microbiologist Professor Stephen Mattingly of the University of Texas, such preliminary chemical analysis 'is the first step in quantitative analysis in college chemistry. I can remember my lab instructor sending me back to the bench because the recovery mass of my unknown did not agree with the known value.'[19]

That the area from which the radiocarbon sample was taken carried significant amounts of microbiological contamination was positively confirmed in 1994 during a Round Table held at the University of Texas. During this meeting, tiny segments from the area Riggi had unofficially appropriated for himself when he trimmed the radiocarbon samples in 1988 were studied under a microscope by a number of scientists.[20] These included specially invited guest Professor Harry Gove, inventor of the AMS radiocarbon dating method used by all three laboratories. The general agreement, with which Gove concurred, was that a substantial amount of contaminating microbiological coating was indeed present.[21]

So the issue became not one of whether the sample taken in 1988 was contaminated, but whether the contamination was sufficient to skew the Shroud's dating by thirteen hundred years. As was generally agreed, 60 per cent contamination was required for a cloth genuinely of the first century to appear to be from the fourteenth. But surely the presence of 60 per cent contamination would be very obvious?

Not so, according to Professor Mattingly, and to his counterpart at the University of Queensland, the late Dr Tom Loy.[22] As argued by both, the micro-organisms' transparency could cause them to be almost invisible. To prove his point, Mattingly made sample cloths that he had artificially prepared with such a coating so that Gove could satisfy himself of the scientific truth of this. Mattingly takes up the story:

> I went back and forth with Harry Gove about this and finally sent him two linen samples. One was un-contaminated and the weight was determined and included with the sample. A second sample with near identical uncontaminated weight was coated with enough bacteria (previously killed by heat) to represent 60% of the dry weight of the linen sample. The contaminated sample was more yellow in colour and had stiffness as previously noted in regard to the Shroud. Harry never corresponded with me again. What he did with the samples, I have no idea. Being a scientist I think he realized that I was correct and he saw no further need to argue with me.

Professor Harry Gove died, at the age of eighty-six, on 18 February 2009, just one month after an uncharacter-istically unanswered email enquiry from me concerning this very question. So, sadly, we lack any confirmation of his final thoughts on this issue.

What can be said with confidence is that, excellent tool of archaeology though radiocarbon dating undeniably is, the dating results it provides can be subject to substantial errors due to factors known and unknown that may affect the object being dated. There are plenty of examples of this in archaeology. When Egyptologist Rosalie David asked for separate carbon datings for the wrappings and body tissue of an Egyptian mummy in the Manchester Museum collection, the wrappings were dated a thousand years younger than the body they had wrapped, with no evidence of any secondary rewrapping.[23] Samples from bog body Lindow Man were sent to three different radio-carbon dating laboratories. Harwell's date was AD 500, Oxford's AD 100 and the British Museum's 300 BC – a discrepancy of eight hundred years, yet all three laboratories claimed accuracies to within a hundred years.[24] The reason for the discrepancy has never been determined. Many more examples could be cited.

As admitted by Dr Willi Wolfli, director of the Zurich AMS laboratory (one of the Shroud three), in a paper written in 1985, after his laboratory had made an error of a thousand years specifically because of a contaminated sample, 'The existence of significant indeterminant errors can never be excluded from any age determination. No method is immune from giving grossly incorrect datings when there are non-apparent problems with the samples originating in the field. The results illustrated [in this paper] show that the situation occurs frequently.'[25]

The vast majority of the finds archaeologists submit to radiocarbon dating laboratories usually have experienced nothing even remotely resembling the long-term handling to which the Shroud has been subjected. Nevertheless, as Hong Kong archaeologist William Meacham has pointed out, members of his profession are very familiar with such 'rogue dates'. Mostly they quietly reject them and go on to seek out some alternative samples, maybe even find a fresh laboratory. In Meacham's own experience of the dating of more than a hundred carbon 14 samples from Neolithic, Bronze Age and Early Historical sites, '78 were considered credible, 26 were rejected as unreliable and 11 were problematic'.[26]

The big problem with the Shroud dating, and one that virtually never happens with any other datings, was that there was no one equivalent to an archaeologist to act as an interpreter (and potential rejecter) of the results. The Church rightly declared that it had no competence in the matter (despite much popular supposition to the contrary, the Shroud neither is nor ever has been something that Roman Catholics are required to believe in). The British

Museum's Dr Michael Tite declined to see himself in this role, again quite rightly, because he was simply another radiocarbon dating scientist, not an archaeologist.

This left the role to be filled by the three radiocarbon dating laboratories – primarily, because with characteristic forcefulness he made himself their chief spokesman, by Oxford's Professor Edward Hall. Unabashedly atheist, and neck-deep in his multi-million-pound propaganda war against Harwell, Hall positively relished the opportunity to be judge, jury and executioner of the Shroud. He lectured on the subject to the British Museum Society. His laboratory's role in proving it to be a fake was proudly displayed at his offices as among his finest achievements. There was no way he was going to modify the conclusiveness of the laboratories' findings with any cautionary proviso of the kind his Zurich counterpart Professor Wolfli had volunteered.

Professor Edward Hall died in 2001, and his successor Dr Michael Tite has retired from what has been renamed the Oxford Radiocarbon Accelerator Unit, now moved to rather more salubrious premises in Oxford's South Parks Road. The unit's present director, Professor Christopher Bronk Ramsey, a professed Christian, was a young student at the time the Shroud was radiocarbon dated and has none of Hall's Old Etonian braggadocio. Characteristically soft-spoken, he has declared himself very willing to make his laboratory's facilities available for any sensible, properly scientific fresh approach to the subject.

When interviewed by me in May 2009, he professed himself open-minded on the issue of the Shroud's authenticity. However, he staunchly maintained that the preliminary treatments AMS laboratories routinely use on samples submitted to them for dating should have removed all possible microbiological contaminants, even though the microbiologists Stephen Mattingly and Tom Loy had both insisted to the contrary. In the light of his expressed doubts that carbon dating could have produced an error as large as thirteen centuries, he was asked how he thought someone of the fourteenth century might have faked the Shroud. He suggested that it was perhaps 'an unintentional copy. Something which originally had something on the cloth that has come off from something which was originally much more marked and crude . . . Like a curtain with a backing cloth, with light shining through it, so you can envisage the pattern from the original pattern on the front showing through it . . .'[27]

Perhaps the most balanced note of caution was published in the prestigious *Journal of Research of the National Institute of Standards and Technology* in 2004. During a thorough evaluation of carbon dating, the author studied the case for dating of the Shroud, and his conclusions were as follows:

> . . . a wave of questioning has followed – not of the AMS method, but of possible artifacts that could have affected the carbon 14 result . . . the question of non-contemporaneous organic matter – whether from incompletely removed carbon contamination from 'oil, wax, tears, and smoke' that the cloth has been exposed to, or from bacterial attack and deposit over the ages. Apart from the effects of such factors on the Shroud, the issue of organic reactions and non-contemporaneous contamination of ancient materials can be a very serious and complex matter, deserving quantitative investigation of the possible impacts on measurement accuracy.[28]

As non-scientists, it seems that we have expected rather too much from the radiocarbon dating method – that it might magically put back the Shroud to the first century

AD in a way that had seemed impossible from the available pre-fourteenth-century historical evidence. But during the last two decades the whole issue of the evidence for the Shroud's history prior to the fourteenth century has in turn been undergoing some profound and exciting new developments. And it is to this, the ultimate means of dismissing the radiocarbon dating 'verdict' of 1988 as just a 'rogue date', that we will now turn.

Chapter 8

'The Saviour's Likeness Thus Imprinted'

... this [the Shroud] could not be the real shroud of our Lord having the Saviour's Likeness thus imprinted upon it, since the holy Gospel made no mention of such an imprint, while, if it had been true, it was quite unlikely that the holy Evangelists would have omitted to record it or that the fact should have remained hidden until the present time.

Pierre d'Arcis, Bishop of Troyes, France, 1389

There is no mystery about the Shroud's history back to the year 1453. As was earlier noted, it was a possession of the Savoy family, rulers of a domain that included parts of what are today north-western Italy and south-eastern France. It is well documented that the dukes of Savoy staged elaborate showings of the Shroud going back to the time of the first-known Savoyard owner, Louis I of Savoy, who died in 1465.

Louis I of Savoy and his Cypriot wife Anne de Lusignan

were a devout couple with a constant retinue of Franciscan friars who acquired the Shroud in 1453 after some obscure dealings with an elderly French widow, Margaret de Charny. Twice widowed, and with no children from either marriage, Margaret was looking for a suitable family to whom to bequeath the Shroud, and the pious and up-and-coming Savoys seemed the ideal choice.

Margaret had inherited the Shroud from her father Geoffrey de Charny, whom for simplicity we will call Geoffrey II. A royal official known as a *bailli*, roughly equivalent to a regional governor, Geoffrey II in turn had inherited the Shroud from his father, Geoffrey I de Charny, who had died in 1356.

It is this mid-fourteenth-century period of the Shroud's history that causes the problem. The works of two early twentieth-century historians, the medieval specialist Ulysse Chevalier and the erudite Jesuit scholar Revd Herbert Thurston in England, seem to be particularly damning.[1] Though both men were staunch Roman Catholic clerics and directly contemporary with the discoveries of Secondo Pia and Yves Delage, neither had any time for 'modern' evidence favouring the Shroud's authenticity. As historians they were unwaveringly certain that the Shroud was a fourteenth-century fake. Indeed Thurston easily matched Professor Hall's forthrightness over the radiocarbon dating findings with his own similarly upbeat conviction concerning history's 'verdict' on the Shroud: 'The probability of an error in the verdict of history must be accounted, it seems to me, almost infinitesimal.'[2]

From the standpoint of the historical documents that Thurston and Chevalier had consulted, such an attitude was entirely understandable. As they determined, very little was known about the Shroud in the time of Geoffrey I. But from around the year 1389, nine years before Geoffrey II's death, there exists a formidable memorandum from Pierre d'Arcis, Bishop of Troyes, written in clerical Latin to the then Avignon Pope Clement VII, which seems to provide all the crucial details.[3] According to d'Arcis, within his own diocese a grave scandal was going on at the de Charny family seat at Lirey, about twelve miles from Troyes. The canons of the collegiate church there, founded thirty-six years earlier by Geoffrey I de Charny, were, it was said, exposing for veneration, with papal authorization, a cloth carefully described as a 'likeness or representation' of the '*sudarium* of Christ'.

There can be no doubt that this was our Shroud. D'Arcis's description of it specifically stated it to bear 'the twofold image of one man, that is to say, the back and the front'. And while the Lirey canons were describing it only as a likeness, it was being shown on a lofty platform flanked with torches, and with great ceremony. And they were making it known privately that it was the actual shroud in which Christ had been wrapped in the tomb – a claim that was attracting multitudes of pilgrims.

To Bishop d'Arcis's considerable annoyance, although these showings were happening within his diocese, they were outside his jurisdiction. Because of the way the Lirey church had been founded, as a collegiate church, Geoffrey II de Charny was legally entitled to go above the local bishop's head to obtain his permission for the showings directly from Cardinal de Thury, the papal legate.

This was a time notorious for abuses relating to relics, and d'Arcis, as an upright churchman, was concerned to stamp out anything of this kind that might be happening on his doorstep. According to his own account, his enquiries indicated that this was not the first time the de Charnys had shown the cloth. A burst of similar showings 'thirty-four years or thereabouts' previously (so during the time of Geoffrey I) had seen the same cloth openly claimed to be the true shroud of Christ. Again according to

1a The scene in Turin Cathedral's old sacristy, 21 April 1988. On the back pew, Oxford radiocarbon dating scientist Professor Edward Hall (*centre*) and representatives of other laboratories watch as a sliver of the Shroud is removed to be apportioned between them. In the foreground are (*l to r*) Cardinal Anastasio Ballestrero, Archbishop of Turin, Dr Michael Tite of the British Museum, and Professor Luigi Gonella, Cardinal Ballestrero's chief scientific adviser.

1b Turin's present archbishop, Cardinal Severino Poletto, inspects the Shroud in the high-tech conservation container that is its present-day home.

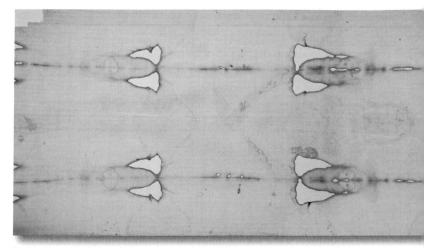

2a The Shroud full length, as it appears today. The triangular-shaped marks are from the chapel fire which nearly destroyed the cloth in 1532.

2b The Shroud's present-day repository. The conservation case seen in plate 1b lies beneath the ivory-coloured fabric covering.

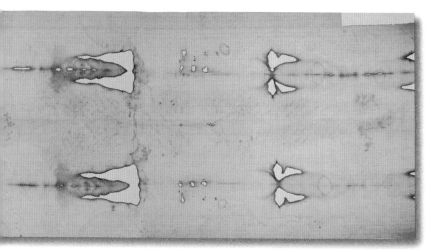

3a (*above*) Artist's copy of the Shroud, dated 1516, preserved in the Church of St Gommaire, Lierre, Belgium. Showing the cloth's appearance before the fire damage of 1532, clearly visible are the four very distinctive sets of triple burn-holes sustained in some earlier fire-damage incident.

3b (*right*) Close-up of one of the four sets of triple burn-holes. These match one another in descending order of penetration, indicating that the Shroud was folded in four at the time the damage occurred.

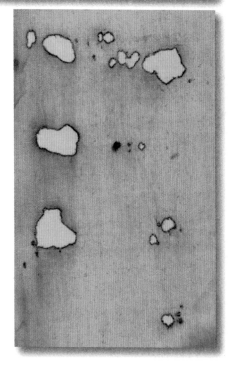

4a Amateur photographer Secondo Pia with a large plate camera similar to the one he used to take the first photograph of the Shroud in 1898, revealing the hidden 'negative'.

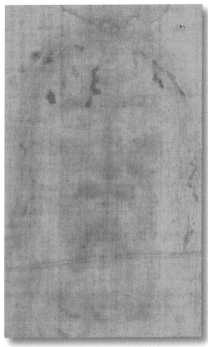

4b The Shroud face, natural appearance.

4c (*above left*) American scientist Dr Emily Craig's attempt at replicating the Shroud negative by a dry transfer technique, 1994.

4d (*above right*) Professor Nicholas Allen's attempt at replicating the Shroud negative, 2007.

4e Italian scientist Luigi Garlaschelli's attempt at replicating the Shroud negative (2009), using a body lightly coated in acid.

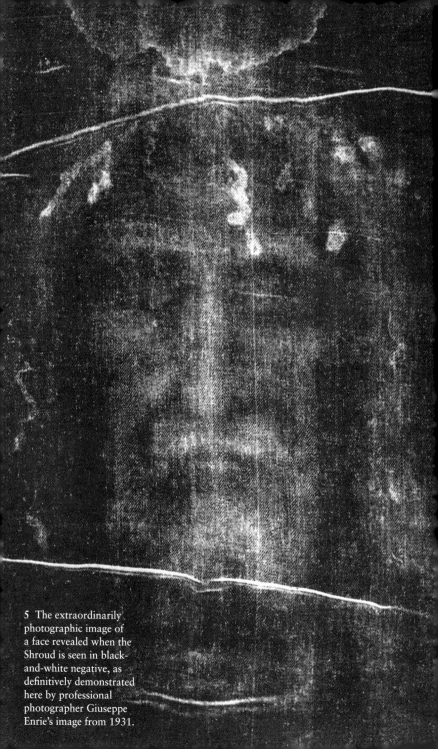

5 The extraordinarily photographic image of a face revealed when the Shroud is seen in black-and-white negative, as definitively demonstrated here by professional photographer Giuseppe Enrie's image from 1931.

6 The Shroud front-of-the-body imprint seen in negative:

(1) Swelling below right eye

(2) Trickles of blood on forehead and hair

(3) Multiple flesh wounds as from severe whipping

(5) Elliptic stab wound in chest

(6) Trickles of blood down forearm from wound in left wrist

(7) Trickles of blood down right forearm

(8) Bloodstains from wound in the feet

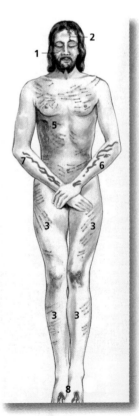

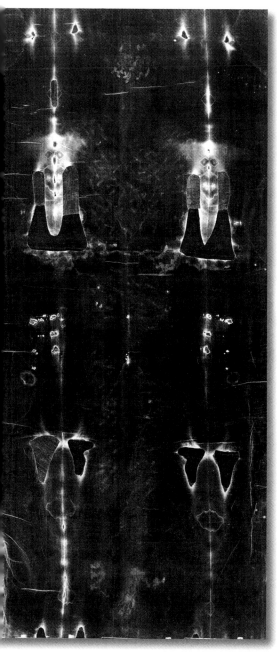

7 The Shroud back-of-the-body imprint seen in negative:

(2) Trickles of blood around back of head

(3) Multiple flesh wounds as from severe whipping

(4) Abrasions across the shoulders

(8) Bloodstains from wounds penetrating feet

(9) Spillage of blood from ankle wound

(10) Spillage across small of back from wound in the chest

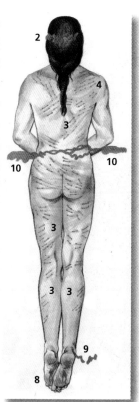

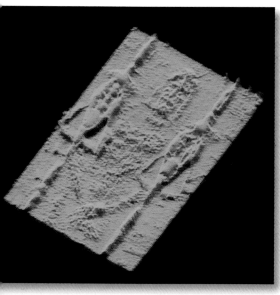

8a (*left*) The Shroud image's three-dimensional characteristics, as revealed by the VP-8 Image Analyzer in February 1976. Here the face and body appear in sculpted relief, framed by the two lines of scorches from the chapel fire of 1532.

8b (*below*) British television cameraman David Crute filming the Shroud in high definition for a BBC TV documentary, January 2008.

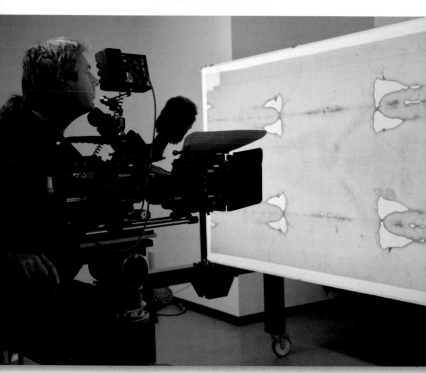

d'Arcis, the then Bishop of Troyes, Henri of Poitiers, had similarly investigated the matter and been told by 'many theologians and other wise persons . . . that this could not be the real shroud of our Lord having the Saviour's Likeness thus imprinted upon it, since the holy Gospel made no mention of such an imprint, while, if it had been true, it was quite unlikely that the holy Evangelists would have omitted to record it or that the fact should have remained hidden until the present time'.

In his memorandum, d'Arcis continued with a sentence that to many is the *coup de grâce* to the Shroud's authenticity, powerfully reinforcing the carbon dating result of 1988. According to Herbert Thurston's translation of d'Arcis's original Latin, 'Eventually, after diligent inquiry and examination, he [Henri of Poitiers] discovered the fraud and how the said cloth had been cunningly painted, the truth being attested by the artist who had painted it, to wit, that it was a work of human skill and not miraculously wrought or bestowed.'

'Game, set and match' though this statement might appear in favour of the Shroud being a fake – it is often quoted as such – this is not, in fact, the case. Latin lacks the definite article, and Thurston's 'the artist' can quite legitimately be replaced by 'an artist'. Similarly, while *depingere* means 'to paint', it can also mean 'to copy', the phrase thereby becoming translatable as 'the truth being attested by an artist who had copied it'. This may not be the first understanding either we in the twenty-first century or anyone of the fourteenth century might glean from d'Arcis's Latin, but we should allow for the possibility that d'Arcis may quite deliberately have intended some ambiguity of meaning, particularly given that he failed to provide any substantiation to such a damning piece of information.

Yet, going back six hundred years, we do not have any

readily apparent reason to cast doubt on the integrity of the two bishops of Troyes involved in the issue. Pierre d'Arcis had held the see of Troyes for some twelve years before the 1389 controversy. Prior to that he had enjoyed a reputable legal career, and possessed no apparent sinister motive for sticking his neck out in this case. Rather the reverse, as by his own admission the Pope appears to have been displeased with his attitude, even threatening him with excommunication. Henri of Poitiers was one of the worldlier churchmen of his time. He doubled as a military commander and kept a mistress busy with some prolific childbearing. But he had plenty of contemporary respect.

Still, in spite of the plethora of documents, one cannot avoid the feeling that all is not quite as it may seem, that there was something a lot deeper to the affair than meets the eye.

One peculiarity was that Geoffrey II married a niece of Henri of Poitiers some time after the alleged fuss about the first spate of showings. Although Geoffrey was very young when the first showings took place, he cannot fail to have later learned something of what went on, certainly if it had been as big a 'scandal' as d'Arcis claimed. So why did he bring the Shroud out again in 1389? By that time he was a well-respected official of France's king. It could surely not have been 'avarice' (the specific charge laid by d'Arcis) that led him to bring out a relic whose authenticity had apparently already been seriously challenged. And why, before commencing the showings, did he so meekly agree to call the Shroud merely a 'likeness or representation', yet actually present it on these occasions with all the reverence and ceremony pertaining to the real thing?

In the case of Clement VII, why as Pope, after he had received Bishop d'Arcis's powerfully worded memorandum, did he not immediately suppress the showings as d'Arcis requested? Instead, Clement apparently upheld the expositions on the understanding that the de Charnys should continue to describe the cloth as only a 'likeness or representation'. And, most intriguingly, he insisted on two occasions that Bishop d'Arcis should remain 'perpetually silent' about the matter.

Not least, why, during the fifteenth century, did Margaret de Charny, who had withdrawn the Shroud from the Lirey church when there was danger from English invasions, resist several determined legal appeals by the church's canons for its restoration to them, only then to go to great lengths to find some suitable noble family (i.e. the Savoys) to take it off her hands? Were such actions the ones of someone knowingly perpetrating a forgery?

Particularly pertinent to this matter is the character of the first-known owner, Geoffrey I de Charny. From everything that is known about him he simply fails to fit the mould of someone intent on a cynical money-making deception. A cadet member of the de Charny family, the relatively minor estates he acquired were gained largely from his two marriages, and from his extremely distinguished career as a professional soldier. In this career he achieved, among both his fellow Frenchmen and his adversaries, the highest reputation for outstanding personal bravery, and for chivalry – a subject on which he wrote the keynote book of the age.[4] On his epaulettes he wore the motto 'Honour Conquers All'. Awarded the coveted role of *porte-oriflamme*, bearer of France's sacred banner of Saint Denis, he died a hero's death holding this aloft to the last while battling the English at Poitiers on 19 September 1356. Fourteen years after the battle, when France had recovered from its resounding defeat that day, he was given a hero's funeral, at royal expense, in the Church of the Célestins in Paris. The poetry he left behind[5] shows him to have been a man of profound, almost melancholic, religious piety. Was this the kind of man who would have been the knowing promulgator of a religious fake?

Curiously, there exists no record of the Poitiers searches from which d'Arcis seemed to be quoting. Indeed d'Arcis actually let slip one key piece of information which seriously undermines his allegations. This was his statement that the inquiry in question had taken place 'thirty-four years or thereabouts' earlier. Which gives rise to the question, why, as a relatively recent successor to Henri of Poitiers, could he not put his hand on a single document from the inquiry that would have given him an exact date to quote to his Pope? As recently pointed out by American scholar Alan Friedlander, 'The account of Henri of Poitiers' investigation forms the centrepiece of Pierre's memorandum. It was crucial to the argument he sent to the pope. Yet he included no word, phrase or statement directly from that investigation. He did not append for the pope's attention any portion of Henri's *informationes*. He failed to quote a line verbatim from Henri's findings.'[6]

The only relevant properly contemporary record we do have actually seems to contradict d'Arcis. This is a letter from Bishop Henri dated 28 May 1356 that speaks warmly of Geoffrey I de Charny's 'sentiments of devotion' and eulogizes over the recent completion of the Lirey church that Geoffrey had founded three years earlier.[7] This surely indicates that no 'scandalous' showings of a fake relic could so far have taken place. Similarly, Bishop d'Arcis specifically mentioned crowds being drawn from foreign countries, and the Shroud having been 'long exposed' in his predecessor's time. But within four months of Bishop Henri's eulogies Geoffrey was lying very dead on the field of Poitiers, having spent much of the interim moving around with the French army as it sought engagement with the invading English army of Edward the

Black Prince. So there was not time for these expositions.

The only possible inference is that d'Arcis possessed no harder documentary evidence to back up his allegations than we possess today. At best he was peddling hearsay. At worst, with a cathedral acutely short of funds and with no major relic to draw cash from pilgrims, he was highly jealous of the income the Lirey church's canons were attracting with their Shroud and was attempting to sabotage them.

Even with this allowed for, however, the de Charnys' guilt has seemed to be independently demonstrated by a variety of factors. Not least, they failed to make any attempt to explain how they acquired the cloth. If the Shroud was genuine, for them to have explained how it came into their hands would surely have put an end to any bishop's fulminations against them. Yet the sum total of any 'provenance' information we have from them, and only from Margaret de Charny at that, is a totally in-adequate one-word mention that the Shroud was *conquis* by her late grandfather, '*conquis*' meaning anything from merely 'gained' to 'acquired by conquest'.

The de Charnys were, it needs emphasizing, not the sort of family who would be expected to have such a fabulous relic, bearing in mind that relics of the Passion were quite literally worth a king's ransom. Their apparent hasty relinquishing of showings of the Shroud after the inquiry by Bishop Henri of Poitiers has seemed like clear evidence of their guilt. Also, from 1389 and for the next sixty years they would describe it in official documents only as a 'likeness or representation' of the shroud of Christ, seemingly as if they themselves were not convinced of its genuineness.

To visit today the village of Lirey where these events took place can only reinforce the unlikelihood that any genuine shroud of Christ could have ended up there.

Tucked away in rolling French countryside south of Bouilly, a little over an hour's drive from Paris, Lirey is a very easily missed hamlet notable only for a few ancient timbered dwellings. Long gone is Geoffrey de Charny's wooden collegiate church where the canons so controversially exhibited the Shroud; an uninspiring nineteenth-century successor stands in its place. There never was any ducal chateau in which might have dwelt the sort of great magnate who might have acquired the genuine Shroud by some legitimate means.

Whatever the answer concerning Henri of Poitiers' inquiry, however, Bishop d'Arcis made two very pertinent points that have reverberated down the centuries: 'This could not be the real shroud of our Lord having the Saviour's Likeness thus imprinted upon it, since the holy Gospel made no mention of such an imprint', and 'if it had been true, it was quite unlikely that the holy Evangelists would have omitted to record it or that the fact should have remained hidden until the present time'.

There can be no denying d'Arcis's point that the 'holy Evangelists' or gospel writers failed to mention any imprint on the burial linens that Peter and John had found mysteriously abandoned in Jesus's empty tomb that first Easter Sunday morning. But as we have already seen, there is a great deal that the gospel writers, for reasons often best known to themselves, omitted to tell us. The writer of the John gospel actually acknowledged this omission in his gospel's last verse.[8] Likewise we have already seen that there has to have been something immensely powerful about that empty tomb scene that morning, not all of which the gospel writers properly described or explained.

Neither the gospels nor any other book of the New Testament tell us anything about what happened to the burial cloths, irrespective of whether or not they were imprinted with 'the Saviour's Likeness'. When the rock-cut chamber that is credibly believed to have been Jesus's tomb[9] was unearthed in the early fourth century on the orders of Roman emperor Constantine the Great's mother Helena, what was reputedly the wood of Jesus's cross was discovered still inside, likewise the nails with which Jesus was pinned to the cross, and even the title-board bearing his 'crime'. However sanguine a view we might take of these objects' authenticity,[10] they undeniably went on to become revered Christian 'relics' whose history and whereabouts were well documented by Dark Ages standards. But there was no mention of any shroud among this 'discovered' collection.

So imprinted or not imprinted, what could have happened to Jesus's shroud in the wake of that first Easter Sunday morning? It seems very unlikely that Jesus's disciples would simply have thrown it away. This said, however, we do need to bear in mind that they were Jews, and there were two things about it that would have caused considerable disquiet. First, as a burial cloth, the shroud was 'unclean', and therefore not the kind of object any of them would have felt comfortable taking around with them to demonstrate that Jesus had risen from the dead. These were men and women who had seen the resurrected Jesus, in the flesh, with their own eyes. That surely was testimony enough, without the need for any 'prop'. Second, if whatever had been left behind in the tomb was indeed our imprint-bearing Shroud, its imprint would have been a source of embarrassment in its own right. Jews abhorred any kind of representational image. The second commandment of Moses specifically forbade any such thing: 'You shall not make yourself a carved image or any likeness of anything in heaven above or on earth beneath.'[11] It would have aroused many deep-seated taboos that may well have precluded any mention of it.

However, there is a flaw in d'Arcis's assumption that the

shroud being exhibited by the de Charnys could not be the real shroud because there had never been a historical mention of an imprinted shroud. Nearly true, except for one reference, which d'Arcis probably would not have known about anyway. This came from an ordinary French soldier, Robert de Clari, who said that in August 1203, shortly before his fellow Crusaders captured and sacked Constantinople, he had seen in the city a church 'which was called My Lady St Mary of Blachernae, where there was the *sydoine* in which Our Lord had been wrapped, which every Friday stood upright, so that one could see the figure of Our Lord on it'.[12] Then, in the very next breath, de Clari added, 'No one, either Greek or French, ever knew what became of this *sydoine* when the city was taken.'

That is not exactly a lot of documentary evidence to support twelve hundred otherwise unknown years of history – except for one thing. Robert de Clari did at least establish that the idea of Jesus imprinting his likeness existed as early as 1203 – that is, over half a century earlier than the earliest date (1260) attributed to the Shroud by radiocarbon dating. And puzzling and disquietingly isolated though Robert's description of the *sydoine* might be, there can be little doubt that he saw *some* actual cloth of this description being shown in Constantinople.

Were there any other clues to a shroud bearing the likeness of Jesus?

Back in the 1960s, an Ampleforth-educated Benedictine monk, Father Maurus Green, made the most exhaustive possible search among early documents for possible references to the Shroud's existence during those so-called 'silent centuries'. He specifically called his study 'Enshrouded in Silence',[13] for the very good reason that he could find hardly anything to satisfactorily refute Bishop d'Arcis's charges. The apocryphal Gospel of the Hebrews contained a vague mention of Jesus after his resurrection giving his *sindon* – seemingly his burial shroud – to the 'servant of the priest' before his appearance to James.[14] In the fourth century St Nino, the apostle of Georgia, thought that Pilate's wife had kept the burial linens for a while.[15] About AD 570 the so-called Piacenza pilgrim described a cave convent on the banks of the river Jordan in which was 'said to be the *sudarium* which was over the head of Jesus', though whether this was imprinted, and whether it was a face cloth or something full-length, was by no means clear.[16] Some time during the early 680s Bishop Arculf of Perigueux, shipwrecked off the Scottish island of Iona, told his rescuer, the local monastery's abbot Adamnan, that he had seen in Jerusalem 'the *sudarium* of Our Lord which was placed over his head in the tomb'.[17] Arculf said this had been brought to Jerusalem only shortly before. It was apparently about eight feet long, so its measurements do not appear to match our Shroud, though it was certainly not a face cloth either. But Arculf mentioned no imprint. Around the twelfth century a *sindon* was listed as among the relics preserved in the collection of the Byzantine emperors in Constantinople. But the mentions of this included no accompanying information concerning whether it bore an imprint or how and when it might have been acquired by the imperial collection, details that were normally well documented.

So it appears that Bishop d'Arcis's point remains fundamentally valid: that if our imprinted Shroud were genuinely that of Jesus, it surely cannot have 'remained hidden' until the time of Bishop Henri. If authentic, it has to have been somewhere throughout the long centuries prior to its unlikely appearance in mid-fourteenth-century Lirey. But where?

The solution to this puzzle seems to lie in focusing on, to use Bishop d'Arcis's own words, its being a piece of cloth 'having the Saviour's Likeness thus imprinted upon it'. Earlier we saw that the 'crown of thorns' was something singular to Jesus. Equally singular to him is the idea that he imprinted his likeness on a cloth. In the entirety of human history, there is absolutely no one else who has ever had that idea attributed to them – not Mohammed, not Buddha, not any saint, only Jesus. Why?

If such an idea had originated only in the mid-fourteenth century, then probably we would have to attribute both it and the Shroud itself to that same un-believably brilliant mid-fourteenth-century 'cunning painter' whom Bishop d'Arcis and the radiocarbon dating laboratories would have us believe in. But this is far from being the case. Rather than being silent about any cloth

Fig. 15 LATE FIFTEENTH-CENTURY WOODCUT depicting the traditional western story of St Veronica holding up her veil imprinted with Jesus's features.

'having the Saviour's Likeness thus imprinted upon it', the long centuries prior to the fourteenth century are actually full of stories and images of this kind. The difficulty is finding one's way through what is a bewildering jungle of data to determine when, how and, most pertinently, with what they originated.

For westerners, the most familiar example of the genre will probably be the famous Veronica cloth. This is popularly associated with the story of a woman called Veronica wiping Jesus's face with her veil as he struggled with his cross through Jerusalem's streets on his way to be crucified. According to the story, Jesus's 'Likeness' became miraculously imprinted on Veronica's veil. Dozens of medieval and Renaissance artists depicted the scene, and thousands of Roman Catholic churches have it included among their 'Stations of the Cross', leading many to suppose the story must be in the gospels (fig. 15).

In fact the story in this form dates no earlier than the late Middle Ages, seeming to have been invented to spice up 'miracle play' dramatizations of the Passion story. In a twelfth-century version[18] there was no woman called Veronica, though at that time the canons of St Peter's, Rome were already keeping under close guard a cloth that was supposed to be the Vera Icon or 'True Likeness' of Jesus. Reputedly this likeness was imprinted not during Jesus's carrying of the cross but when he wiped his face after the 'bloody sweat' in the Garden of Gethsemane. A popular attraction for pilgrimages to Rome during the Middle Ages, this cloth can be traced historically no earlier than the eleventh century.[19] It seems to have been an official 'copy' for the western world of something that was altogether older and more mysterious being preserved at that time in the Byzantine east, in Constantinople.

It is this older and more mysterious eastern cloth which now commands our attention. Even putting a name to it is not easy. In depictions of it in numerous churches across Greece, Serbia, Russia and elsewhere it is mostly labelled the 'Mandylion', a word that is not even Greek but (via Arabic) obscurely related to the English-language 'mantle'. Among the very few modern-day scholars to have paid much attention to it the cloth has mostly been called the Image of Edessa, after the very oriental city of Edessa in which it spent most of its known history, and this is the label we will adopt here.

In an extraordinarily complex, difficult-to-handle subject of study, one of the most central facts about this Image is that its imprint of Christ's likeness was, as the Greeks called it, *acheiropoietos* – 'not by hand made'. That is, it was not the work of any artist, this notably being the exact (and ostensibly unique) characteristic so powerfully attributed to the Turin Shroud.

The second central fact is that this cloth had a history as an actual historical object extending from the time of Jesus only up to the Crusader sack of Constantinople in 1204. It mysteriously vanished during that sack – at the very same time, we may note, as the disappearance of the *sydoine* so singularly spoken of by Crusader Robert de Clari. So could it (likewise the *sydoine*) have been one and the same as the cloth we today know as the Turin Shroud? If this were correct, then the latter's missing history would be filled at a stroke, and Bishop d'Arcis's doubt that knowledge of it could have 'remained hidden' up to his time very readily explained.

In the early centuries the widespread popular understanding of the Image's origin was that its imprint of Christ's likeness had been created while Jesus was alive, not dead – though, as we will discover, ideas notably differed concerning how and when this imprinting might have occurred. Likewise the general run of artists' 'copies' of this cloth – which equally notably differ widely from one another (pl. 22b–d, 25b & 26a) – suggests that it was only a hand-towel-size piece of fabric, not anything on the scale of the Shroud. Accordingly they show only the face of Jesus, not a full body.

But as with so much else relating to our Turin Shroud, all is not necessarily as it might seem. The entirety of the known history of the Image of Edessa resides in what we today call the country of Turkey, where Max Frei's pollen findings indicated the Shroud had been at some time in the course of its history. In Turkey, centuries of wilful Islamic destruction and neglect have taken a huge toll on anything and everything Christian. However, in recent decades, in some of the country's remoter areas, some of the very oldest depictions of the Edessa cloth have been coming to light. Though rarely even noticed by the thousands of tourists who now flock to the region, some of these are to be found in the long-abandoned rock-cut churches and chapels dotted around one of the world's weirdest landscapes, the tufa pinnacles and 'fairy chimneys' near Göreme in Cappadocia (pl. 22a). One of the least visited of these churches, the so-called Sakli or Hidden Church, can be reached only by a steep, rocky track which requires the help of a guide even to find it. Above one of its arches is a painting of a wide piece of cloth imprinted with a sepia-coloured, front-facing face of Jesus with a striking semblance to the equivalent area of the Shroud we know today (pl. 22b & 23a). This particular painting dates no later than the mid eleventh century – a full two centuries earlier than the earliest date attributed to the Shroud by radiocarbon dating. And for the artist who created the painting, the original cloth he was depicting was already very old.

Independently, British-born linguist Mark Guscin, in the wake of years of intensive sleuthing among the libraries of the Eastern Orthodox monasteries at Mount Athos and elsewhere, has recently completed the first professional translations into English from manuscripts containing copies of some of the earliest key documents concerning the Image of Edessa. These documents have lain mostly untouched and unread for many decades. As Guscin's recently published studies have revealed, there was actually a lot more to this cloth than has been widely supposed about it, even among those of the Eastern Orthodox Church.

Could this once intensely revered 'lost' piece of cloth, readily corresponding to Bishop d'Arcis's description of our Shroud as having 'the Saviour's Likeness thus imprinted upon it', truly have been the cloth we today know as the Turin Shroud, simply long unrecognized as such?

The Byzantine world revelled in mystique and mystery. The Byzantines' ancestors, the ancient Greeks, believed in gods who exacted terrible punishments on mortals who beheld their divinity. Unlike the cloth we today know as the Turin Shroud, which has been frequently displayed before huge crowds throughout the last six hundred years, the Edessa cloth was surrounded with the densest clouds of the most hallowed secrecy. Penetrating that secrecy and mystery will not be an easy task. But that is the mission on which we are now about to embark.

Chapter 9

'Blessed City'

Blessed is the town in which you dwell, Edessa, mother of the wise. By the living mouth of the Son has it been blessed by the hand of his disciple. That blessing will dwell in it until the Holy One reveals himself.[1]

St Ephrem of Edessa, Syriac ecclesiastical writer, *c*. 370

A COMMUNITY OF exiles from Iraq living in the quiet, multicultural town of Fairfield near Sydney, Australia, might seem a long way removed from anything to do with the Turin Shroud. But this devout Christian community have an assembly room called the Edessa Hall, and their church, the Assyrian Church of the East, belongs neither to the Eastern Orthodox nor to the Roman Catholic rites, but is allegedly older than both. To this day these 'Assyrians' speak a Syriac language very close to that spoken by Jesus and his disciples, and according to their firmly held tradition a disciple of Jesus called Addai[2] brought Christianity to their forebears when they were living in the then pagan city of Edessa. Eastern Orthodox tradition, in turn, tells us that very soon after Jesus's crucifixion this same disciple brought to them from Jerusalem the mysterious Christ-imprinted cloth which would become known as the Image of Edessa. Edessa's king Abgar was converted, along with many of his citizens, and Edessa duly became the world's first, and oldest, Christian city.

That tradition – in fact an amalgam of two traditions, Assyrian and Orthodox – almost immediately provokes a hornets' nest of controversy among scholars. But before we get embroiled in this, one of the likeliest and most understandable of layman questions is, 'Where on earth is Edessa?'

The Edessa of our 'Image of Edessa' (not to be confused with its namesake in Macedonian Greece, or with Odessa in the Ukraine) is today known as Şanliurfa in south-eastern Turkey (pl. 14a). A very Islamic city just a few miles from the country's borders with Syria and Iraq, Şanliurfa is not exactly likely-looking as the world's first and oldest Christian city. Via the magic of Google Earth you can easily visit it on your home computer. Zoom in to a few close-ups of its ancient central area and the minarets of Islamic mosques are easy enough to spot, but you will not find a single Christian church. Not unexpectedly, it is not one of the popular destinations for American evangelical tour groups.

Yet as ever with the subject of the Shroud, appearances can be deceptive. If we were able to Google Earth back in time to the sixth century we would see a city bristling with dozens of Christian churches, representing three different denominations each with its theology school. Some of these churches were already several centuries old, enjoying a universal reputation for being the oldest in the Christian world. We would find pilgrims from foreign lands visiting its Christian shrines, one of them containing the bones of Jesus's disciple St Thomas, brought all the way back from India. We would note hundreds if not thousands of hermit-like monks camped outside the city's walls. In the minds of sixth-century Christians far and wide, Edessa was a very special city, blessed directly by Jesus himself.

But the starting point of our interest must be Edessa as it would have looked in Jesus's time. Originally known as Orhay in the local Syriac language, it was swept up by Alexander the Great's conquests and the Macedonian Greeks changed its name to Edessa. In the first century it was a prosperous mercantile city that greatly benefited from being at the crossing point of two well-established caravan routes, one heading east across Asia to India and ultimately China, the other heading south to Jerusalem and ultimately Egypt. Men in baggy trousers and turbans traded silks and spices in its market-places. They spoke much the same Syriac language as that of Jesus and his disciples. Yet they were not Jews by religion, instead worshipping the local pagan deities Bel and Nebo, so they had no qualms about figurative images.

Politically, Edessa lay as a small 'buffer' state between the mighty Roman and Parthian Empires, its rulers Arab petty kings of the Aryu or Lion dynasty installed by the Parthians after their rolling-back of Alexander the Great's conquests. Edessa's ruler contemporary with Jesus was King Abgar V (AD 13–50). And closely linked to the story of Abgar's conversion by Jesus's disciple Addai we find the origins of the Image of Edessa.

The oldest surviving history of Christianity was written early in the fourth century by Bishop Eusebius of Caesarea,[3] and in this Eusebius tells how Abgar V of Edessa, then suffering from an incurable disease, heard of the miracles Jesus was performing[4] and sent to Jerusalem a messenger bearing a letter addressed to Jesus, asking him to come to his city to heal him. Jesus declined, saying he needed to stay in Jerusalem to await his fate, but he blessed Abgar for his show of faith and promised that after being 'taken up' he would send one of his disciples to Edessa to cure him and bring him the Christian message.

According to Eusebius, Abgar's and Jesus's 'actual letters' were still preserved in the Public Record Office in their original Syriac, which he had translated for his readers' benefit. And joined with the two letters was another Syriac document, also dating from the time of the Abgar dynasty, telling what happened next. Although these original documents have inevitably disappeared, in the early nineteenth century some early Syriac manuscripts were discovered, one called the *Doctrine of Addai*,[5] which despite having some anachronistic later interpolations essentially checks out with Eusebius's account of what followed, suggesting that both were based on genuine earlier texts.

Apparently, the disciple Jesus had promised to send to Edessa was the very same Addai whom the Assyrian Church of the East still reveres as its founder, his name given by Eusebius in its Graeco-Roman form as Thaddaeus. On Addai's arrival, Abgar sent for him, whereupon according to the *Doctrine of Addai* version, 'When Addai came up and went to Abgar, who was accompanied by leading members of his court, on his going towards him a wonderful vision was seen by Abgar in the face of Addai. At the moment that Abgar saw the vision, he fell down and worshipped Addai. Great astonishment seized all those who were standing there before him, because they did not see the vision which was seen by Abgar.'

Immediately following this mysterious 'wonderful vision', Abgar reportedly declared his belief in Jesus, and Addai cured him of his disease. Addai was then allowed to preach the Christian message before an assembly of

Edessan citizens. Many were converted, including members of the city's Jewish community and even some of Edessa's pagan priests. According to both Eusebius and the *Doctrine*, all this happened in AD 30, the year of Jesus's crucifixion, and therefore well over a decade before St Paul started his missionary journeys.

Although neither Eusebius nor the *Doctrine* manuscript explained what the 'wonderful vision' was that had been seen solely by Abgar, later Eastern Orthodox tradition unhesitatingly identified this as the Christ-likeness imprinted cloth known as the Image of Edessa. Among the manuscripts in the Mount Athos monasteries studied by Mark Guscin were early copies of the Eastern Orthodox Church's tenth-century official *Story of the Image of Edessa*, which makes precisely this identification. According to this, Addai 'placed the Image on his own forehead and went in thus to Abgar. The king . . . seemed to see a light shining out of his face, too bright to look at, sent forth by the Image that was covering him.'[6] A typical depiction of this scene, though the Image is not shown actually on Addai's forehead, can be found in a manuscript of the eleventh century originally created at the Stavronikita monastery on Mount Athos, but which is today in Moscow (pl. 14b).

So what about that hornets' nest of controversy? Immediately needing to be made clear is that, even without its tenth-century Image of Edessa component, the Abgar story is one that historians have long viewed with the greatest scepticism. As far back as the fifth-century Pope Gelasius (pontificate 492–6), the letters supposedly exchanged between Abgar and Jesus were declared to be apocryphal, an assessment most modern-day scholars regard as fully justified. For instance, the second sentence of Jesus's letter, as quoted by Eusebius – 'It is written of me that those who have seen me will not believe in me' –

alludes to St John's gospel chapter 20 verse 29. It thereby presupposes that John's gospel had been written in Jesus's lifetime, which of course it could not have been. Bluntly, the letters seem to be fakes, albeit very early ones.

Furthermore, although historically there was an Abgar V of Edessa directly contemporary with Jesus, historians have long doubted that any king could have been converted to Christianity so early in the religion's existence without some independent record having been made. But apart from 'church propaganda' such as the *Doctrine of Addai*, no such record survives. Everything of Edessa's once famous Record Office, along with its equally once famous churches, was destroyed following the Turks' capture of the city in 1144. Abgar's coinage, which might have shown some religious affiliation, bore no likeness of him, having been issued in the name of his Parthian overlords. And in the *Annals* of the Roman author Tacitus, virtually our only surviving 'historical' source, Abgar V features unflatteringly only as a 'deceitful ruler', a trickster who favoured the Parthians rather more than the Romans.[7]

Abgar V was part of a dynasty of rulers bearing this same name, and one successor slightly more favoured by historians as the Abgar whom Addai converted (and who therefore may have been the true recipient of the Image of Edessa/Shroud) is Abgar VIII,[8] who reigned from 179 to 212. In its entry for the year 201, the *Chronicle of Edessa*[9] included a very detailed description of a lethal flood in Edessa during which the floodwaters 'destroyed the great and beautiful palace of our lord king and removed everything that was found in their path – the charming and beautiful buildings of the city, everything that was near the river to the south and north. *They caused damage, moreover, to the nave*[10] *of the church of the Christians* [my italics]. This is one of those tiny nuggets of information indicating that Christianity genuinely must have arrived very early in Edessa, to the extent of its having an officially recognized Christian church building as early as AD 201. As such this is a world first for Edessa, yet historians all too often sit on their hands over acknowledging this.

A second nugget is Abgar VIII's coinage. In a recent article describing the evidence for Abgar VIII's conversion to Christianity as 'extremely flimsy', distinguished Oxford Syriac scholar Professor Sebastian Brock remarked that 'important . . . in this connection is the negative evidence of the coins of the kings of Edessa, none of which bear any hint of a Christian symbol'.[11] When I pointed out to him that on several examples of Abgar VIII's coins, some of them housed in London's British Museum,[12] there is an unmistakable Christian cross on the king's head-dress (pl. 15a), Professor Brock very graciously acknowledged, 'It certainly looks as if I was too categorical.'[13] Abgar VIII, who issued his coins in close liaison with the Romans, seems to have dared to be open about his Christian affiliations only during the reign of Emperor Commodus, whose wife/mistress Marcia had Christian leanings. As the earliest-known instance of a monarch displaying the Christian cross symbol on his head-dress, this was another Edessan world first.[14] It also sets Abgar VIII's adoption of Christianity back in time to no later than AD 192, because Commodus died in that year.

A third nugget is an archaic-looking sculpted stone lion (pl. 15b) that stands forlornly in the open-air, outdoor section of Şanliurfa's present-day museum, typically with no accompanying explanatory information.[15] Judging by the hole drilled in the animal's mouth it clearly once served as a city fountain; but our interest is in what stands on top of its head: an unmistakable sculpted Christian cross, an all-too-rare sight in present-day Şanliurfa. In Syriac, the word for 'lion' is *aryu* – the name of Edessa's ruling dynasty. This fountain has to have stood in Edessa when the city was ruled by a Christian king of the Abgars' Aryu dynasty, a line that ended for ever when the Romans took over in AD 215.

We can therefore say with some confidence that Christianity arrived in Edessa while the city was ruled by members of the Abgar line, that one of these kings definitely adopted Christianity, and that this most likely happened before AD 192, because of the Abgar VIII/Commodus coin. But was Abgar VIII the first or the second of his dynasty to adopt the new religion? That is, was the Abgar of the story of the Image of Edessa's arrival in the city Abgar VIII, for whose acceptance of Christianity we have some definite supportive evidence, or was it Abgar V, Jesus's direct contemporary, as attested by Eusebius and the *Doctrine of Addai* manuscript, but otherwise unsubstantiated?

Strongly favouring the latter is the fact that the known circumstances of Abgar VIII's reign and its immediate aftermath simply do not 'fit' the *Doctrine of Addai*'s account of events after the 'wonderful vision' episode and King Abgar's conversion. According to the *Doctrine*, Addai went on to make many converts in Edessa and its surrounds, among these Aggai, maker of the royal head-dresses, before dying a peaceful, natural death in the city. Addai was then greatly honoured by Abgar by being buried in the same great sculpted mausoleum 'in which those of the house of Aryu, the ancestors of the father of king Abgar, were placed'. Abgar then died himself, and 'years after' there was a reversion to the old pagan religion by 'one of his sons'. When this son called upon Aggai to renounce his Christianity and make him a head-dress bearing the old pagan symbols, Aggai refused, whereupon the son ordered Aggai's legs to be broken, resulting in his death.

Abgar V is known to have died in AD 50, and to have been followed by two sons, first Ma'nu V, then on his death in AD 57 a younger son who became Ma'nu VI. If Christianity had achieved significant success in Edessa under Abgar V and Ma'nu V, it is easy to understand aggrieved priests of Edessa's supplanted old religion persuading Ma'nu VI to turn back the clock. Not only do these circumstances fit those recounted in the *Doctrine of Addai*, they also make sense of how the Shroud, if it was brought to Edessa very soon after the crucifixion, could well have disappeared again very quickly – indeed, before a single gospel had been written – as our later information indicates happened.

The circumstances of Abgar VIII's reign were quite different. After this Abgar's death only one son succeeded, Abgar IX, whom the Romans almost immediately seized and deposed, thereafter making Edessa a Roman *colonia*. The monarchy was never reinstated, so there was neither the time nor the right circumstances for any successor of Abgar VIII to instigate persecutions. And the *Doctrine of Addai* gives no hint of any Roman involvement in this particular story, even though Edessa has plenty of later traditions of Christian martyrdoms that did take place while the city was under Roman mastery.

Also needing to be pointed out is that Addai, albeit as the disciple of Christ responsible for evangelizing Edessa, actually has rather more historicity than many modern commentators are prepared to allow. As early as AD 190 the Church father Clement of Alexandria, in his book *Outlines*, alluded to the existence of Addai's tomb in Edessa as part of a listing of the burial places of Jesus's disciples. Clement, who lived between *c*. 150 and 215, would hardly have included Addai in such a list if this individual had been contemporary with his own time. The site still exists,[16] on a spectacular mountain-top location some six miles out of present-day Şanliurfa, reachable by the roughest and most winding of tracks (pl. 16a). The bones of Addai and Abgar are of course long gone, historically recorded to have been transferred in the year 494 to the safety of a church inside the walls of Edessa due to Persians raiding the surrounding countryside – and just as well, for today the site consists of little more than mounds of rubble.

But the overriding point is that western Christianity's New Testament may lack any mention of whoever might have taken charge of Jesus's Shroud after the crucifixion, just as Bishop d'Arcis insisted so stridently, but in the world of eastern Christianity the disciple Addai was not only associated with the bringing of a Christ-imprinted cloth to Edessa at some time earlier than AD 192, he was also sufficiently flesh-and-blood and non-legendary for the whereabouts of his physical remains to be known and reliably recorded.

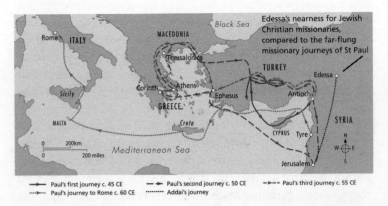

Fig. 16 THE MISSIONARY JOURNEY OF ADDAI TO EDESSA, seen on the dotted line at the right, compared with the far-flung missionary journeys of St Paul. Edessa's proximity for Jesus's disciples, the fact that its citizens spoke their same language, and its location on a major trade route, all make it very credible as a location for first-century Christian missionary activity.

That Addai's Image-bearing missionary journey (fig. 16), even though it did not gain a mention in western Christianity's canonical gospels, happened in the first century rather than the second is further indicated by any glance at a map of the missionary journeys of St Paul. Every one of Paul's journeys started from Antioch, modern-day Antakya in south-eastern Turkey, from which he ventured five hundred miles westwards to Ephesus, a further five hundred miles westwards to Malta, and ultimately even further, to Rome. In contrast to these far-flung destinations, Syriac-speaking Edessa lies only 180 miles to Antioch's east, and on a direct trade route from both Antioch and Jerusalem. Is it really likely that throughout Christianity's first 150 years the first Christians should have ignored Edessa as a target for their missionary activities?

That they did not is further indicated by the chronicle of one of Edessa's further-flung neighbours, the small border kingdom of Adiabene, whose capital was Arbela, today the large Iraqi city of Arbil. Arbela's ancient lineage of bishops began with one Pkhida, who can reliably be dated to the year 104. And according to Arbela's chronicle it was Addai who converted Pkhida to Christianity, thereby again indicating that Addai belonged to Abgar V's first century rather than Abgar VIII's second. As has been pointed out by the Estonian-born American scholar Arthur Voobus, if Christianity had reached as far as Adiabene by the year 100, there can be 'no doubt' that in Edessa 'the Christian faith had been established before the end of the first century'.[17]

Being as sure as we can be that the Image arrived very early in Edessa is important, because the next difficulty we face is that almost as quickly and mysteriously it vanished, and in circumstances sufficiently dire that all living memory of its hiding place became lost, arguably as a result of drastic persecution of that first Christian community, just as had happened to Addai's successor Aggai.

That the Image certainly did not stay around is evident from two key facts. First, when it was dramatically re-discovered in the sixth century (the circumstances of which we will learn in the next chapter) it had clearly been very purposefully hidden away, and had remained that way for a very long time. Second, when Christianity was re-established in Edessa (as it arguably was from Abgar VIII's reign onward), there was no sign of the Image. Instead, what seems to have remained in Edessa through the next several centuries was a deep sense that the city had been specially blessed by Jesus, his supposed letter to King Abgar, fabrication though it may have been, being very central to this, and acting as some kind of substitute.

For however unconvincing this letter may have been to Pope Gelasius in Rome, and may be to modern scholars, its fame spread right across the world, and with quite remarkable potency. Numerous papyrus and parchment copies have been found as far afield as Egypt dating from the fourth through to the thirteenth centuries, some of them having magical protective properties associated with them. Versions inscribed on stone have been found in northern Anatolia, at Philippi in Macedonian Greece, and at Kirk Magara near Edessa (fig. 17). In England it was included in a service book of the Saxon era, in a position of honour immediately after the Lord's Prayer and the Apostles' Creed.[18] From the variations in text between one early example and another – reflected also in the variations between manuscript renditions – there is a very strong sense that there never was one master 'authorized' version. Versions from the late fourth century on, for instance, enigmatically take on the extra sentence 'Your city shall be blessed and no enemy shall ever be master of it.'

Had the Christ-imprinted cloth Image of Edessa been around late in the fourth century the one person who would undoubtedly have let us know all about it was a highly observant lady pilgrim whom historians mostly label Egeria, in the absence of any certain knowledge of her real name (the single manuscript of her travels, when it was found by chance in Italy in the late nineteenth century, had lost both its beginning and its end, leaving only the middle). Having journeyed all the way from western Gaul or Spain, Egeria arrived in Edessa some time between the years 384 and 394. If anything as interesting as the Image of Edessa had been in evidence in the city, there can be no doubt that this intrepid lady would have sought it out, and given us a full description.

With engaging chattiness, Egeria tells us that she visited Edessa's newly built church containing the remains of St Thomas, recently brought from India. Hosted by the local bishop, she went on to the still extant palace of the Abgar dynasty, where she viewed stone sculptures of Abgar and his son 'Magnus' (i.e. Ma'nu).[19] Next on her itinerary was

Fig. 17 'BLESSED ARE YOU, ABGAR . . .' Jesus's blessing to King Abgar, as it is likely to have been displayed on the gate of Edessa, from a copy of the inscription in Greek, found by a grave in the Kirk Magara district of Şanliurfa. This was transcribed by German scholars at the beginning of the twentieth century, and published in 1914, but has long since disappeared.

Edessa's famous fish-pools (pl. 15c), the only tourist attraction from her time that still survives in Şanliurfa. Last stop was the city's gate, where the bishop read out to her Jesus's letter to Abgar – its text with the added protective sentence seems to have been inscribed on the gate's brickwork – after telling her a long introductory story about how it had magically protected Edessa from a Persian army's attempt to capture the city.[20] But Egeria made no mention of any Image being kept in the city. And for well over a century after Egeria's time other prolific contemporary writers, among them the famous St Ephrem of Edessa, were also silent on the subject. It was as if it had never existed.

One important point to be observed is that even though this 'silent' pre-sixth-century period was much closer to the time of Jesus than our own time, there prevailed an essentially universal lack of any awareness of what Jesus had looked like. One of the notable omissions on the part of the gospel writers was the provision of any detail of Jesus's physical appearance. And because of the already mentioned Jewish abhorrence of images it is most unlikely that anyone ever painted a portrait of Jesus in his lifetime.

So when, in the reign of the Roman emperor Constantine the Great, Christianity became an official religion of the formerly pagan Roman Empire in which representational images abounded, people very understandably began asking what Jesus had looked like. And despite the strong disapproval of some traditionalist churchmen such as the earlier mentioned Bishop Eusebius,[21] representational images of Jesus gradually began to creep in.

Probably the earliest known example is a poorly preserved mid-third-century fresco found at Dura Europos featuring Jesus healing the paralytic man. Curiously, it depicts him as young, beardless and with short hair.

Altogether better preserved is an outstanding fourth-century mosaic from a Roman villa in Dorset, England, today displayed in the British Museum (fig. 18). The face is again vague, youthful and beardless, only identifiable as Jesus because of the monogram behind his head.

In the fifth century it is much the same story. Despite the occasional equally vague bearded exceptions,[22] the youthful, beardless, Apollo-like type predominates, as in numerous depictions of Jesus's miracles on sarcophagi preserved in the Vatican museums and in the Museum of Archaeology in Istanbul (pl. 17a). The general lack of any clear idea of what Jesus had looked like is explicitly confirmed by St Augustine, from this same century. Augustine

Fig. 18 PORTRAIT OF JESUS, FOURTH CENTURY. Detail of mosaic from a Roman villa at Hinton St Mary, Dorset, England, showing the beardless likeness of Jesus typical at this early time.

described the portraits of Jesus extant in his time as 'innumerable in concept and design', and flatly stated, 'We do not know of his external appearance, nor that of his mother.'[23]

Readily apparent, then, is that throughout exactly the same near five-century period of the Christ-imprinted Edessa cloth's mysterious absence from the historical record, its whereabouts unknown, there prevailed a corresponding lack of any recognized authority, textual or visual, for what the human Jesus had looked like.

But all this was about to change with a very remarkable rediscovery.

Chapter 10

Rediscovered – as 'King of Kings'

When Orthodox think of Christ Crucified, they think not only of his suffering and desolation; they think of him as Christ the Victor, Christ the King.
Bishop Kallistos Ware, *The Orthodox Church*, 1963

FOLLOWING THE WEAKENING of the old Roman Empire, and its reinvention of itself as the Constantinople-controlled Byzantine Empire, Edessa remained one of this 'new' empire's border outposts. But as a prosperous commercial centre it presented a tempting prize for predatory neighbours, and rarely was the danger more acute than in 503, when Persian King of Kings Kavadh[1] turned up before its walls, riding an elephant at the head of a large and fearsome army. Remarkably, Edessa survived. In the words of a contemporary chronicler, 'The whole plain was filled with them [Kavadh's Persian army], and all the gates of the city were open, but the Persians were unable to enter it because of the blessing of Christ. Fear overcame them, and they remained at their own positions . . .'[2]

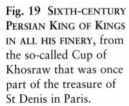

Fig. 19 SIXTH-CENTURY
PERSIAN KING OF KINGS
IN ALL HIS FINERY, from
the so-called Cup of
Khosraw that was once
part of the treasure of
St Denis in Paris.

Whatever we may think of the protective powers of
Jesus's letter of blessing to King Abgar, something very
powerful saved Edessa from the Persian army on that
occasion.[3]

Some forty years on, in the late summer of 544, there
was a virtual re-run when Kavadh's favourite son and
successor Khosraw[4] (fig. 19) appeared before Edessa's
walls with an even more impressive army. Though there
was now a fierce struggle, the Persians building a huge
mound of timber to scale Edessa's walls, and the Edessans
digging a tunnel beneath their walls to undermine this, the
result was much the same. Khosraw, like his father before
him, was obliged to leave almost empty-handed.

But on this later occasion, as we learn from a near-
contemporary chronicler, the agent of Edessa's divine
protection was significantly different. According to
Evagrius, who was a schoolboy at the time of Khosraw's
attack, at the moment when all seemed lost for the
Edessans they

brought the divinely created Image, which human hands had not made [*acheiropoietos*], the one that Christ the God sent to Abgar when he yearned to see him. Then, when they brought the all-holy Image into the channel they had created and sprinkled it with water, they applied some to the pyre and the timbers. And at once the divine power made a visitation to the faith of those who had done this, and accomplished what had previously been impossible for them: for at once the timbers caught fire and, being reduced to ashes quicker than word, they imparted it to what was above as the fire took over everywhere.[5]

No longer was Jesus's letter to Abgar being accredited with protecting Edessa. Now, and without a word of explanation, the Image of Edessa was suddenly at large again as an extant historical object, and possessing the same power as the letter. Indeed superior to it, because from now on it completely took over the letter's protective role, as if it had been the true agent all along.

Furthermore, the Image was quite unmistakably being regarded as the Christ-imprinted cloth which had been brought to King Abgar five centuries earlier, even though the earlier extant documents had spoken only of a 'wonderful vision' seen by Abgar alone. And for the first time its imprint was being specifically described as 'not by hand made' – the literal meaning of the Greek word *acheiropoietos* used by Evagrius. Although this is the first instance of the Image of Edessa being described in this way, it would certainly not be the last, and other later descriptions, using different words, convey much the same meaning. In a nutshell we have exactly the same 'miraculous' Christ-imprinted property that has seemed so special about the Turin Shroud being attributed to an otherwise still very mysterious piece of cloth present in sixth-century Edessa, at a date a full seven hundred years before the earliest date attributed to the Shroud by radiocarbon dating, and with a link directly back to the first century AD.

So what had happened? How and when had the Image of Edessa come to light after having gone unmentioned for so long? Typical of the difficulties the subject presents, no immediately contemporary description of this event survives. Only the tenth-century *Story of the Image of Edessa*, the texts of which Mark Guscin researched so assiduously in the monasteries at Mount Athos, provides a glimmer of an idea.

The *Story*'s relevant section begins by explaining how after Abgar V had been converted to Christianity he ordered the image of a pagan god that had been over his city's gate to be taken down and replaced by the Image of Jesus. After his death, when Abgar's second son reverted to paganism, the son ordered the pagan image to be restored, and that of Christ destroyed. However, Edessa's bishop of that time managed to pre-empt this. In the words of the tenth-century writer, 'Given that the place where the Image was kept was shaped like a cylindrical semi-circle, he [the bishop] showed great foresight and lit a lamp in front of the Image and put a tile on top of it. He then sealed the surface off with gypsum and baked bricks, finishing the wall off on the same level.'[6]

The *Story* then moves us on five hundred years to its version of the dramatic events of Khosraw's siege of Edessa in 544. Reputedly the moment of greatest danger came when Edessa's citizens became aware of the Persians digging underneath their walls:

The city's inhabitants were at a loss and had absolutely no idea what to do ... At night the bishop – Eulalios – had a vision ... telling him to get the image of Christ that had not been made by human hands and parade it in a procession, and the Lord would certainly demonstrate his wonders. The bishop answered that he did not even know if the Image existed, and if it did whether it was there or anywhere else. The one who had appeared to him . . . told him that it was hidden away above the city gate in such and such a place, and in such and such a way. The bishop was encouraged by the clarity of the vision and went to the place in solemn procession. He searched and found the sacred image unharmed, and the lamp that had not gone out after so many years. Another likeness of the first like-ness had been formed on the tile that had been placed in front of the lamp for protection, and it is still kept in Edessa even today.[7]

Oil from the lamp ignited, incinerating the Persians burrowing beneath Edessa's walls, and when Bishop Eulalios processed around the top of the walls 'holding the Image outstretched in his hands' a change of wind caused the flames from the Persians' burning wood pyre to blow back in their faces, panicking them and precipitating Khosraw's ultimate withdrawal.

Now, tall stories abounded in the Byzantine world and we have no need to believe that the Image saved sixth-century Edessa quite so single-handedly. Nor is there any Eulalios from this time to be found in the surviving lists of Edessa. But what the *Story*'s writer tells us about where the Image was found, it having apparently remained in this same location throughout five centuries, has an impressive factuality to it. Indeed, that the Image's rediscovery was a real-life sixth-century archaeological discovery – one that seems to have happened some time between 503, when the letter of Jesus was still Edessa's protection, and 544, when the Image had evidently taken over this role – is made very credible.

According to the *Story*'s information, the Image had been kept 'hidden away above the city gate' in a place

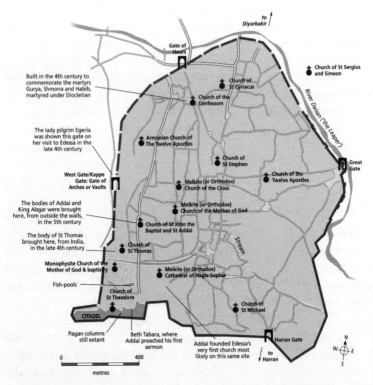

Fig. 20 EDESSA IN ITS HEYDAY AS A CHRISTIAN CITY, showing the main locations of relevance to the story of the Image of Edessa. Only the citadel remains.

'shaped like a cylindrical semi-circle'. Various icons depict a discovery-place of this kind (pl. 17b), and later depictions of the Image right across the Orthodox world are frequently located over entranceways (see, for example, pl. 23a & 26a), so the tradition of this location was clearly a strong one. Sixth-century Edessa had at least four gates set into its walls (see map, fig. 20), and the western gate was specifically known as the Kappe Gate, meaning Gate of Arches or Vaults. This seems to fit the *Story*'s 'cylindrical semi-circle' description very well. Vestiges of this gate were still extant at the time of my first visit to Şanliurfa in the 1970s, though hardly anything remains today.

Also making this Kappe Gate a credible candidate for the Image's hiding-place over the five centuries (and therefore potentially the Shroud's) is its commanding height above the broad flood-plain below. Due to some occasional very erratic behaviour on the part of Edessa's river Daisan ('the Leaper'), the city suffered several serious floods during the centuries the Image was immured above the gateway: in 201 (as earlier noted), and again in 303, 413 and 525. To this day vestiges of the line of Edessa's western walls, in which the Kappe Gate was set, can be seen rising above Şanliurfa's Haleplibahce Street, way above any conceivable flood height. As remarked in the *Story*, whoever chose this location to hide the Image had indeed shown 'great foresight'.

But why would anyone, in any era, choose such a strange hiding-place? The clue lies in the 'tile' found with the Image. Like the Image itself, this tile became a historical object in its own right, called by the Byzantines the Keramion (fig. 21), and specifically noted by the tenth-century *Story* writer as 'still kept in Edessa even today'. It is thereby known to have been a piece of ceramic with Jesus's face represented on it in the identical front-facing disembodied manner of its companion. Back around the first century it was common practice in Parthian tributary states, and indeed elsewhere in the pagan world (fig. 22a), for relief sculpture heads of gods to be set up over gateways, or their near vicinity. Examples can still be seen at Parthian Hatra, one of Edessa's near neighbours (fig. 22b). Typically for Parthian art, such heads were rigidly frontal. Because the area of the face on the Shroud happened to exhibit this same rigid frontality, it would have been perfectly natural for Abgar, who had none of the Jewish qualms over religious images, to order a ceramic version of his new 'god' set up over his city's gateway.

Accordingly, it would have been this same ceramic, or tile, version of Jesus's face, rather than the Image itself, as described in the *Story*, which Abgar's second son ordered to be removed from above the gate when he reverted to paganism and began persecuting Edessa's Christians. Whoever carried out this removal may have simply turned the tile around so that its 'face' side was turned inwards to the cavity behind. The clay oil lamp reportedly found in the same cavity suggests that this operation was carried out at night. And someone seems to have had the idea of using this same cavity to hide the Image/Shroud until the persecutions of Edessa's Christian community had blown over. By daybreak the gateway's brickwork would have been sealed up with mortar, no evidence of any Christ portrait remaining. If this was indeed how and where the

Fig. 21 TYPICAL DEPICTION OF THE KERAMION, OR HOLY TILE. From a late-twelfth-century church at Episkopi in the Mani, Greece.

Shroud lay hidden between the mid first century and some time in the first half of the sixth century, it would certainly have enjoyed near hermetically sealed conditions[8] throughout.

But why should we believe that this Image of Edessa cloth was our Shroud? The main clue lies in a quite extraordinary change in how artists portrayed Jesus's likeness, which happened very soon after the Image of Edessa cloth came to light. We noted in the last chapter how right up until at least the end of the fifth century the portrayals of Jesus lacked any authority, most representations depicting him beardless. As evidenced by St Augustine's remarks, there was a general lack of any awareness of what he looked like. But in the art of the sixth century there occurred a remarkable transformation in the way Jesus was depicted.

Just two of several surviving examples will serve to illustrate this. The first is a 'Christ Pantocrator' icon painted in encaustic – a wax technique, the recipe for which became lost after the eighth century – that is preserved in the remote monastery of St Catherine in the Sinai desert (pl. 17d). The second is a relief portrait of Christ on a silver vase that was found at Homs in Syria, and is now in the Louvre in Paris (pl. 17c). Firmly datable to the sixth century, both are authoritative, definitive versions of the distinctive likeness that today we instinctively recognize as Jesus Christ. And if we compare these front-facing likenesses with the face as visible on the Shroud before any discovery of the hidden photographic negative, there is a very uncanny resemblance: the same frontality, the same long hair, long nose, beard, etc. It is as if someone has studied the Shroud's facial imprint and for public consumption has very carefully crafted an interpretative official likeness from this in the guise of Christ Pantocrator – the 'King of All'.

Fig. 22 HIDING PLACE OF THE IMAGE OF EDESSA. (a) Second-century image of a god set above a fountain, Sagalassos, Turkey, and (b) mythological heads on the Parthian palace at Hatra, Iraq, illustrative of the type of pagan decoration that would have been popular in the time of the Abgar dynasty. (c) An artist's reconstruction of the hiding place of the Image of Edessa cloth, and its companion Holy Tile, or Keramion, above the West Gate of Edessa. The tile may have been turned around to face inside the cavity. Or it may have been laid protectively on top of the Image, as indicated here.

Now, historically no one tells us directly anything of what happened to cause this new and unprecedented confidence in what Jesus had looked like. If the Shroud was one and the same as the Image of Edessa, the latter was regarded as far too holy for anything so unseemly as a public showing. We may recall that on its arrival in Edessa back in the first century King Abgar was the only person privileged to view it, and on its rediscovery in sixth-century Edessa it seems to have been kept with similar reserve and secrecy.

But thanks to a remarkable manuscript discovery made at St Catherine's Monastery in 1975, which included a cache of early Georgian manuscripts whose contents were translated only subsequent to 1994, we now know that there was in the sixth century a specific movement to disseminate this very likeness, and that this movement stemmed from Edessa and its environs.

Tradition in Georgia, the former republic of the old Soviet Union, has long held that some time around the mid 530s twelve Assyrian monks left Mesopotamia and travelled north to found several monasteries in Georgia. Present-day tour groups to Georgia can follow in these missionary monks' footsteps, and in Georgia's capital Tbilisi there is a very badly worn sixth-century Christ Pantocrator icon, the Anchiskhati – an almost exact counterpart to the one at St Catherine's Monastery in the Sinai – which is thought to have been brought to Georgia by this mission (fig. 23a).[9]

The quite remarkable new insight from one of the recently discovered Georgian documents from Sinai[10] is what it tells of the activities of two of these Assyrian monks, Theodosius from Edessa and Isidore from Edessa's sister city Hierapolis. Theodosius is specifically described as 'a deacon and monk [in charge] of the Image of Christ' in Edessa. As Georgian scholars recognize,[11] this Image can

be none other than our Image of Edessa, thereby confirm-
ing Evagrius's information that this was an extant
historical object by this time,[12] one evidently sufficiently
important to have its own 'carer'. Theodosius's companion Isidore was apparently responsible for a tile image[13]
belonging to Edessa's sister city Hierapolis.[14] Both monks
travelled to Georgia specifically to paint interpretative
versions of their charges for the newly founded churches
there.

Never before have we been afforded a glimpse of who
lay behind the rash of Christ portraits that appeared in the
sixth century. It is quite evident from the Georgian document that they were Assyrian artist-monks from Edessa
and its environs who saw themselves as missionaries or
icon evangelists for the newly revealed 'divine likeness'
that had been so recently rediscovered in Edessa. Even
little details such as their travelling in twos puts one in
mind of the way Jesus in person sent out his disciples five
hundred years earlier. Because of the awesomely divine
nature of the imprint they had been highly privileged to
witness directly, their artworks had a lot of magic
associated with them: one of them reputedly appeared
miraculously on a Georgian church wall accompanied by
the message in Syriac lettering, 'I am revealed by He whom
you see here'.

With the exception of the Tbilisi icon, none of these
sixth-century missionaries' artworks would appear to have
survived in Georgia, where in the sector that is now north-
eastern Turkey the once superbly frescoed churches lie in
appalling ruin and neglect (pl. 18b). However, other parts
of the then Christian world provide us with a slightly
better glimpse of the monks' 'icon-evangelizing' activities.
One example is a little-known sixth-century Christ face in
a roundel painted on the underground walls of a spring,
the Holy Spring of Nicodemus, at Salamis in Cyprus.

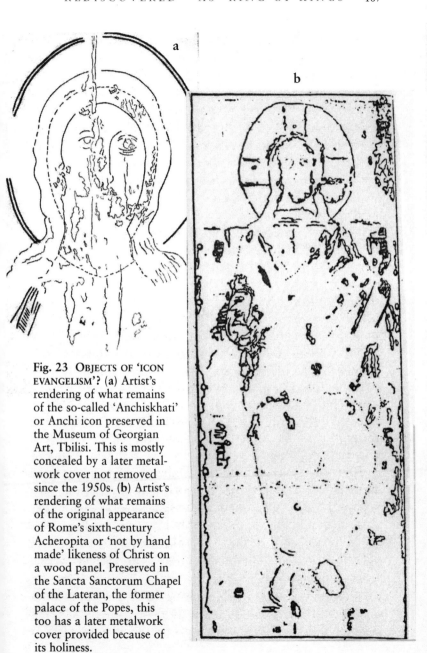

Fig. 23 OBJECTS OF 'ICON EVANGELISM'? (a) Artist's rendering of what remains of the so-called 'Anchiskhati' or Anchi icon preserved in the Museum of Georgian Art, Tbilisi. This is mostly concealed by a later metalwork cover not removed since the 1950s. (b) Artist's rendering of what remains of the original appearance of Rome's sixth-century Acheropita or 'not by hand made' likeness of Christ on a wood panel. Preserved in the Sancta Sanctorum Chapel of the Lateran, the former palace of the Popes, this too has a later metalwork cover provided because of its holiness.

Though this was found back in the 1930s[15] and it and its accompanying inscriptions carefully copied by an artist (pl. 18a), the Turkish invasion of northern Cyprus in 1974 subsequently rendered it inaccessible. However, Cyprus archaeologist Marc Fehlmann has recently managed to re-photograph it (pl. 19b),[16] and a fascinating element is its inscription *menuete ho Theos* – 'God is revealed' – immediately recalling the 'I am revealed' inscription of the lost Georgian example described in the manuscript discovered at St Catherine's, Sinai.

Very likely another work of this same 'icon-evangelizing' group was a reputedly *acheiropoietos* mosaic head of Christ in the St John Lateran Basilica, Rome, the original sixth-century appearance of which has sadly been almost entirely lost as a result of clumsy nineteenth-century restoration (pl. 18c). However, just like the Georgian examples, this powerful likeness was said to have appeared 'miraculously' in the basilica's apse[17] – very credible, because a thirteenth-century restoration found it to have been created on its own independent bed of travertine marble, enabling it to be clandestinely installed overnight.[18] Another example in the sanctuary of the Church of the Holy Sepulchre in Jerusalem – now lost, but its appearance preserved in depictions on pilgrim flasks[19] – is likewise said to have appeared miraculously. A lot of magic was associated with these images – arguably a spin-off from the revelatory original that had inspired them.

Reinforcing the association of these very special Christ likenesses with the Image of Edessa is the sixth-century mosaic fragment of the same face (pl. 19a) that has recently come to light in Şanliurfa, as described in this book's introduction. Comparison of this fragment with the sixth-century Cyprus wall-painting reveals a compelling similarity. It is impossible to be sure what surrounded the face on the Şanliurfa mosaic, but the likelihood is a roundel, as on another strikingly similar version on a sixth-century icon that was at St Catherine's in Sinai but which is now in the City Museum of Eastern and Western Art in Kiev (pl. 19c).

At this period there seem to have been no depictions of the Image of Edessa in the later 'popular' form of a disembodied face on a landscape-aspect cloth (as seen in plates 22, 25b and 26a). Instead there were at least three interpretative versions: a 'pure' roundel version, in the form of the face disembodied within a circle, as on the Kiev icon (pl. 19c), and on a recently discovered wall-painting in the church of the Holy Cross, Telovani, Georgia (fig. 24);

Fig. 24 ROUNDEL FORM OF DEPICTING THE IMAGE OF EDESSA, from a wall-painting of the late eighth century found in 1989 in the Church of the Holy Cross, Telovani, Georgia. The accompanying Georgian inscription specifically describes it as the 'Holy face of God'.

the 'head and shoulders Pantocrator' version, which might be either within a roundel or on a rectangular icon, as on the Homs vase and the St Catherine's Pantocrator icon respectively (pl. 17c & d); and a full-length Christ Enthroned, of which one example is a sixth-century icon in Rome, specifically called the Acheropita (fig. 23b), which was carried in procession to invoke its protective properties whenever Rome was under some kind of threat[20] – exactly like the Image of Edessa in 544. Such versions were entirely in keeping with Byzantine Orthodox thought: as has been pointed out by present-day bishop Kallistos Ware, even in death Christ had to be perceived as 'Christ the Victor, Christ the King, reigning in triumph'.[21]

But we still come back to a key question. Even if we accept that the Image of Edessa was a piece of cloth bearing Christ's imprint, so far that is all we have heard about it, that it bore the face of Jesus. Why should we believe it was a cloth of the fourteen-foot dimensions of the Shroud? And what evidence do we have that this Edessa cloth actually was the Shroud?

In the case of the Image of Edessa's dimensions, one important indicator is to be found in one of the very first documents to provide a 'revised version' of the King Abgar story in the wake of the cloth's rediscovery. The document in question is the *Acts of Thaddaeus*, dating either to the sixth or early seventh century. Although its initially off-putting aspect is that it 'explains' the creation of the Image as by Jesus washing himself, it intriguingly goes on to describe the cloth on which the Image was imprinted as *tetradiplon* – 'doubled in four'. It is a very unusual word, in all Byzantine literature pertaining only to the Image of Edessa, and therefore seeming to indicate some unusual way in which the Edessa cloth was folded.

So what happens if we try doubling the Shroud in four? If we take a full-length photographic print of the Shroud, double it, then double it twice again, we find the Shroud in eight (or two times four) segments, an arrangement seeming to correspond to what is intended by the sixth-century description (fig. 25). And the quite startling finding from folding the Shroud in this way is that its face appears disembodied on a landscape-aspect cloth exactly corresponding to the later 'direct' artists' copies of the Image of Edessa.

In the *Story of the Image of Edessa*, the Image is

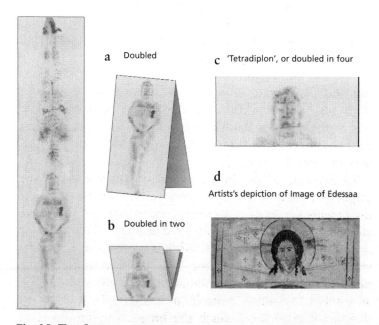

a   Doubled

c   'Tetradiplon', or doubled in four

d   Artists's depiction of Image of Edessaa

b   Doubled in two

Fig. 25 THE SHROUD FOLDED *TETRADIPLON*, OR DOUBLED IN FOUR. When the Shroud is folded in the 'doubled in four' manner of texts describing the Image of Edessa, the face appears disembodied on a landscape aspect cloth in the exact semblance of artists' depictions of the Image of Edessa. Here the Shroud is reconstructed with the damage from the fire of 1532 and from the 'triple burn-hole' incident both removed.

specifically described as mounted on a board. So a folding for presentation purposes in this 'doubled in four' way actually makes a great deal of sense. It reduces the Shroud's extremely awkward fourteen-foot length into a manageable and presentable twenty-one inches by forty-five inches, and displays by far the most meaningful section of the cloth, the face. And if we think of the face as seen in this way in the dim lighting conditions of a church interior – conditions in which, as we know from surgeon Dr Pierre Barbet, the different colour of the bloodstains does not show up – it is easy to understand how the face might have been supposed to be of a watery origination, exactly as envisaged in the sixth-century *Acts of Thaddaeus* account.

Moreover, not only does this document use the word *tetradiplon*, thereby indicating the Image of Edessa to have been on a large cloth, in the very next sentence it also uses the word *sindon* for the Image – the very same word all three synoptic gospel authors used for Jesus's burial shroud. Nor is the *Acts* alone in this. Both Mark Guscin[22] and Georgian scholar Dr Irma Karaulashvili[23] have pointed out three other documents from this same early period which do exactly the same. This is not to suggest that the Image of Edessa had necessarily yet been recognized as Jesus's burial shroud; it may well not yet have been unfastened from its 'doubled in four mounting'. For our present purposes the documentary confirmation that the Image of Edessa was a large cloth, and not the hand-towel size often envisaged, is enough.

This said, however, even from much this same early time there is actually one further even more compelling indicator that the Image of Edessa was one and the same as our Shroud. The seventh century saw another wave of Pantocrator-type depictions of Christ, which we have shown to be based on the Image of Edessa. One of these

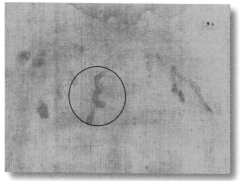

9a (*above*) Detail of 'crown of thorns' bloodstains on the forehead. Circled is the stain in the shape of a reversed '3' that has particularly impressed medical specialists.

9b (*above right*) Danish physician Dr Niels Svensson, one of many present-day medical specialists convinced of the Shroud's authenticity.

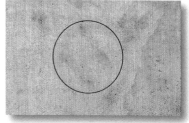

9c Two of the hundred-odd dumb-bell-shaped marks (circled) visible on the Shroud, indicative of a savage whipping.

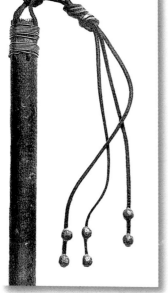

9d Reconstruction of a Roman whip called the flagrum, tipped with dumb-bell-shaped pellets.

9e Roman coin of *c*. 100 BC showing a flagrum in use against a naked man in a gladiatorial contest.

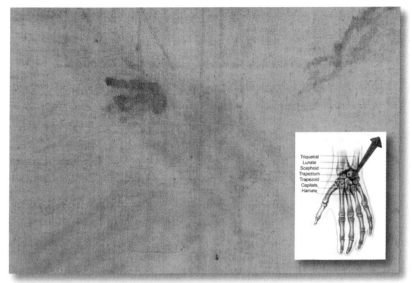

Triquetral
Lunate
Scaphoid
Trapezium
Trapezoid
Capitate
Hamate

10a Bloodstains on the Shroud as from nailing through the wrists.

10b (*inset*) Anatomical view of the hand, showing the apparent point of penetration of the nail through the wrist bones.

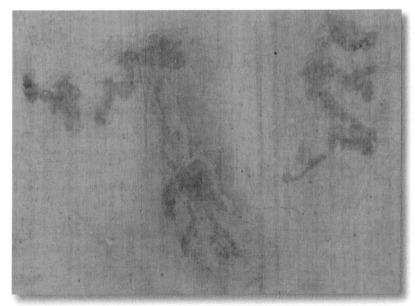

10c Bloodstains on the Shroud as from nailing through the feet. Note the blood at left which seems to have come from the area of the ankle or heel, and spilled directly on to the cloth when the body was laid in it.

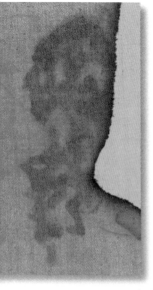

11a (*far left*) Bloodstain as from bladed weapon plunged into the chest.

11b (*left*) Example of Roman lancea described in the gospel of John as used to check Jesus was dead.

11c (*below*) Identical wound on 'Dying Gaul' statue, Rome.

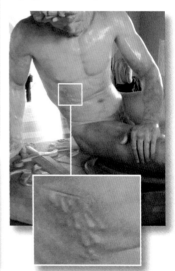

11d (*right*) The lance-wound seen in negative, showing actual position in chest.

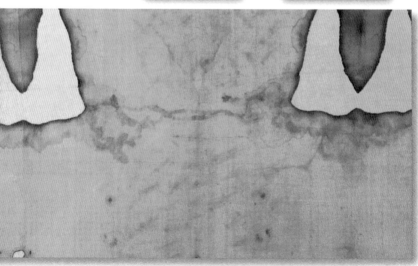

11e Back-of-the-body bloodstains apparently spilled from the lance-wound in the chest.

12a (*left*) High-definition close-up of the area of the nose on the Shroud (see area within rectangle in **12b** *below*), showing the lack of any pigment-type substance composing what the eye sees as the body imprint. Note the distinctive weave, also a bloody tinge to the bottom of the moustache.

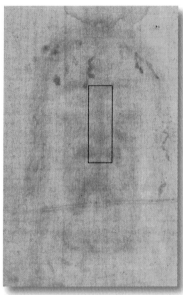

12c Highly magnified area of bloodstain, showing the presence of blood particles.

13a (*above*) Conservator Dr Mechthild Flury-Lemberg working on the Shroud during the major conservation initiatives undertaken in 2002.

13b (*right*) The Shroud's previously inaccessible underside, exposed by Dr Flury-Lemberg's work for the first time in more than four centuries. The body image is all but invisible on this side, but the bloodstains are still apparent. Laid out on the white rectangles are the patches sewn on to the Shroud to hide the damage sustained during the 1532 fire.

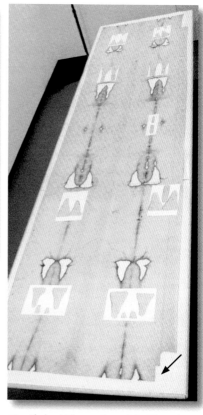

13c (*below*) The section of the Shroud removed for carbon dating in 1988. Arrowed in 13b is the site from which it was removed.

14a (*above*) Şanliurfa, formerly the very Christian city of Edessa, as it looks today. Prior to 1144, many of the world's earliest Christian churches lay in the area immediately behind the mosque in the foreground, including a Hagia Sophia cathedral regarded as one of the wonders of the world.

14b (*left*) King Abgar receives his 'wonderful vision' (from an eleventh-century manuscript).

15a (*left*) Coin of Abgar VIII featuring a Christian cross on the king's tiara.

15b (*below left*) Sculpted lion fountain with a Christian cross on the lion's head. The lion denoted the Aryu or Lion dynasty to which the Abgars belonged.

15c (*below right*) Şanliurfa's fish-pools, a tourist attraction since as early as the fourth century AD.

16a Ruins at Deyr Yakup, a now lonely hill outside Şanliurfa, thought to have been the location of the mausoleum of Edessa's kings, also of Jesus's disciple Addai.

16b Linguist Mark Guscin, author of the definitive book on the Image of Edessa, in the midst of investigating one of Şanliurfa's dustier landmarks.

16c (*below*) Just one of the high-quality sixth-century mosaics Dr Mehmet Önal is currently excavating at the Haleplibahce site in Şanliurfa. Such mosaics provide a tantalizing glimpse of the undoubtedly superior examples which decorated Edessa's long-destroyed Hagia Sophia cathedral.

can be found in the little-visited St Ponziano catacomb in Rome's Transtevere district (pl. 20b). It is of exactly the same type as the Pantocrator icon at St Catherine's Monastery in Sinai that we earlier established as having been painted under the influence of the Image of Edessa. However, it features one highly important extra detail: on the forehead between the eyebrows there is a starkly geometrical shape resembling a topless square. Artistically it does not seem to make much sense. If it was intended to be a furrowed brow, it is depicted most unnaturally in comparison with the rest of the face. But if we look at the equivalent point on the Shroud face (pl. 20c) we find exactly the same feature, equally as geometric and equally as unnatural, probably just a flaw in the weave. The only possible deduction is that fourteen centuries ago an artist saw this feature on the cloth that he knew as the Image of Edessa and applied it to his Christ Pantocrator portrait of Jesus. In so doing he provided a tell-tale clue that the like-ness of Jesus from which he was working was that on the cloth we today know as the Shroud.

Seven decades ago Frenchman Paul Vignon identified another fourteen such oddities frequently occurring in Byzantine Christ portraits (fig. 26), likewise seemingly deriving from the Shroud.[24] Among these is a distinctive triangle immediately below the topless square. But like a Man Friday footprint of the Shroud's existence six centuries before the date given to it by carbon dating, the topless square alone is enough.

So, identification with our Shroud notwithstanding, where was the Image kept in Edessa following its rediscovery? In 525, Edessa suffered the worst of a succession of catastrophic floods, thirty thousand of its citizens reportedly dying in the disaster. The Byzantine emperor at the time was the septua-genarian Justin I. Very soon to succeed him, however, was his brilliant nephew Justinian, builder of Constantinople's Hagia

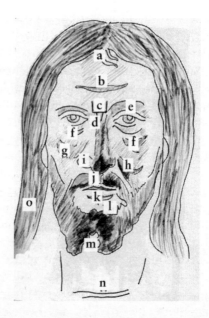

Fig. 26 THE VIGNON MARKINGS. The French scholar Paul Vignon noticed these and other markings recurring in Byzantine portraits of Jesus, seemingly deriving from features visible on the Shroud: (a) Two strands of hair; (b) Transverse streak across forehead; (c) Topless square between eyebrows; (d) 'V' shape at the bridge of the nose; (e) Raised right eyebrow; (f) Heavily accentuated 'owlish' eyes; (g) Accentuated left cheek; (h) Accentuated right cheek; (i) Enlarged left nostril; (j) Accentuated line between nose and upper lip; (k) Heavy line under lower lip; (l) Hairless area between lower lip and beard; (m) Forked beard; (n) Transverse line across throat; (o) Left sidelock of hair lower than right. In some instances the marks in the artworks appear in mirror reverse.

Sophia. And Justinian quickly spared no expense on Edessa's reconstruction. He completely re-routed the river Daisan in an attempt to ensure no such calamity could ever happen again, and rebuilt the city's walls (it is possible that it was in the course of this work that the Image came to light hidden in the old Kappe gateway). And he constructed a magnificent new cathedral, Edessa's own Hagia Sophia.

Although no trace of this cathedral can be seen in present-day Şanliurfa, several praise-filled descriptions of it exist. Surrounded by water – the two mosques in Şanliurfa's fish-pools area convey something of the idea – it was built entirely of stone, and capped with a dome along very similar lines to that of its Constantinople equivalent. Its textured marble was specifically likened to the 'Image not by hand made' (yet another sixth-century reference to the Image),[25] and its interior walls were covered with mosaics. Although the latter were long ago destroyed, something of their quality may be glimpsed in the near-contemporary floor mosaics, with secular themes, of a Byzantine governor's villa currently being excavated in Şanliurfa's Haleplibahce Park area (pl. 16c).[26] The Hagia Sophia cathedral was the Image's illustrious new home after all those centuries secluded in the cavity, its role akin to the Temple of Jerusalem containing the Ark of the Covenant.[27] The adulatory Syriac hymn describing it glowingly referred to it as containing 'the very essence of God'.

But however secure things might have seemed now that the Image was housed in a state-of-the-art new cathedral, trouble was never far away. Between 609 and 626 the Persian menace rose again in the form of new 'King of Kings' Khosraw II, who seems to have gained control of Edessa suspiciously easily and bloodlessly. This was probably due to the help of the city's powerful Monophysite faction, bitter enemies of the Byzantine Empire's official 'Chalcedonian' Orthodoxy. Indicative of this, Khosraw ordered all Edessa's Orthodox churches and monasteries, which would have included the Hagia Sophia cathedral, to be handed over to the Monophysites. Following this, Persia's governor Cyrus began to strip them of their gold and silver, shipping it all back to Persia.

Enslaved contingents of Edessan citizens were about to follow when in the nick of time the Byzantine Empire's vigorous new ruler Heraclius managed to recapture the city.

But hardly had Heraclius restored Edessa's churches and monasteries to their rightful owners when a new and altogether longer-lasting danger appeared before Edessa's walls – the newly invented religion of Islam. Despite Justinian's lavishing of so much expense on Edessa, there had never been much love lost between its eastern Syriac-speaking inhabitants and their Greek-speaking 'protectors' in distant Constantinople. Accordingly, when the Islamic Arabs arrived, again there was a suspicious lack of serious opposition among the Edessan citizenry. And as part of a general toleration the Islamic Arabs accorded to Christianity at this time, the Image's security seems to have been assured.

For a brief period, therefore, the Image enjoyed a strange new status quo during which all three of Edessa's frequently quarrelsome Christian denominations – the third was the Nestorians – were kept in reasonable order by their new Islamic overlords. Intriguingly, it is from this period that a Nestorian bishop's letter survives that refers to Edessa as 'the city of the Mighty Lord . . . the holy place which the Omnipotent God chose from among all the countries of the world . . . as a sanctified throne for the Image of his adorable face and his glorified incarnation'.[28]

As has been suggested by that bishop's present-day successor,[29] the words 'and his glorified incarnation' could indicate an awareness of a full body imprint on the Image of Edessa, as on our Turin Shroud. But if the Image of Edessa was indeed our Shroud, it was due for a very big shock – quite literally.

Chapter 11

Trial by Fire

*... the Image of the Lord made without ink on cloth ...
was not effaced or burnt when it was tested by fire and
water before the great multitude.*

Thirteenth-century Arabic document found in the Wadi'n-
Natrun, Egypt

IT WAS EASTER SUNDAY, 3 April 679. Deep underground,
two great plates of the earth's crust shifted positions. As
no one could have been aware before the age of science,
Edessa and its near neighbour Batnae[1] had been built
rather too close to what seismologists call the North
Anatolian Fault. Batnae, at the earthquake's epicentre, was
razed to the ground, and even though Edessa was fifteen
miles away many of its buildings collapsed, including large
parts of the Hagia Sophia cathedral.[2] Although we are not
told so directly, somehow amid the rubble the Image
survived.

Edessa, as noted in the last chapter, was now under
relatively benign Islamic Arab control. Although the
Islamic religion was less than half a century old, already
deep hatreds had developed between the followers of Ali

(effectively, today's Shi'ites), who wanted Islamic leader-ship to stay within the family of the Prophet Mohammed, and what are today the Sunnis, who preferred a leadership based on ability. Belonging to the latter faction was the astute Caliph Mu'awiyah, notable for his tolerance towards Christians, whose territories included Syria, the Holy Land and Egypt.

At the time of the earthquake, Mu'awiyah was encamped close to Edessa, where he dreamed of the Shi'ite faction's downfall and his own empire's success. Somehow this dream and the need to rebuild Edessa became con-nected in his mind. A former governor of the Edessa region, he was almost certainly aware of the magical pro-tective properties associated with the Image, but whatever the reason, and most unexpectedly for a Muslim ruler, he ordered Edessa's cathedral to be rebuilt.

Although we are not told directly what happened to the Image while this rebuilding was going on, inevitably the cloth had temporarily lost its home. And as the German historian Dr Karlheinz Dietz of the University of Würzburg has pointed out,[3] it was at this time that the shipwrecked Bishop Arculf of Perigueux, mentioned back in chapter 8 (see page 153), told Iona monastery's abbot Adamnan of the *sudarium* of Christ that he had seen in Jerusalem while on his pilgrimage to the Holy Land. This story's leading figure, indeed the only historical individual directly named in it, happens to be none other than Caliph Mu'awiyah, referred to in Adamnan's Latin as 'Mavias, the king of the Saracens'.

According to Arculf's story, as related by Adamnan, when two different factions began bickering over the *sudarium*'s ownership, Mu'awiyah was called upon to adjudicate. Mu'awiyah duly told those then in possession of the *sudarium*, 'Give the sacred linen cloth which you have into my hand.'[4] Adamnan's rendering of Arculf's tale continues:

In obedience to the king's [i.e. Caliph Mu'awiyah's] command, they bring it [the *sudarium*] from its casket and place it in his bosom. Receiving it with great reverence, the king ordered a great fire to be made in the square before all the people, and while it was burning fiercely, he rose, and going up to the fire, addressed both contending parties in a loud voice: 'Now let Christ, the Saviour of the world, who suffered for the human race, upon whose head this *sudarium* . . . was placed in the Tomb, judge between you by the flame of fire, so that you may know to which of these two contending parties this great gift may most worthily be entrusted.'[5]

In the typical manner of the many tall stories from this era, the *sudarium* was described as fluttering above the fire 'like a bird with outstretched wings' before safely descending in the direction of the group that Arculf called the 'believers', who accepted it 'with great honour, a gift to be venerated as sent to them from heaven'. They then 'covered it up in another linen cloth and put it away in a casket of the church'.

According to Arculf, the *sudarium* he saw involved in this 'trial by fire' staged by Mu'awiyah was 'about eight feet long' – clearly, then, a body-length piece of cloth, though only half that of the Shroud when fully extended. But what if the Shroud were 'doubled' at the time? As we may recall from chapter 6, the Shroud was quite definitely folded in half down its length and half across its width at the time of the otherwise unknown incident during which it sustained the triple burn-holes (pl. 3b). To an outside observer it would have seemed about eight feet long when folded in this way. And because the holes occur so centrally in this folding arrangement (fig. 11), they have every semblance of the result of thrusting through the cloth three times something like a sputtering pitch-soaked red-hot poker. Such pitch-soaked hot iron rods were classic implements of Dark Ages 'trial by fire' ordeals.[6]

Obviously it would have appeared to any spectator at the ordeal as if the Shroud had sustained major damage, as indeed it had. However, if the cloth was then returned to the *tetradiplon*, or 'doubled in four', folding arrangement reconstructed of the Image of Edessa (fig. 25), it would have appeared completely unharmed: none of the triple burn-holes are visible when the Shroud is folded in this way. A document in Arabic found in the Wadi'n-Natrun in Egypt specifically describes the Image of Edessa going through such an ordeal and emerging unscathed: 'Hail Abgar, who was worthy to behold the Image of the Lord made without ink on cloth, the image of the worker of miracles. It was not effaced or burnt when it was tested by fire and water before the great multitude.'[7]

So if it was indeed our Shroud, alias the Image of Edessa, which underwent Mu'awiyah's 'trial by fire', it could well have been adjudged to have passed this Dark Ages 'radiocarbon test' with flying colours.

Whatever the truth of this scenario, obviously it remains highly speculative that the *sudarium* of Christ that Arculf saw in Jerusalem and the Image of Edessa, alias our Turin Shroud, were one and the same object. However, Arculf does explicitly state that the cloth had come to Jerusalem only about three years previously,[8] which immediately raises the question of where it had been beforehand.

As Dr Dietz has pointed out, the best calculation of when Arculf was in Jerusalem, based on a variety of different pointers, is between 681 and 683 – exactly the period when the Image of Edessa had temporarily lost its home. Also, if there was any time a dispute might have broken out concerning the Image's ownership in Edessa, the period immediately after the earthquake has to have been that time. After Arab Muslims took over the Holy Land in

638, fully respecting Christianity's holy places, it was Jerusalem, not Rome or Constantinople, which the Muslims recognized as Christianity's spiritual centre. Historically, this particular period happens to be a very obscure time with regard to what was happening in the Jerusalem Patriarchate responsible for the minority Chalcedonian or Orthodox faction in Edessa. Because of the Hagia Sophia cathedral's collapse, it is very possible that they ordered the Image to be brought back to the spiritual home it had left six centuries before – something that would certainly have upset the Edessans every bit as much as if the present Pope, as owner of the Shroud we know today, ordered that henceforth it should be housed in Rome, not Turin.

Earlier we noted that there were no fewer than three Christian denominations in Edessa at this time, of whom the Monophysites were dominant, by far. When the Persians briefly controlled Edessa they had actually put the Monophysites in charge of the Hagia Sophia cathedral and its prized Image. Accordingly, it would scarcely be surprising for them to have tried to exert some pressure for the Image to be handed over to them now that the Orthodox no longer had even a fitting home for it.

One of the most powerful Christians of this time, not only in Edessa but throughout Mu'awiyah's dominions, was a Monophysite business magnate by the name of Athanasius bar Gumaye. Born to one of Edessa's well-to-do families, early in his career Athanasius had been appointed tutor to a brother of the future caliph Abd al-Malik. When his charge became Governor of Egypt he acted as his right-hand man, in the course of which, according to one contemporary report, 'He collected gold and silver like pebbles and he had four thousand slaves, all bought out of his own purse, besides grand houses, villages, various estates and gardens worthy of a king.'

After twenty-one years in Egypt, where he built two churches, Athanasius returned to his native Edessa, where he owned no fewer than three hundred shops and inns. And it was some time around the year 700 – that is, less than two decades after the Mu'awiyah 'trial by fire' incident, by which time the cathedral had been repaired, but its adherents were being very heavily taxed to pay for it – that we are told by a near-contemporary chronicler:

The Edessans owed part of the taxes which they had to pay and had nothing with which to pay it. A crafty man . . . advised the collector of taxes, 'If you take the Image[9] they will sell their children and themselves rather than allow it [to be removed].' When [he] did this, the Edessans were in consternation . . . They came to the noble Athanasius and asked him to give them the 5,000 dinars of the taxes, and to take the Image to his place until they repaid him. He gladly took the Image to his place and gave the gold. Then he brought a clever painter and asked him to paint one like it. When the work was finished and there was a replica as exactly as possible like [the original] because the painter had dulled the paints . . . so that they would appear old, the Edessans after a time returned the gold and asked him for the Image. He gave them the one that had been made recently and kept the old one in his place. After a while he revealed the affair to the faithful [Monophysites], and built the wonderful shrine of the Baptistry. He completed it at expense great beyond reckoning, spent in honour of the Image, because he knew that the genuine Image . . . had remained in his place. After several years he brought it and put it in the Baptistry.[10]

This is a quite extraordinary piece of information. As if our trying to identify the Image of Edessa with Turin's Shroud is not complex enough already, here we have a near-contemporary chronicler telling us of a duplicitous act of substitution involving the Image of Edessa which occurred in the early eighth century. It would seem that Athanasius used the opportunity of redeeming the Image from the Arab tax collector who had been holding it in pawn to fob off the Orthodox community with a cleverly painted fake, and keep the original for his own Monophysite community. If the story is true – and as we will see in the next chapter, Athanasius's deception does seem to become manifest two centuries later – we have the confusion that for the next two centuries the cloth that was being kept in the superbly restored Cathedral of Edessa as the official Image of Edessa was actually a fake, while another version being kept by the Monophysites in their lavishly upgraded Baptistry was the true original.[11]

Six hundred miles away in Constantinople, the icon evangelism that had been initiated with such vigour in the sixth century was reaching a new high point. In 692 the then twenty-three-year-old Byzantine emperor Justinian II had presided over a council of the eastern churches, the Quinisext. Among this council's canons, or decrees, was one officially approving what had already long been happening in Christian art: the representation of Jesus in human form, rather than depicting him symbolically as a lamb in deference to concerns over breaking the second Commandment. In the words of canon 82, 'In order to expose to the sight of all, even with the help of painting, what is perfect, we decide that henceforth Christ our God must be represented in his human form instead of the ancient lamb.'[12]

Justinian II's demonstration of his personal support for this policy was to issue gold *solidi* coins with his own likeness, standing, on the reverse (or 'tails' side), and the Image of Edessa-inspired Christ Pantocrator likeness on the obverse (pl. 20a). These coins are inscribed 'Rex

Regnantium', or 'King of those who Rule', indicative of exactly the same Orthodox spirit associated with the Image of Edessa likeness presented in the last chapter. It is the earliest-known representation of Jesus's likeness on any coin. These coins were not only superb works of art, they were also clever psychology on the part of a ruler conscious of his growing unpopularity because of the heavy taxes he was imposing to fund his ambitious building projects.[13]

However, if by so publicly representing his rule as being under the aegis of Christ's image on the Image of Edessa Justinian II had hoped he would gain some kind of personal divine protection, he was doomed to disappointment. Only three years after the Quinisext Council's declaration, he was deposed by a popular uprising. In an attempt to make sure he could never regain his throne, the rebels cut off his nose, it being an imperative that Byzantine emperors should be of unblemished appearance.

In the event, in 705 a noseless Justinian did regain his throne. And in an act of apparent spite against what he presumably perceived to be the Image of Edessa's failure to protect him, he chose for his post-705 coin issues a different variety of Christ portrait, a frizzy-haired type.[14] Almost certainly this was based on the Image of Edessa's rival as a 'not by hand made' image of Jesus, the so-called Image of Camuliana,[15] which unlike the Image of Edessa had actually been brought to Constantinople, in 586.

Justinian lasted only six more years before losing his throne again – this time along with his head. As if now to underline the hollowness of the Image of Edessa's protective powers, around 717 Edessa suffered a further serious earthquake, the Hagia Sophia cathedral again among the buildings damaged.

Six years after that there was completed in Edessa a book of Orthodox homilies, written in Syriac, today preserved in the British Museum's collection as Oriental Manuscript 8606.[16] Its scribe recorded as part of its 'publication' details, 'This book was written and completed in the blessed city Urhai in the time of office of . . . Jannai, priest and abbot[17] of the House of the Image of the Lord, and Jannai priest and archivist[18] of the same church.' As is generally recognized, the House of the Image of the Lord can only mean the Hagia Sophia cathedral, where the Image of Edessa was normally kept, so appearances were being kept up that the Orthodox possessed the true original.

Within months of the Edessan scribe writing these words, in Constantinople the Byzantine emperor Leo the Isaurian, motivated by exactly the same objections to religious images that we noted earlier had hindered the early development of Christ portraiture, began to put into reverse everything that had been achieved by nearly two centuries of icon evangelism. From the beginning of his reign Leo abandoned the policy of displaying the Image of Edessa-inspired likeness of Christ on the imperial coinage, and from 726 onwards – in the teeth of objections from Constantinople's Patriarch, who was quickly deposed – he began a campaign of systematic destruction of all religious images. One of the first to go was the great depiction of Christ, almost certainly one of the Image of Edessa-inspired Pantocrator variety, over Constantinople's Chalke Gate, the entranceway to the emperors' Great Palace. When those assigned to destroy this image began to carry out their task, bystanders reputedly attacked them for sacrilege, resulting in the first spilled blood. But the destruction continued. A similar image of Christ over the throne in the imperial throne-room – almost certainly another of the Image of Edessa-inspired variety – likewise disappeared. Even reputedly 'not by hand made' rivals to the Image of Edessa, such as the Image of Camuliana, were not spared, because we never hear of the latter again. Right across the Byzantine Empire icons were burned, mosaics destroyed, murals whitewashed over. Quite how much was lost is impossible to calculate, but the destruction was undoubtedly widespread, akin to what would later happen to so many great works of Roman Catholic medieval art during the English Reformation.

Inevitably there were objectors. Particularly vociferous was the monk John of Damascus who, while clearly demonstrating his understanding that the Image of Edessa had been created by Jesus in life, in the manner of the 'Jesus washing himself' story, specifically quoted it as an argument for artistic representations of Jesus having divine approval: 'Abgar, king of the city of Edessa, sent an artist to paint the Lord's image but could not do so because of the shining brilliance of his face. The Lord therefore placed a *himation* on his divine and life-giving face and wiped his own imprint on to it.'[19] Notably, the word used for the cloth, *himation*, just like *sindon*, indicated something of full garment-size proportions rather than the pocket-handkerchief-size cloth some scholars have insisted the Image of Edessa to have been.

Meanwhile Edessa, being beyond Byzantine juris-diction, might have seemed immune from all the image-smashing but for the fact that even before Emperor Leo had begun his destruction the Islamic caliph Yazid, a successor to Mu'awiyah, had ordered much the same throughout his domains. Fortunately Caliph Yazid's edict does not seem to have carried the same force as that of Leo the Isaurian. Nor do his successors seem to have been enthusiastic about pursuing the policy. Certainly in 787 a certain Leo the Reader of Constantinople when he visited Edessa was able to report, 'When I, your unworthy servant, went to Syria with the royal commission, I came to Edessa and saw the holy image that was not made by human hands, held in honour and venerated by the faithful.'[20] By no means clear from Leo the Reader's remarks is whether he saw just the Image's casket (which would have been the norm) or the Image itself. Even in the case of the latter, there of course remains the uncertainty over whether this was the true original or Athanasius bar Gumaye's cunning replica.

A document written in Constantinople in the tenth century allows us a unique glimpse of the extraordinary awe that surrounded the presumed Image while it was kept in a cathedral so highly regarded that even Arab Muslim geographers recognized it to be one of the wonders of the world.[21] According to the document,[22] twice a week only the Orthodox clergy opened the doors of the cathedral's special sanctuary housing the Image to allow the many pilgrims who congregated in Edessa to have their much-coveted glimpse of the casket containing the cloth. Even so, the sanctuary was protected by a special grille. On only two occasions during the year was the Image taken out of this sanctuary, both during Lent.

When this happened, its casket would be set on a throne and carried in a most elaborate procession, flanked by golden sceptres, and followed by twelve ritual fan bearers, twelve bearers of censers filled with sweet-smelling incense, and twelve torches. Every element of the ceremonial had its own special symbolism and meaning. At the procession's culmination, the Image's casket would be carried up the nine steps of the cathedral's apse, leading to a special altar-throne. There it would be wrapped in white linen and placed on the throne as King. It would then be carried, still enthroned, into the cathedral's sanctuary where only Edessa's bishop was allowed to kiss its white linen covering and wrap it in another cloth, this time of purple, after which it was returned to its own sanctuary. Never during the ceremony were the ordinary faithful allowed any proper sight of the cloth. As summed up by the tenth-century text, 'no one was allowed to draw near or touch the holy form with his lips or eyes. Thus did fear of God increase their faith and make people shudder with greater fear for the revered object.'[23]

Back in Byzantium, the passion for iconoclasm unleashed by Leo the Isaurian lasted over a century, but eventually burned itself out. In 843 a pro-images policy was reintroduced, and we are told that in the throne-room 'once again the image of Christ shines above the imperial throne and confounds the murky heresies; while above the entrance is represented the Virgin as divine gate and guardian'.[24]

The Image of Edessa-inspired likeness of Christ above the Great Palace's Chalke gateway was restored. The Image of Edessa-inspired likeness of Christ 'Rex Regnantium' was restored on the imperial coinage – somewhat crudely copied from the old coinage. Still, for now at least, the Image of Edessa itself lay out of reach, waiting for the day when Byzantine armies would be strong enough to reach out beyond the empire's frontiers and again penetrate into the eastern territory it once held.

Chapter 12

Baghdad Surrender

When the Image and the letter of Christ were about to leave Edessa, thunder, lightning and a terrible rainstorm suddenly struck, by some chance or providence.

Story of the Image of Edessa

IT WAS SPRING IN THE YEAR 943, exactly a century after religious images had been restored in Constantinople. In Edessa, the city's Muslim emir looked out with dismay over his walls at a sight his messengers had warned him to expect, and which had now become hard reality: an eighty-thousand-strong Byzantine army led by John Curcuas, a hugely successful Armenian-born commander-in-chief with the reputation of a Montgomery or a Rommel.[1]

Throughout the last year Curcuas had taken his army well into the north-western edges of Arab-held territory, raiding and capturing towns and villages and taking prisoners by the tens of thousands. Weakened by large-scale losses to Shi'ites in the south, and with a succession of ineffective caliphs based in Baghdad, the Arab world lacked both the forces and the resolve to defend this part of its dominions. Clearly Edessa's fall to Curcuas's army would only be a matter of time.

But to the emir's utter astonishment, Curcuas made no attempt to attack. Instead he offered him an immediate deal. He said that he was prepared to spare the city and release two hundred high-ranking Muslim prisoners he was holding all for just one thing: the safe hand-over of the Image of Edessa.

Faced with such an unprecedented demand, the emir was at a loss to know what to do. Keenly aware of the strong opposition he could expect from his city's Christian community, he sent an envoy riding post-haste to his superiors in Baghdad to seek their advice. Extraordinarily, Curcuas allowed him the time for this, merely conducting a few low-key raids in the surrounding territories in the interim, even though he could hardly have failed to realize that Arab-held Mesopotamia lay virtually at his mercy and he could easily make some impressive strategic gains.

All of which makes quite apparent that Curcuas's demand was no off-the-cuff whim. As the army's commander-in-chief, his orders had come directly from Byzantium's septuagenarian and now ailing emperor Romanos Lecapenos. Sharing Curcuas's Armenian birth and army background, Romanos had risen through the ranks, first to admiral, then to increasingly powerful positions within the all too typically hapless Byzantine court of a quarter of a century earlier. Although illiterate and uncouth, he had arranged for his daughter Helena to marry the then under-age hereditary emperor Constantine Porphyrogennetos, 'to the purple born'. Thereby enabled to assume executive command of the Byzantine Empire, Romanos had given himself the title Emperor and never relinquished it. In the event, his strong leadership benefited the empire both politically and economically. And his usurped son-in-law Constantine, whom Romanos had kept on at the court rather than quietly murdering him in the normal Byzantine manner, had at least been able to while away his time indulging his tastes for literature and the arts.

But in his old age Romanos's thoughts had turned increasingly to religion, particularly after the loss of his most competent son, Christopher. Fearful of divine judgement on his injustices towards his son-in-law, he seems to have decided upon the mission to capture the Image of Edessa for Constantinople as a way of improving his own image for posterity. The last time that truly great relics of Christ had been brought to Constantinople was back in the days of Constantine the Great, when Constantine's mother Helena had excavated Jesus's tomb in Jerusalem and brought back with her pieces of the wood of the cross, and other items. Romanos had named his own daughter after this Helena, likewise one of his sons Constantine. For him to wrest from Muslim territory Edessa's fabled imprint-bearing cloth and whisk it off to Constantinople to join the world's most definitive collection of Christ's relics would surely be a most fitting and honourable end to his reign. Hence John Curcuas's very special mission.

When the Edessa emir's emissary arrived in Baghdad with the news of Curcuas's extraordinary demand, the then caliph, al-Muttaqi, duly convened his qadis (his chief legal advisers) to consider the problem. Their debate was prolonged, with some strong stands taken.[2] All sides exhibited quite remarkable respect for the Image of Edessa – hardly what might be expected among the highest echelons of image-abhorring Islamic society had the Image genuinely been just the 'some old icon' modern-day historians often suppose.

One faction strongly recommended outright rejection of Curcuas's demand, arguing that the 'Mandil', as they called the Image of Edessa, had been in Islamic territory from time immemorial. No Byzantine emperor had ever previously laid claim to it. It would be quite unacceptable for Muslims to surrender it to the Greeks. But then the highly revered old vizier Ali ibn 'Isa spoke up. His thoughts were for the Muslim prisoners who might be set free by the Image's surrender. 'Let the Muslims be freed from their imprisonment,' he urged. 'Release them from their confinement among the unbelievers. Pluck them from the sufferings and the sorrows that they are currently enduring.' And it was Ali ibn 'Isa's argument that won the day, on the grounds that the lives of Muslim captives outweighed all other considerations. Edessa's emir was duly instructed that he should surrender the Image in return for the two hundred high-ranking prisoners held by Curcuas, but with the extra proviso that Byzantium should issue a special decree promising Edessa and its near neighbours perpetual immunity from any future Byzantine attack.

Curcuas duly agreed to the deal – he had already captured another thousand prisoners while waiting for Baghdad's answer – but that was just the easy part. It was quite unthinkable that a crude soldier should go into Edessa and lay his bloodstained hands on so priceless a Christian relic. Furthermore, it needed someone with some first-hand knowledge of the Image to ensure that the Edessans were surrendering the true original, not just palming them off with a faked substitute.

Early in 944, some nine months after Curcuas had first appeared beneath Edessa's walls, Abraham of Samosata, a local bishop who had happened to visit Constantinople at the time of these events, became the agreed official scrutineer and receiver of the Image on behalf of Emperor Romanos Lecapenos. As the *Story of the Image of Edessa* makes clear, Abraham insisted on inspecting no fewer than three Edessan versions of the Image, including one kept by the Nestorians, before he was satisfied that he had what the *Story* specifically calls 'the true unpainted image'.[3] Whether this was the one that was kept by the Monophysites, or the one held by his own fellow Orthodox, we shall never know. Almost as an afterthought, Abraham apparently also took Jesus's letter to Abgar.

Now the problem was how to get these two sacred treasures, both of them popularly believed to have magically protected Edessa throughout the last nine centuries, physically out of the city and into the protective custody of Curcuas's army encamped outside. As the *Story* records,

a revolt arose among the faithful in Edessa and great unrest took hold of the city, as they did not want to let their sacred objects and their homeland's safeguard be taken away. In the end the Saracen leader [i.e. the Muslim emir] had them handed over by persuading some, forcing others and frightening yet others with threats of death. When the Image and the letter of Christ were about to leave Edessa, thunder, lightning and a terrible rainstorm suddenly struck, by some chance or providence. The ones who had resisted before were stirred up again and claimed that the divine will was clear from what had happened – God did not wish these most holy objects to be removed.[4]

The sacred objects' bearer party apparently comprised not only Abraham, Bishop of Samosata, but also his Edessan counterpart, the latter's most senior aide, other leading Christians and a representative of the emir. Eventually the emir managed to get this whole group and their highly revered charges out of the city and on their way, now under the protection of Curcuas's massive army escort.

The decision seems to have been already taken that a sea journey would be far too risky, particularly with the

Arabs now having significant sea-power capabilities. So there now stretched before them an arduous six-hundred-mile march overland across the near entirety of what is today Turkey.

The first hurdle, at only a day's march distance, was crossing the river Euphrates just before Bishop Abraham's own Samosata, a town which no longer exists because it was flooded in 1989 as part of Turkey's Ataturk Dam project.[5] On the party's arrival at the ferry's crossing point there was some fresh and equally fierce opposition from locals who had followed them there. However, according to the *Story* – and again we come across one of those tall tales in which the Dark Ages delighted – this time divine intervention favoured the Byzantines: 'The boat that was intended to ferry the bearers across the Euphrates was still moored on the Syrian side, while the rioters were still in the grip of tumult. Yet as soon as the bishops who were carrying the divine Image and the letter had boarded, suddenly, with no rowers, no helmsman and no other vessel to tow it, their boat set off for the land on the other side, guided only by the will of God.'[6]

After a few days spent recovering in Samosata, the bearer party and their army escort proceeded onwards, until by early August they had reached Anatolia's north-west shore by the river Sagaris. Only a boat-ride across the Bosphorus strait stood between the Image and its new home, Constantinople.

Today's Istanbul, despite the impressive modernization it has undergone in recent years, conveys all too inadequate an impression of the Constantinople 'Queen of Cities' as it was at that time. In the tenth century, Constantinople lay at the eastern end of Europe much like a fairy-tale palace in a wilderness of hovels. As a centre of art, culture and commerce it had no peer, having preserved intact all the knowledge and experience of the old Roman

Empire. Trade poured into it from all quarters, by land, by sea and by river. Its palaces, churches and shrines were the envy of the world. Male and female citizens alike wore perfumes and decked themselves with jewellery. Christ himself was believed to have guided its founder, the first Christian emperor Constantine the Great, in determining the fourteen-mile circuit of its walls. In its Great Palace (fig. 27, pp. 218–19), a once huge complex of colourful buildings that have now almost completely disappeared save for a few mosaics and a sea-wall, the emperor ruled in god-like luxury amid an interminable daily round of ceremonies, with visitors prostrating themselves before him whenever they were granted the honour of an audience.

As soon as word arrived at this Great Palace that the Image was now just across the Bosphorus, Romanos sent his very able chief minister Theophanes, together with leading members of the Senate, to call upon the bearer party while they were resting overnight at a local monastery. Early the next morning this reception committee together with the bearer party then carried the Image to a second monastery, where 'the casket that concealed the miracle-working Image was reverently placed in the same church of the monastery, and when it was uncovered from the casket they saw and venerated it with due reverence'.[7]

The fifteenth of August is celebrated with great solemnity both in western and eastern churches as the Feast of the Assumption of the Virgin Mary, and during that day Emperor Romanos, his two sons Stephen and Constantine, and rightful emperor Constantine Porphyrogennetos, all of them designated as 'emperors', had been at Constantinople's Church of the Virgin at Blachernae, to take part in the ceremonies connected with the feast. Located in the far north-western corner of the city, the Blachernae church had very special significance for the Byzantines because it contained the reputed robe of the Virgin Mary, widely believed to have protected Constantinople from foreign adversaries in the seventh and ninth centuries.[8] Every Saturday evening a 'miracle' would happen in this church: the curtain veiling its sacred icon of the Virgin would be lifted in the air by a supernatural breeze – the kind of 'magic' in which the Byzantines revelled.[9] (We may also note that it is the same church where two and a half centuries later the French Crusader Robert de Clari would see the *sydoine* bearing Jesus's full body imprint.)

According to the *Story*, 'The bearers of the holy objects arrived in the evening [at the Blachernae church], and the casket holding the Image and the letter was placed in the upper chapel . . . The emperors went up to the casket, and greeted and venerated it, although they did not open it. They then conveyed it to the royal galley with honour, due escort and many lighted lamps, and so came with it to the Palace. They placed it there in the divine chapel called Pharos . . .'[10]

This Pharos chapel,[11] a fabled shrine within the Great Palace sadly long lost to us, already contained the collection of illustrious relics of Jesus's Passion that Constantine the Great's mother Helena had brought back from Jerusalem six centuries earlier. Its contents included large pieces of wood from the true cross, a section of the 'King of the Jews' title that had been on the cross, the nails that held Jesus on the cross, the crown of thorns, the lance that pierced Jesus's side, his seamless coat, and a phial of his blood, each resplendent in their own glittering reliquaries. These would become the constant companions of the Image of Edessa and letter for the next two and a half centuries. But first there were some out-and-about adventures planned for the new arrivals.

According to the *Story*, the very next day, the sixteenth,

Emperor Romanos was too ill to take part. However his two sons, together with his son-in-law Constantine, solemnly entered the Pharos chapel. Accompanied by much singing of psalms, they picked up the sacred casket and carried it to the nearby waterside, the Sea of Marmara, where their galley was waiting. This was then rowed along the walls to the city's westernmost part where the imperial party with its revered burden disembarked. There waiting to receive them were the city's Senate, its Orthodox Patriarch and 'the whole body of the clergy'. All now proceeded to walk the short distance north to Constantinople's then magnificent, relief-studded Golden Gate, an entranceway traditionally associated with emperors returning in triumph. The bearer party purposefully strode through it, in the words of the *Story*, carrying 'the box holding the precious and sacred objects as if it were another Ark of the Covenant or something even greater'.[12]

The cortège then made its way right into Constantinople's centre, 'believing that in this way the city would be made holier and stronger, and would be kept unharmed and unassailable for all time'. According to the *Story*'s author, it was impossible to describe in words just how much 'intercession, prayer to God and thanksgiving took place throughout the city as the divine Image and the sacred letter' moved through the vast crowds of people who had 'gathered together from all over, like huge waves'.[13]

Then it was on into the magnificent Hagia Sophia cathedral where 'all the clergy' paid their special respects while both the Image and the letter were temporarily placed 'in the innermost recesses of the sanctuary'. From there the colourful procession went on to the Great Palace, a ten-minute walk away, where the Image was placed on the imperial throne in clear symbolic recognition of its

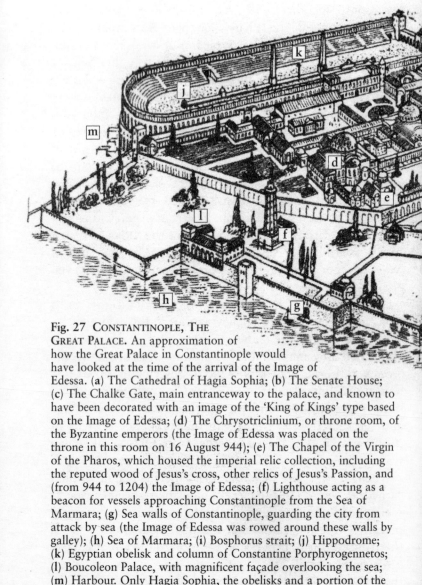

Fig. 27 CONSTANTINOPLE, THE GREAT PALACE. An approximation of how the Great Palace in Constantinople would have looked at the time of the arrival of the Image of Edessa. (a) The Cathedral of Hagia Sophia; (b) The Senate House; (c) The Chalke Gate, main entranceway to the palace, and known to have been decorated with an image of the 'King of Kings' type based on the Image of Edessa; (d) The Chrysotriclinium, or throne room, of the Byzantine emperors (the Image of Edessa was placed on the throne in this room on 16 August 944); (e) The Chapel of the Virgin of the Pharos, which housed the imperial relic collection, including the reputed wood of Jesus's cross, other relics of Jesus's Passion, and (from 944 to 1204) the Image of Edessa; (f) Lighthouse acting as a beacon for vessels approaching Constantinople from the Sea of Marmara; (g) Sea walls of Constantinople, guarding the city from attack by sea (the Image of Edessa was rowed around these walls by galley); (h) Sea of Marmara; (i) Bosphorus strait; (j) Hippodrome; (k) Egyptian obelisk and column of Constantine Porphyrogennetos; (l) Boucoleon Palace, with magnificent façade overlooking the sea; (m) Harbour. Only Hagia Sophia, the obelisks and a portion of the façade of the Boucoleon Palace remain extant.

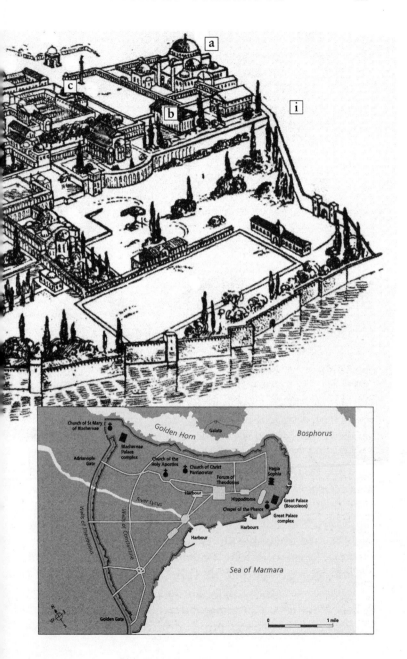

nature as Rex Regnantium, King of those who Rule. Here there was yet more singing of psalms and hymns before finally it was returned to the chapel of the Pharos, where 'it was consecrated and placed on the right towards the east for the glory of the faithful, the safety of the emperors and the whole city together with the Christian community'.[14]

Amid so much ceremony and self-evident excitement it is difficult to determine when and where, if at any point at all, anyone meaningfully saw the Image removed from its casket in a way that could enable proper study. Nevertheless, that this actually happened is confirmed by an independent contemporary account, not part of the *Story of the Image of Edessa*. According to this, 'A few days beforehand, when they [the imperial party] were all looking at the marvellous features of the Son of God on the holy imprint, the Emperor's sons [i.e. Stephen and Constantine] declared that they could only see the face, while Constantine his son-in-law said he could see the eyes and the ears.'[15]

Given the extraordinary efforts that had been made to obtain the Image, several historians have expressed puzzlement that it should have appeared so indistinct to the few who were allowed to view it directly. As the eminent Cambridge historian Sir Steven Runciman remarked, 'It is possible that the young Lecapeni [i.e. Emperor Romanos's two sons Stephen and Constantine] were drunk, though in that case it is curious that Constantine [i.e. the rightful emperor], who was notoriously fond of stimulants, should have missed the opportunity for drinking too.'[16]

If the Image of Edessa was genuinely one and the same object as today's Shroud of Turin, no such explanation is of course necessary. The Shroud's watery-looking impression and its uncertainty of detail would readily explain Romanos's sons' perception difficulties. Although we might question how Constantine Porphyrogennetos, even with his strong artistic interests, saw 'eyes' on the imprint, this perception corresponded to the then universal idea that the Image had been created by Jesus in life. This idea was notably shared by several of the artist copyists of the Shroud during the sixteenth and seventeenth centuries, despite their full awareness – almost certainly not yet shared by Constantine – that they were looking at the imprint of a dead body. It is almost entirely thanks to the discovery of the photographic negative that we know the eyes to have been closed in death.

Whatever our stand may be on identifying the Image of Edessa with the Shroud, there were some watchful tenth-century commentators who saw serious political and spiritual implications for Romanos's sons' having difficulty perceiving Christ's features on the cloth, compared with their legitimately royal brother-in-law. Not long before the Image had arrived in Constantinople, Emperor Romanos, as part of his crisis of conscience, had made out a will giving son-in-law Constantine Porphyrogennetos precedence in the succession over his two sons. This rebuff ultimately drove Stephen and Constantine to stage a coup, on 20 December 944, in which they arranged for their frail and elderly father to be seized and shipped off to house arrest in a monastery on the island of Prote out in the Sea of Marmara. However, they had not taken sufficient account of Constantinople's citizens' feelings on the matter. When rumours of the two brothers' actions began to circulate around the city a mob quickly gathered outside the palace gates demanding the appearance of rightful emperor Constantine Porphyrogennetos. Constantine had to be hauled bare-headed from his beloved library, and when he emerged the climate of public opinion was so strong in his favour that Romanos's sons were quickly obliged to acknowledge him as senior emperor.

For a while Stephen and Constantine were allowed to rule alongside their brother-in-law, but still they plotted, attempting to lure him to a fatal breakfast-party. At last, heeding the pleas of his ever-watchful and loyal wife Helena, on 27 January 945 Constantine had the unpleasant pair arrested. Much to the delight of the Byzantine populace, who had learned to detest them, they were dispatched first to the same island of Prote to which they had sent their father – who gave them more than a piece of his mind – then separately on to full prison in more outlying locations.

So, within months of the Image of Edessa's arrival in Constantinople, Constantine VII Porphyrogennetos at last properly succeeded to the imperial throne that had been rightly his throughout the past thirty-three years, ever since he had been a sickly boy of seven. For the ever superstitious Byzantines, the two events were not unconnected. At the very time the Image had arrived on the Anatolian shore, while the bearer party were taking their first sight of Constantinople after their long journey, reportedly a madman had cried out, before many high-ranking witnesses, 'Constantinople, take the glory and the blessing, and you, Constantine, your throne!' The madman had promptly been cured of his insanity. The inability of Romanos's sons to distinguish the Image properly was simply another telling portent.

Certainly Constantine, no less superstitious than his fellow countrymen, very quickly gave the Image of Edessa the appropriate recognition for the long-hoped-for turnabout in his fortunes, reinforcing this for all time in a variety of ways. First, he had the date of the Image's 'official' arrival in Constantinople, 16 August 944, instituted on the Eastern Orthodox Church calendar as a permanent annual feast day for the Image. (Remarkably, the Orthodox Church continues to observe this feast day, even though the Image itself has been lost to them for more than eight hundred years.)

Second, and with no less foresight, Constantine commissioned the writing of the *Story of the Image of Edessa*, the single work that more than any other has enabled us to piece together a cogent history for the Image. Preserved in some three dozen early manuscripts – there are copies in Moscow, Paris, Spain, Vienna and the Vatican, with the lion's share scattered around the monasteries of Mount Athos – the *Story* carries in its opening sentence the somewhat unbelievable claim that Constantine himself was its author. In fact, because of Constantine's well-attested bookishness, it is not at all unlikely that he took the closest interest in its compilation. Whatever the scale of his personal involvement, the *Story*'s authority was such that it was adopted completely unaltered for the Menologion, or collection of saints' lives, most assiduously assembled in Constantine's time by Byzantium's renowned editor and bibliophile Symeon Metaphrastes.

The third no less enduring way in which Constantine marked his indebtedness to the Image was via his coinage. He had a special love of the goldsmith's craft and within weeks of his accession he issued a particularly beautiful gold *solidus* coin with his own bearded likeness on one side and the now familiar Rex Regnantium Christ likeness (earlier established as having the strongest association with the Image of Edessa) on the other. Similar Rex Regnantium-type images had been used on the coins of Constantine's near-immediate predecessors[17] during the century since the overthrow of Iconoclasm. In all these, something of the basic, familiar facial likeness is there. But the coins issued by Constantine – the one emperor we know to have seen the imprint on the Image of Edessa directly with his own eyes – exhibit a remarkable change (pl. 21a): nothing other than what appears to have been a deliberate attempt to reproduce in the Christ face features quite uncannily close to the exact imprint that appears on the Turin Shroud.

This characteristic, which first occurred less than a year after the Image of Edessa's arrival in Constantinople, was actually noted over twenty years ago by a Hungarian-born Oxford scholar with a very strong interest in Byzantine coins, Dr Eugene Csocsán de Várallja. As Csocsán de Várallja remarked of Constantine Porphyrogennetos's coin issues, 'Just following the arrival of the Edessa [Image in] . . . 944 . . . a completely new image of Christ appeared on the bezants. On these coins Christ's nose became as elongated as on the Shroud, the angle of his eyebrows changed to match the Shroud eyebrows, and the slightly differing angle of each moustache seems to mirror that on the Shroud. In addition the Christ image took on just as impressionistic a character as on the Shroud.'[18]

Two decades on there is one further feature that can be added to these observations: the very distinctive mark running down from the hairline to immediately above Christ's (spectator's) right eyebrow, just to the right of the nose. It appears too deliberate to be some random blemish, and is in fact repeated on later coins. On the Shroud, in this identical location is the reverse '3'-shaped blood flow that runs from hairline to eyebrow.

Around a year before Constantine VII Porphyrogennetos died (in 959), he sent a letter of encouragement to his troops then campaigning around Tarsus.[19] He told them that he was sending them a supply of holy water that had been consecrated by contact with the relics of Christ's Passion held in his palace. These relics he carefully specified as 'the precious wood [of the cross], the unstained lance, the precious inscription [i.e. the title attached to the cross], the reed which caused miracles, the life giving blood from his side, the venerable tunic, the sacred *spargana*, [and] the *sindon* which God wore . . .'[20]

Constantine made no apparent mention of the Image of Edessa, despite his close identification with it. The *spargana* are usually understood to be the baby Jesus's swaddling clothes, but the intriguing word is *sindon*. This is because there is simply no earlier record of the presence of Jesus's burial shroud in the imperial collection. So did Constantine mean the Image of Edessa, which he had otherwise omitted to mention and, as was earlier noted, was specifically referred to as a *sindon* in several pre-tenth-century texts? Is this evidence that in the time between the Image's arrival in Constantinople in 944 and his writing a letter to his troops in 958, Constantine had become aware that the Image was one and the same as Jesus's burial shroud?

Clearly, if the cloth we now know as the Turin Shroud was the cloth that was brought to Constantinople in 944 as the Image of Edessa, we are now tantalizingly close to evidence that could establish this with an at least reasonable degree of confidence. So, given the significant amount of documentary information we have already noted as having come down to us from Constantine's time, what further clues can contemporary materials reveal for us?

Chapter 13

Not for Common Gaze

If you really want to see what my face looks like, I am sending you this linen cloth, on which you will be able to see not only the form of my face but the divinely transformed state of my whole body.

Message from Jesus to King Abgar (according to the
late-tenth-century Codex Vossianus)

IN 1958, AN EXPEDITION OF American art historians[1] visited St Catherine's Monastery in the Sinai desert to study its unrivalled collection of early Byzantine icons. While there they came across what remained of a panel painting that had originally been in three sections, as a triptych.[2] The central panel, almost certainly a depiction of the Image of Edessa, was lost, leaving just the two wings. The left-hand wing showed the seated figure of the disciple Addai/Thaddaeus, bringer of the Image to Edessa, with the smaller figures of two saints[3] below him. The right-hand wing featured a seated King Abgar holding the Image of Edessa (pl. 21c), and below him another couple of saints.[4]

The now two-part icon still exists at St Catherine's, and its right-hand wing is the more interesting. For as was quickly noted by German-born expedition member Professor Kurt Weitzmann, the features of 'King Abgar' bear the closest resemblance to those of Emperor Constantine Porphyrogennetos as known from his coin portraits (pl. 21d). Likewise, Abgar's costume and crown is that of a Byzantine emperor from this same period. Evidently the icon painter was flatteringly comparing Emperor Constantine's recent acquisition of the Image of Edessa to King Abgar's of the same nine centuries earlier. Weitzmann therefore confidently pinpointed the icon's date to somewhere between the beginning of Constantine's reign as sole emperor in January 945 and his death in 959.

Obviously of prime interest for us has to be how the artist has depicted the Image being held in Constantine's hands. If the panel indeed dates between 945 and 959, it ranks as the earliest known depiction of the Image of Edessa in its direct form as an actual piece of cloth[5] as distinct from its indirect or interpreted versions as Christ Pantocrator, Christ Enthroned, and those of the roundel variety we have seen hitherto.

Immediately apparent is that the piece of cloth we see on this icon, even though it features a disembodied, front-facing face in broad semblance of the equivalent area on the Shroud, does not seem to be on anything like the Shroud's double body-length scale. While the 'face-only' form is consistent with the earlier-advanced 'doubled in four' mode of how the Shroud was presented as the Image, a fringe runs along the bottom edge where we would expect the Shroud's fold line to be. As a result of this, some art historians, understandably encouraged by the radio-carbon dating findings of 1988, have used this icon to dismiss any possibility that the Image of Edessa could have been one and the same as the Shroud.[6]

Yet one of the most persistent features of our inquiry has been the deceptiveness of appearances. And in this instance a dismissal of identity purely on these grounds would be particularly over-hasty. As art historians are well aware, most Byzantine artists were Orthodox monks who quite characteristically took little account of representational reality. When in the sixteenth century one such monk looked at a painting he had commissioned from Italian Renaissance artist Titian he expressed his horror at Titian's naturalism: 'Your scandalous figures stand out quite from the canvas: they are as bad as a group of statues!'[7] To Orthodox monks, their paintings were icons of what they represented in almost exactly the same sense as those on a home computer's 'desktop', simply symbols or tokens for the altogether greater and more interesting realities that lay behind them.

Thus it is instructive to see the St Catherine's Sinai depiction of the Image of Edessa in the context of five equivalent examples that were created during the following century. Two of these are manuscript illuminations, one in the Greek Patriarchal Library, Alexandria (pl. 22c), the other formerly in the Stavronikita monastery, Mount Athos, now in Moscow (pl. 14b).[8] The remaining three are mural paintings to be found in rock churches nestled amid the picturesque landscape at Göreme, Cappadocia. The first, among the better preserved, is in the earlier-mentioned Sakli or 'Hidden' Church, only discovered in 1957 after a landslide had blocked off its entrance for five hundred years (pl. 22b). The second is in the dimly lit Karanlik Church (pl. 22d), whose frescoes[9] were beautifully preserved mainly because for centuries after Byzantium's fall it served as a pigeon house: the droppings shielded the paintings from what would otherwise have been certain Turkish vandalism. The third, and poorest preserved, is in the nearby St Catherine Chapel.

The Alexandria, Sakli and Karanlik depictions all have their fringes at the sides, rather than on the top and bottom edges as on the Sinai example. Uniquely among the group, the Alexandria version has a lattice decoration to the cloth's surface, whereas the three Göreme examples all feature various kinds of roundel decorations. The three Göreme murals are also decorated with vertical bands, whereas the two manuscript illuminations are not.

By such significant variations in their details all five depictions essentially cancel themselves out, along with their St Catherine's Sinai panel painting counterpart, for being anything like reliable representations of whatever the Image of Edessa may actually have looked like. It seems improbable that any of the six artists responsible for them had actually viewed at first hand the original Image they were copying, supposedly more directly than anyone had hitherto.

For certainly, just as when it was housed in its old home in the Cathedral of Edessa (or its Monophysite equivalent), in its new home in Constantinople's Pharos Chapel the Edessa Image was regarded as too holy for normal human gaze. Despite all the excitement of the Image's reception in Constantinople, and all the public activities whereby it theoretically bestowed upon Constantinople the same magical protection Edessa had enjoyed, the general Constantinopolitan public seems only to have been allowed to see the closed casket containing the Image, not the Image itself. There were no known public expositions of the Image, such as would later become relatively common in the case of the Turin Shroud. To the Byzantine mind, something of this kind for so holy an object was quite unthinkable. In the words of the hymn that had been specially written into the Byzantine Menaion, or monthly calendar, for the Orthodox Feast of the Image of Edessa instituted by Constantine Porphyrogennetos,

How can we with mortal eyes look upon this Image
Whose celestial splendour the host of heaven presumes not to
    behold?
He who dwells in heaven has now clearly come down to us in
    this revered Image.
He who is seated on the Cherubim visits us this day
Via an Image which the Father has delineated with his
    immaculate hand,
Which he has formed in an ineffable manner,
And which we sanctify by adoring it in faith and love.[10]

Yet, despite its all-powerful, 'too holy for mortal gaze' mystique, there undoubtedly were, as we have already seen, some individuals in Constantine's time who were allowed a direct, close-quarters sight of the Image of Edessa. Bishop Abraham of Samosata had necessarily been accorded this privilege back in Edessa, in order to satisfy himself that it was the genuine article and not some cunningly contrived fake that he was bringing back to his emperor. Constantinople's chief minister Theophanes and leading members of the Byzantine Senate appear to have been granted the same privilege when they acted as reception party for the Image at the time of its arrival. Constantine Porphyrogennetos and his two brothers-in-law definitely saw it, and at close quarters, hence the intriguing account of their reactions described in the last chapter. And there must have been others, as can be gleaned not least from the relative wealth of document-ation concerning it which survives from Constantine's time.

Thus, the *Story of the Image of Edessa* – whether or not it was directly written by or merely commissioned by the emperor – includes some tantalizing indirect snippets of information about the Image's physical appearance, even though it never provides us with any direct description.

In terms of the Image's housing, the *Story* includes

several mentions of its being carried around Constantinople, together with the letter of Jesus to Abgar, in a *kibotos*, which means a coffer or chest. In the description of how the Image had been stored in Edessa the alternative word used was *theke*, carrying much the same meaning. There is therefore a strong suggestion that whether it was being transported long distance or being stored long term, its housing was rather more substantial than might be expected for something that was merely a headscarf-size piece of cloth.

The *Story* also makes fairly explicit that, as a piece of linen cloth, the Image was mounted in some form rather than merely being stored loose. For it relates that after King Abgar had been cured of his disease, he ordered the Image to be 'fixed to a wooden board and adorned with the gold that can still be seen. He had these words inscribed on the gold: "Christ, the God. Whoever hopes in you will never be disappointed." '[11]

The strong inference is that at the time of the *Story*'s composition – understood to have been no later than 16 August 945 – the Image was being preserved in Constantinople in the very same mounting provided for it while it was being kept in Edessa, a mounting possibly dating even as far back as Abgar's time. As noted earlier, this seems to have been *tetradiplon*, or 'doubled in four', style. If the Image was indeed one and the same as the Shroud, this would have left just the face visible, readily accounting for the face-only depictions, such as that on the Sinai triptych, featured earlier in this chapter. Arguably, therefore, even in 945, and even among those few in Constantinople who had seen the Image directly, no one may yet have been aware that the Image was anything more than a face cloth.

This is certainly borne out by the two alternative versions the *Story* provides (both of which were apparently in circulation at the time of its writing) concerning how the Image was formed. According to the first version, the messenger Abgar had sent to take his letter to Jesus had also been instructed 'to take back with him a likeness of Jesus' face'. Perceiving this, Jesus 'washed his face in water and wiped the liquid from it onto a cloth that he had been handed, and arranged in a divine way beyond understanding for his own likeness to be imprinted upon the cloth'.[12] According to the second version, provided by the *Story*'s author because, by his own admission, 'it would not be at all strange if confusion has arisen in the story over such a long time', 'It was said that when Christ was about to willingly undergo suffering he displayed human weakness and prayed in anguish. The gospel tells us that his sweat fell like drops of blood and then it is said that he took this piece of cloth, which can still be seen, from one of his disciples, and wiped off the streams of sweat on it. The figure of his divine face, which is still visible, was immediately transferred on to it.'[13]

From the viewpoint of any possible identification of the Turin Shroud with the Image of Edessa, obviously the second version is at least the more promising. Here, for the first time in textual references to the Image of Edessa, we have the information not just that the imprint was watery-looking, but that it was created at a time when Jesus had apparently been pouring 'streams' of a bloody sweat of the kind described in St Luke's gospel chapter 22:44. The notably tentative suggestion is that this may have happened during Jesus's deeply emotional agony in the Garden of Gethsemane shortly before his arrest.

Besides taking the Image closer to the events of Jesus's Passion than had been understood previously, this new, alternative version reminds us of one of the many oddities we noted much earlier about the Turin Shroud: the seemingly watery consistency and colour of its main blood flows. At the end of the last chapter we saw how the artist who had created the coin die for Constantine Porphyrogennetos's gold coins included the reverse '3'-shaped watery blood flow on the Shroud man's forehead. So, back in the year 945, with the Edessan Image having been mounted still with only its face area visible, it would have been perfectly understandable even for those theoretically 'in the know' in Constantinople to have supposed the Image to comprise Jesus's face alone, and for them to have envisaged it deriving from a bloody sweat which had poured from Jesus while he was still alive, rather than its having been created by his crucified body while it lay in death.

While everyone was under precisely that false impression, before the turn of the second half of the tenth century the Eastern Orthodox Church's propaganda machine seems to have been set in motion to put out this 'official' story around the then Christian world. As we saw earlier, celebration of the Image was written into what became the Menaion, or monthly calendar, with rites for commemorating lives of saints according to their feast days, and the Synnaxarion, the shorter version, both forming an essential part of the apparatus for any Orthodox church or monastery. The monk Symeon Metaphrastes seems to have worked tirelessly generally to upgrade, elaborate and formulate all such service books that the Byzantines used in their elaborate liturgies. Notably, nowhere among this immediate output is the word *sindon* used of the Image of Edessa as it had been in some of the earlier Syriac sources.

And because the Image was obviously no longer the Image of Edessa but of Constantinople, a new name for it gradually crept in. Seemingly adopted from the Arab *mandil*, which had never previously existed in the Greek language, it became known as the *Mandylion*, or to give it its most frequently used title the *agion Mandylion*, or 'Holy Mandylion'.[14] The earlier-mentioned Edessa Image mural in the Karanlik Church at Göreme seems to have been one of the first depictions to carry such an inscription name, albeit in a rather scrambled form.[15]

In all these ways there was created by the very Constantinople clique that was theoretically in the know a very powerful, and still prevailing, perception that this cloth bore an imprint of Jesus's face only. The general understanding also was that this image had been created while Jesus was still alive, and on a piece of cloth of a size that today would be a headscarf – therefore, for us, from more than a whole millennium later, seemingly impossible to equate with the fourteen-foot full-body Turin Shroud.

Yet not very long after 945 some subtle hints began to emerge that all about the Image may not have been quite as plain and above-board as many had assumed. As noted by Mark Guscin during his extensive browsing among the early manuscripts preserved in the monasteries at Mount Athos, in several of the Synnaxarion manuscripts, at the very beginning of the entry for 16 August – that is, the celebration of the Feast of the Image of Edessa – there occurs the following verse:

> In life you exuded your likeness on to a *sindon*,
> In death you entered the final *sindon*.[16]

Although this did not exactly seem much to go on, Guscin also noticed in some of these same Mount Athos manuscripts a change in the request of King Abgar. He was represented as instructing his messenger to bring back to him details not only of Jesus's face and hair, but of his 'whole bodily appearance'.[17] As further noticed by Guscin, a late tenth- or eleventh-century manuscript of the sixth-century *Acts of Thaddaeus*, one of only two of this composition to have survived to our time, differs from its partner in precisely this same piece of information, merely using different Greek words for this purpose.[18]

Supplementing and expanding upon this, back in the early 1990s Rome-based scholar Gino Zaninotto had brought to attention a manuscript preserved at the University of Leiden in the Netherlands, the Codex Vossianus, in which Jesus, in his letter to Abgar, was represented as saying, quite illogically but reflecting a changed understanding that the image was of the full body, not just the face, 'If you really want to see what my face looks like, I am sending you this linen cloth, on which you will be able to see not only the form of my face *but the divinely transformed state of my whole body* [my italics]. When you have seen it you will be able to soothe your burning desire. May you fare well for all time in the wisdom of my Father.'[19]

Because of its Carolingian-style handwriting, the Vossianus manuscript cannot date much later than the end of the tenth century. Furthermore, little more than a century later it finds support from another Latin source, the *History of the Church* written by English monk Ordericus Vitalis in 1130, in which Ordericus recorded that 'Abgar the ruler reigned at Edessa; the Lord Jesus sent him a sacred letter and a beautiful linen cloth he had wiped the sweat from his face with. The image of the Saviour was miraculously imprinted on to it and shines out, *displaying the form and size of the Lord's body* [my italics] to all who look on it.'[20]

Some time around the middle of the eleventh century, which would coincide with the centenary of the Image's arrival in the city, there also occurred in Constantinople an intriguing flurry of fresh interest in the Image. This seems to have begun in 1032 when the Byzantine general George Maniakes, taking advantage of quarrels between Arab chiefs on the empire's eastern borders, managed to capture

Edessa on behalf of the then emperor, Romanos III. Great-grandson of his namesake who had commissioned John Curcuas's Edessa expedition back in 943, Romanos III had become emperor by a very last-minute marriage to Zoe, fifty-year-old daughter of his then dying predecessor Constantine VIII.

According to the contemporary Arab-Christian historian Yahya of Antioch, Maniakes duly brought back with him to Constantinople the purportedly 'genuine' letter of Jesus to Abgar, the Curcuas-borne one of 944 having apparently been a fake.[21] Genuine or otherwise, neither the letter nor the Image brought Romanos any personal protection because within two years, in one of those murky episodes that abound in Byzantine history, he was found dead in his bath. Whether he had suffered a heart attack and drowned or had deliberately been held under the water is unclear. Within hours, Constantinople's Patriarch was summoned to the Great Palace where he was first shown Romanos's corpse, then ushered into the next room. There the newly widowed Empress Zoe was sitting side by side with her lover, a courtier called Michael, both already arrayed in full imperial regalia. They ordered the Patriarch to marry them on the spot, whereupon Michael became Emperor Michael IV of Byzantium.

In a bid to ensure the maintenance of law and order, the forceful Zoe immediately sent off the Image of Edessa and Jesus's letter to Abgar, together with other relics, to a resort on Byzantium's central Black Sea coast, as sureties to persuade the well-respected aristocrat Constantine Dalessenos to come out of retirement and return as chief minister to Constantinople.[22] This unexpected excursion for the Image would almost certainly have been via a state-of-the-art imperial *dromon* or war galley, with as many as a hundred rowers.

Two years later, when Byzantium was in the grip of famine due to a prolonged drought, it is recorded that Emperor Michael IV personally carried the Image of Edessa in procession to the Church of the Virgin at Blachernae to plead for rain. Clearly this was yet another instance of the Image's protective powers being invoked, except that in this instance the Almighty displayed a sense of humour: a sudden violent hailstorm reduced the proceedings to a shambles.[23]

Yahya of Antioch reported seeing the Image of Edessa in Constantinople's Hagia Sophia cathedral[24] in the year 1058. For the reasons stated earlier, it is unlikely that this was a public display of the Image, though it might well have been paraded inside its casket. More probably Yahya was referring to the ceremonial unveiling of a just-completed mosaic of the Image known to have decorated the apex of the arch leading into Hagia Sophia's apse, above a still extant mosaic of the Virgin and Child.[25] Although no such mosaic is visible today – the space is blank (pl. 24b) – in the mid nineteenth century Turkish sultan Abdul Medjid ordered Hagia Sophia's few remaining Christian mosaics to be plastered over, some of which were then lost in an earthquake in 1894. However, in the 1930s American restorers brought to light some mosaics from Empress Zoe's time surviving in the southern gallery, among which the redoubtable Zoe herself is depicted, together with her third and final husband Constantine IX, the pair flanking a magnificent Christ Pantocrator.

That a depiction of the Image of Edessa definitely existed above the arch of the apse is known from an engraving of 1680 (pl. 24a) by the French artist-traveller Guillaume-Joseph Grelot,[26] who made extensive sketches of what he saw in and around Constantinople in the course of his travels in the region. Grelot's sketch, which is of the whole apse, is sadly not detailed enough to tell us much (he himself thought it to be a Veronica). However, if the original was anything like the quality of the surviving accompanying Archangel Gabriel, it would have been a very fine piece of artwork indeed. Accordingly, it may well be no accident that it is from precisely this time on that depictions of the Edessan Image in murals, manuscript illustrations, etc. of the kind discussed earlier in this chapter begin to proliferate.

In Hagia Sophia, Constantinople, the positioning of the Image of Edessa mosaic at the apex of the apse arch was reminiscent of the above-the-gateway location of the original cloth when rediscovered in Edessa five hundred years earlier. It also conveyed the Image's increasingly exalted status in the minds of those responsible for commissioning church decorations.

Five hundred miles away, the very simply executed late-eleventh-century murals in the tiny, obscurely located Sakli or 'Hidden' rock-cut church at Göreme could hardly seem more different from the Constantinople Hagia Sophia's vast scale and one-time decorative magnificence. Yet at Sakli too the depiction of the Image of Edessa crowns an arch, one of three such arches fronting the sanctuary. And particularly fascinating in its case are the figures painted in its immediate vicinity (pl. 23a).

The most striking of these is a woman in front of a tall building, with on the pillar of the arch opposite her a winged angel. As is evident from a near-identical scene painted on an icon at St Catherine's Sinai (pl. 23b), the theme is the Annunciation. The woman is the Virgin Mary, spinning a skein of blood-red wool for the veil of the Jerusalem Temple (the tall building). The angel is the Archangel Gabriel; the moment depicted is his announcement to her that she is being divinely impregnated with the infant Jesus.

The fascinating aspect of the St Catherine's scene is that

at the end of a beam of supernatural light coming down from heaven can be seen a vanishingly faint impression of the embryonic Jesus being *imprinted* into Mary's womb (pl. 23c). Any casual viewer of the Sinai icon might not even notice this tiny detail. Yet in all Byzantine art, no work in its *character* takes us closer to the evanescent, shadowy impressionism of the imprint that is on the Turin Shroud. And on the Sakli mural, despite the scene being exactly the same Annunciation, we find no beam of light, no embryo, only the Image of Edessa set on high between Mary and the archangel. So was the Sakli monk artist deliberately likening the divine imprinting that happened at Jesus's birth to the similarly divine imprinting that would happen at his death?[27] The Temple veil that Mary was spinning was of course that which would be rent at the very moment of Jesus's death – yet more haunting symbolism.

Whatever the validity of this, it is from this same mid-eleventh-century period that we come across an artist providing us with what seems to be an absolutely unique depiction of how the Image of Edessa was being kept during this later leg of its stay in Constantinople. Only very recently brought to the attention of western scholarship is a manuscript called the Alaverdi Tetraevangelion preserved at the Institute of Manuscripts, Tbilisi, Georgia.[28] This comprises the four Christian gospels and the story of the bringing of the Image of Edessa to King Abgar. We know exactly when the manuscript was created because it carries a date, 1054. This corresponds to the period of fresh interest in the Image earlier noted, and also to the known visit to Constantinople of Georgia's King Bagrat IV. Among the few ever allowed to see something of the treasures of the Pharos Chapel were visiting kings and their entourage.

The Georgian artist has accordingly depicted for us

what seems to have been the Image of Edessa's casket as stored in the chapel (pl. 24c).[29] And if this understanding is correct, the scale of the object he has depicted can only further support the ever-growing indications that the Image of Edessa secretly featured the imprint of Jesus's entire body, as on the Shroud. For what we see is not only a box-like lower casket, much like the one in which the Shroud was stored between 1604 and 1998, but rising from this a rectangular gold-covered panel, much larger than anything which might be expected for a mere face cloth. Displayed on the panel's gold cover is what might at first be mistaken for the Image itself, until we notice that the artist has been at pains, by the deliberately unconvincing way he has depicted the cloth's fringe, to show that this is merely a token of the true original cloth which we may infer to be hidden behind the gold covering.[30]

Usefully supporting and supplementing this is a verbal description of the Image's storage arrangements as recorded in Latin by a clearly intrigued visitor to Constantinople around 1090: 'This wonderful linen cloth with the face of the Lord Jesus, marked by direct contact, is kept with greater veneration than the other relics in the palace, and held in such great esteem that it is always kept in a golden case and very carefully locked up. And when all the other palace relics are shown to the faithful at certain times, this linen cloth on which the face of our redeemer is depicted is not shown to anyone and is not opened up for anyone except the emperor of Constantinople.'[31] For a cloth that we know to have been accompanying the wood of the true cross, the crown of thorns, the lance that pierced Jesus's side, a phial of Jesus's blood and much else to be 'kept with greater veneration' has to be accounted a high esteem indeed, nothing less than the ultimate in holiness.

Accordingly, we may ever more confidently suspect that

it was the cloth we today call the Turin Shroud which lay beneath that gold cover. No longer was it mounted 'doubled in four' style. (That mode of folding may have been used for a briefer period than previously thought.) Instead it was most likely draped over a board, so that the frontal half of the body hung one side and the back-of-the-body half the other. This would have minimized the amount of creasing and been much better for the cloth's long-term preservation. From the raking-light photography of the Shroud carried out in recent years some appropriate crease lines can be noted all along what would have been the broad dividing line between the two halves.

Exactly when the discovery of the Image's full body imprint was made we can only guess. But whenever it happened, just as today's Christians make no attempt to rewrite the Book of Genesis to accord with Darwin's Theory of Evolution, so it would not have been in the Byzantine mindset to rewrite what had become Orthodox tradition. Better for the extraordinary secret that lay inside the Image's locked gold case to stay that way.

Nevertheless, as corroboration that from this time on there was a growing, albeit very tacit, awareness of the Shroud's full body imprint, some significant things happen in the field of art. As has been shown by the same Professor Kurt Weitzmann whose team found the Constantine VII–Abgar icon at Sinai, from the eleventh century on what had been a mummy-style mode of depicting Jesus's entombment gradually gave way to a new concept of how Jesus was buried.[32] The Byzantine Greeks called this new mode the Threnos, or Lamentation, its main feature being that Jesus is wrapped in a large cloth readily compatible with today's Turin Shroud.

An early example of this genre can be seen on an ivory panel dated to the eleventh century preserved in London's Victoria and Albert Museum. Another is on a champlevé

Fig. 28 ENTOMBMENT SCENE ON ENAMEL from pulpit preserved in the monastery of Klosterneuberg, near Vienna. The figure of Christ is strikingly similar to that on the Shroud. The work of master decorator Nicholas of Verdun, the enamels on this pulpit were completed no later than 1181.

enamel panel that forms part of the decoration of a magnificent twelfth-century pulpit preserved at Klosterneuberg, near Vienna, a work that is known to have been completed by master decorator Nicholas of Verdun no later than 1181 (fig. 28). As an otherwise inexplicable innovation in art, we see in some of the examples Christ's body depicted in a very stiff attitude, the hands crossed over the loins in the same so-called 'modesty pose' visible on the Shroud. Worthy of note is not only the 'pose' itself, but that the right arm is over the left, with an awkward crossing point at the wrists, exactly as on the Shroud.

Perhaps most compelling of all is a drawing on a page of the Hungarian Pray manuscript preserved in the National Szechenyi Library, Budapest (pl. 25a).[33] Not only do we yet again see the awkward arm crossing, this time, most unusually, Jesus is represented as totally nude, exactly as on the Shroud. Again exactly as in the case of the Shroud, all four fingers on each of Jesus's hands can be seen, but no thumbs. Just over Jesus's right eye there is a single forehead bloodstain. Delineated in red, this is located in exactly the same position as that very distinctive reverse '3'-shaped stain on Jesus's forehead on the Shroud that we noted earlier. Exactly as in the case of the Shroud, the cloth in which Jesus is being wrapped is of double body length type, the second half, as known from other versions of the same scene, extending over Joseph of Arimathea's shoulder. If all this is not enough, the cover of what appears to be the tomb is decorated with a herringbone pattern in which can be seen four holes in an identical arrangement to the so-called 'poker-holes' on the Shroud that we have suggested were sustained during Caliph Mu'awiyah's 'trial by fire' experiment back around 680.

However, just when the Shroud/Image of Edessa identification might seem all but established, we encounter a setback. During the twelfth century, various visitors to Constantinople, when writing about their experience, provided lists of the relics they had been told were kept in the imperial collection. Whereas previously there had been no mention of Constantinople possessing Jesus's burial linens, now, without any indication of how these had suddenly appeared, they were listed, and with no reference to any imprint. But also listed, as if a separate object, was what sounds like the Image of Edessa.

The first instance of this occurs in a list written in Latin by an English pilgrim around 1125. After mentioning 'the holy handcloth[34] . . . which Christ sent to King Abgar of Edessa' which 'has the face of the Savior without painting',[35] the pilgrim went on to list 'the crown of thorns, the mantle, the scourge, the cane, the sponge, the wood of the Lord's cross, the nails, the lance, blood, the robe, the girdle, shoes, the linen cloth and *sudarium* of the entombment'.[36/37] A similar duplication occurs in the writings of an Icelandic abbot, Nicholas Soemundarson, who visited Constantinople in 1157. Soemundarson mentioned both 'the *sudarium* which was over his head', with no indication that this bore an image, and 'the *mantile* which our Lord held to his face, on which the image of his face was preserved'.[38]

While such duplication is disquieting (though in any event it is unlikely that either pilgrim saw the relics they listed), it is easy enough to account for. Tourist confusion is not uncommon in any century. Even in modern times visitors to Turin viewing a photographic copy of the Shroud sometimes suppose they are seeing the true original. But there may well be a deeper explanation. If the Byzantines wanted now to acknowledge that they had Jesus's burial cloth, as they seem to have done, they could not simply tear up as so much waste paper the Image of Edessa's tradition that Jesus had created it in life, as an image just of his face. As stressed earlier, this would not have been in the Byzantine mindset. Their solution would have been to use a copy of the Image as the 'face only' relic and gradually suggest that there existed an imprint on the burial linens likewise.

There is certainly a hint of this in how, in the earliest years of the thirteenth century, we find Nicholas Mesarites, custodian of the Pharos Chapel relic collection, referring to what is undoubtedly Jesus's burial shroud (whether imprinted or not imprinted). First, he described this as proof of Jesus's resurrection: 'In this chapel Christ rises again, and the *sindon* with the burial linens is the clear proof.'[39] Then, in his second reference to this same shroud, he remarked intriguingly, 'The burial *sindon* of Christ: this is of linen, of cheap and easily obtainable material, still smelling fragrant of myrrh, defying decay, because it wrapped the mysterious, naked dead body after the Passion.'

Which immediately raises the question, what was it about this shroud that caused Mesarites to claim it as proof of Jesus's resurrection, also to be so confident that Jesus's dead body had been laid in it naked? Was he actually talking about the Christ-imprinted cloth of Edessa, all along supposed to have borne Christ's face only, but now gradually assuming its true identity as Jesus's burial shroud?

Ironically it would take an ordinary soldier – and a despised westerner – at last to reveal the Image's true identity, and in very short order. For in the early 1200s Frenchmen and Venetians joined forces for what history now knows as the Fourth Crusade. The objective was to attack Muslim strongholds in the Holy Land. However, the plan introduced by one leader, Boniface de Montferrat, was to stop off at Constantinople to help a Byzantine prince, Alexius, regain the Byzantine throne on behalf of his father Isaac II, the legitimate emperor, who had been deposed, blinded and imprisoned in a coup back in 1195.

When the Crusader fleet and army arrived at Constantinople and began a determined attack on the city the emperor who had usurped the throne, Alexius III, fled to Thrace. With relative ease the Crusaders were then able to take the city, in the course of which the blinded Isaac II was released from prison and reinstalled on the Byzantine throne as joint-emperor with his 'rescuer' son, now Alexius IV.

But all along the Crusaders had been expecting that a

duly grateful Constantinopolitan citizenry would thereupon generously fund their otherwise punitively expensive
expedition to the Holy Land. Accordingly, while the
Crusader leaders and Byzantium's new emperors
negotiated just how much this was going to cost
Constantinople's citizens, and how they were going to
extract it from them, the Crusader army was free to stroll
Constantinople's streets.

It was like a teenager who has been brought up on
hamburgers and chips being let loose in a fine dining
restaurant. Rough Frankish soldiers, vulgar in their
manners, ill disciplined and unwashed, found themselves
striding through streets more beautiful than anything they
had ever seen before, rubbing shoulders with highly
cultured eastern Greeks whose perfumes and jewels, worn
by men and women alike, struck them as decadent and
effeminate.

And it is at this point that we turn again to the aforementioned writings of Robert de Clari (see chapter 8), the
ordinary soldier with the Fourth Crusade who wrote a
book about the marvels he had seen during his time
wandering around Constantinople. One of those marvels
was Jesus's shroud at the Church of the Virgin Mary at
Blachernae. To recall Robert's exact words: 'There was
another church which was called My Lady St Mary of
Blachernae, where there was the *sydoine* in which Our
Lord had been wrapped, which every Friday stood
upright, so that one could see the figure of Our Lord on
it.'

It is such a short passage yet, as an indisputable piece of
writing from the year 1203, absolutely revelatory. For
suddenly everything comes together. The *sydoine* in
Robert's Old French is undoubtedly the *sindon* that
wrapped Jesus in the tomb, so recently, and so
mysteriously, claimed to be among the relics of Jesus

preserved in Constantinople's Pharos Chapel. But Robert tells us that this *sindon* has an imprint of Jesus's figure on it. And the only relic of Jesus known to have any such imprint on it is the Image of Edessa – the cloth which all along we have been suggesting was Jesus's shroud. The reference to its being 'upright' also precisely matches the mode of mounting the cloth, as if Jesus was standing upright, that earlier we inferred from the Georgian Alaverdi manuscript illumination.

And this cloth's identification with the former Image of Edessa also makes sense of the fact that Robert tells us his viewing occurred at the Church of the Virgin Mary at Blachernae. We already know that the Image of Edessa and the *sindon*, irrespective of whether they were or were not one and the same object, were normally kept in the Pharos Chapel of the Great Palace. But as we noted earlier in the instance of the drought in the time of Emperor Michael IV, whenever Constantinople felt itself to be in some danger, the custom was to carry the Image in procession from the Great Palace to Blachernae to invoke its legendary powers of protection. There can be little doubt that this was what Robert de Clari so uniquely witnessed: a very special showing of the cloth itself, not just inside its casket, to persuade Constantinople's citizens that they had nothing to fear from the rough Crusaders in their midst, that Christ and the Virgin Mary were on their side protecting them just as they always had throughout similar dangers during Constantinople's long history.

But this time things were different. Relations between the Byzantines and the Crusaders worsened, and suddenly the latter, still unpaid for their services, found Constantinople's gates firmly shut against them. Whereupon all the fine intentions for a crusade to the Holy Land turned to greed for possession of the glittering city in which they had so recently walked.

A determined attack began, all the more vigorous because it was for their own ends. The Byzantines resisted as best they could, but long ago they had lost interest in war, and their leadership was no match for their opponents' ingenuity. After managing to breach the walls not far from the Blachernae area, the Crusaders poured into the city.

All the hatred, all the envy at the opulent Byzantine world that so outshone their own towns and villages was released as the Crusaders burst into houses, palaces and churches and began to claim the spoils of war. Nicholas Mesarites described the scene: 'war-maddened swordsmen, breathing murder, iron-clad and spear-bearing, sword-bearers and lance-bearers, bowmen, horsemen, boasting dreadfully, baying like Cerberus and breathing like Charon, pillaging the holy places, trampling down on divine things, running riot over holy things, casting down to the floor the holy images of Christ and His holy Mother and of the holy men who from eternity have been pleasing to the Lord God'.[40]

It is without doubt one of the most shameful episodes in western European history. Pope Innocent III, when he heard the news, was horrified that a Christian army should have abused fellow Christians in this manner – and rightly so. But by the time the messenger reached him, the damage had long been done. It took only three days for Constantinople, the queen of cities, to be wrecked in a manner from which she was never fully to recover. Among much else, many of the priceless bronze statues from Greek antiquity were melted for coinage.

Somewhere in all this confusion the Image of Edessa (alias our Shroud) mysteriously disappeared. Very mysteriously. Robert de Clari's words leave no doubt that this was something that had interested him, and he had made specific enquiries, but 'No one, either Greek or French, ever knew what became of this *sydoine* when the city was taken.'

Chapter 14

The Templar Secret

In 1287 a well-born young Frenchman called Arnaut Sabbatier was admitted into the Order of the Knights Templar ... During his initiation ceremony, after he had taken monastic vows of poverty, obedience and chastity, the preceptor led him to a secret place, to which only the brothers of the Temple had access. There he was shown a long linen cloth on which was imprinted the figure of a man and instructed to venerate the imprint by kissing its feet three times.

Vatican archivist Dr Barbara Frale, describing her discovery of an as yet unpublished document concerning the Templars' initiation ceremony, *Osservatore Romano*, 5 April 2009

UP TO NOW, despite the many uncertainties, we have had as a source of strength that there was in existence an undoubted object known to history as the Image of Edessa, and that this could have been our Shroud. It is ironic that this object should disappear within months of the surfacing of the most positive evidence of identity, but such is the stuff of history.

From 1204 there stretches a period that has been termed 'the missing years', a gap of some 150 years from the disappearance of what was unquestionably the Image of Edessa during the Crusaders' sack of Constantinople to the appearance of what is undoubtedly the Turin Shroud in France in the 1350s, in the possession of the knight Geoffrey de Charny.

In many ways a gap of this kind is to be expected. If the theory is correct that the Image of Edessa and today's Shroud are one and the same, there *had* to be a period of discontinuity for the Image's true identity to be forgotten and for the Shroud to emerge, as it does in the fourteenth century, with apparent total ignorance as to its earlier whereabouts. Had this break not occurred, the mystery that has been so difficult to unravel throughout would have been no mystery at all. The words of Robert de Clari are worth repeating: 'No one, either Greek or French, ever knew what became of this *sydoine* when the city was taken.'

But how do we explain, and if possible satisfactorily resolve, this daunting break?

One thing seems definite: the cloth did not stay in the hands of the Byzantine Greeks. At the outset it is possible, indeed likely, that it was some Byzantine with access to the Pharos Chapel who perhaps slipped in and secreted it away. If he – or she – did so, they did not retain it long, perhaps being captured, or pawning it at an early stage for much-needed cash. Certainly the cloth does not appear to have reached the now seriously fragmented new Byzantine mini-kingdoms formed by royal-blood survivors in the Balkans and at what is today Trabzon on Turkey's Black Sea coast. No mention of it has been found in their records, such as they are. Nor did it come to light when in 1261 a Byzantine emperor was able to walk back into Constantinople, following the Crusader withdrawal from the city. In the fifteenth-century accounts of the fall of Constantinople to the Turks, the Virgin's robe played a part in the Greeks' unsuccessful attempts to protect their city. But there is no mention of anything answering the description of the Image of Edessa.

Not that the Greeks, or their fellow Orthodox neighbours, forgot the cloth that they had guarded so closely for more than two and a half centuries. Far from it. It was portrayed on countless icons and mural paintings in Greece, Cyprus, Macedonia, Serbia, Bulgaria, Georgia, Russia, those parts of Turkey not yet fallen to the Turks, and more. Some of these examples, such as the wall-painting in the Church of St Euthymios at Thessaloniki in Greece, included the same association of the Image with the Annunciation that we saw back in the eleventh century at the Sakli Church in Cappadocia.[1]

There is also more than a hint of attempts to make up for the previous secrecy and reticence, for some of these newer depictions more closely resemble the Shroud than anything seen hitherto.

Thus, in the narthex of the jewel-box-like Hagia Sophia at Trabzon, constructed around 1260 on the orders of the Trebizond emperor Manuel I Komnenos, we find over the doorway a mural of the Image of Edessa which, despite the serious damage to its right-hand half, can be seen to bear the closest resemblance to the equivalent facial area on the Turin Shroud (pl. 25b). Unlike the late-eleventh-century Göreme murals, with their widely variegated roundels imaginatively added to the cloth's surface, on this Trabzon example we see a largely plain cloth with just three very simple stripes at the edge, and a disembodied front-facing face that is very close in coloration to that on the Shroud.

Around 1265, a thousand miles to the west, the Serbian king Stefan Uros I commissioned an extensive scheme of murals for his new and picturesquely sited monastery of Sopocani. Over a main doorway – as usual evoking the location of the Image's discovery – Stefan's Serbian artists painted a depiction of the Image of Edessa (pl. 26a) that has almost exactly the same characteristics as those at Hagia Sophia, Trabzon.

Indeed, a notable feature of both these depictions is the way that, more distinctively than previously, the artists showed the Christ face set on the cloth in a landscape aspect rather than a portrait aspect – (i.e. as shown in the sketch on the left rather than on the right in fig. 29). The particular significance of this mode of setting is that it is totally at variance with a virtually universal artistic convention whereby whenever artists create portraits they set the face on an upright rectangle rather than a horizontal rectangle. The simple principle is that it is visually unappealing to set a face on a landscape-shaped background, particularly a virtually plain one, and, even more important, it is wasteful of available space. So for the artist copyists of the Image of Edessa to have depicted the Christ face in this highly unorthodox manner strongly suggests that they were trying to reproduce a curiosity of the original – and the equivalent area of the Shroud has precisely this appearance (pl. 26b).

Fig. 29 LANDSCAPE ASPECT VERSUS PORTRAIT ASPECT. Artistically it makes more sense to present a face 'portrait' style, as on the right.

It is also notable that it is precisely in Trebizond's and Serbia's 'Byzantine' kingdoms during the same thirteenth-century era that we find something even more compelling in its apparent reminiscence of the 'lost' Shroud. Displayed in the Museum of the Serbian Orthodox Church in Belgrade is a five-foot-long red silk liturgical cloth called an *epitaphios* (pl. 27a), directly symbolic of Jesus's shroud, and of a kind that continues to be used to this day in churches of the Eastern Orthodox rite all around the world for their ceremonies on Good Friday and Easter Saturday. In some places the Orthodox monks process with the *epitaphios* on their heads (pl. 27b), reminiscent of how Addai presented the Image of Edessa to King Abgar. It is typical of such *epitaphioi* to have as their central feature a beautifully embroidered figure of Christ lying in death, very similar to those arguably Shroud-inspired Lamentation or Threnos scenes that made their appearance in Byzantine art back in the twelfth century, as noted in the last chapter. On some examples the embroiderers have intriguingly incorporated a Shroud-like herringbone pattern (pl. 27c).

But the particular fascination of the embroidered image of Christ in death on the Serbian Orthodox Church Museum *epitaphios* is that, because of its strict full-body frontality, it is even more compellingly reminiscent of the Shroud's frontal image than anything we have seen hitherto. On it we see the same long-haired, long-nosed, bearded face. We see the same crossed hands. And although the thumbs are depicted, the way the long fingers of the lower hand parallel those on the Shroud is particularly striking.

The Slavonic inscription on this *epitaphios*, 'Remember, Lord, the soul of your servant Milutin Ures', enables us to date it to the reign of Milutin II Uroš, ruler of Serbia between 1282 and 1321. Thirteenth-century Serbia's

Nemanjic dynasty were individuals of quite unusual piety, and this particular example was most likely ordered by Milutin's widow Simonis, daughter of the then Byzantine emperor Andronikos II, for her husband's intended last resting place at the Bankska Monastery.

Not the least of the *epitaphios*'s interests for us is that it was not a single, one-off piece of embroiderer's art, but one of a type. Princeton University's Art Museum has another example, in poorer condition and very crudely 'restored' in places, but with exactly the same frontality. Donated by American philanthropists Mr and Mrs Sherley W. Morgan, it was bought by the couple in 1930 from an Istanbul dealer who told them it was from the dramatically sited Soumela Monastery near Trabzon.[2] A third example of the same genre is at the monastery of Pantokrator, Mount Athos.[3] In the case of the Milutin *epitaphios*, such is its presence that when British actor Sir Laurence Olivier visited the museum where it is displayed, he reportedly spent a long while standing before it in silent prayer, following which he recited a speech from *Hamlet*. In the case of all three works it is very difficult to understand how they could have been created without some knowledge, however indirect, of the image on the cloth that we know today as the Turin Shroud.

Aside from these tantalizing visual suggestions of some distinctly powerful memory lingering after 1204, there really is no other evidence for the Image of Edessa/Shroud having stayed in Byzantine hands. So where did it go? The surprising fact is that it certainly did not *seem* to make any appearance in the west, at least in any recognizable form.

This said, there are a couple of hallowed icons that are sometimes claimed as the true Image. One, long preserved in the Church of St Bartholomew of the Armenians in Genoa,[4] is thought to have been a gift from the Byzantine emperor John V Palaeologos (reigned 1341 to 1376) to

Genoese captain Leonardo Montaldo, who bequeathed it to St Bartholomew in 1384. Painted partly on a cloth pasted on to a wooden panel housed inside a silver-gilt frame beautifully decorated with scenes from the Edessa cloth's history, the icon carries on its frame, in Greek letters, the label *to agion mandylion* – the Holy Mandylion. However, such an inscription gives it no more credence for being the true original than a Louvre Museum postcard of the *Mona Lisa* labelled 'Mona Lisa'. Both the cloth and the wood panel it is pasted on have recently been carbon-dated to the thirteenth century.[5]

The second icon, kept since 1870 in the Matilda Chapel of the Vatican but whose previous home was Rome's Church of San Silvestro in Capite, is housed in a frame of seventeenth-century Italian workmanship. Otherwise it is very similar in size, aspect and character to the Genoa example, again having been painted on cloth pasted on to wood. It formed part of an exhibition of Vatican treasures that toured the USA, Canada and Germany between 2003 and 2005.[6] There is no record of its existence before the sixteenth century. The Genoa icon seems to have been created as one of the gifts that John V Palaeologos used for trying to secure western military and financial aid against the Turks – what one writer has dubbed 'icon diplomacy' – and the Matilda Chapel icon was most likely of similar date and origin.

A third western contender, favoured by Cambridge historian the late Sir Steven Runciman,[7] is a *sanctam toellam* or 'holy towel' bearing a picture, referred to in a document of 1247.[8] This records that Constantinople's then impoverished Latin emperor Baldwin II sold it to St Louis, King of France, who duly installed it in his beautiful Sainte Chapelle in Paris. The difficulty with this identification is that this cloth was regarded as the most minor of the relics the saintly king purchased in 1247.

Louis's far more significant acquisition was Jesus's crown of thorns. An engraving of the Sainte Chapelle's collection made in 1790[9] likewise shows the *toellam* as a very insignificant object compared to the rest. Identification with the once-so-fabled Image of Edessa therefore seems highly unlikely, and whatever it was, it was destroyed in 1792, during the widespread spoliation of France's sacred treasures that accompanied the French Revolution.

But if none of these three was the true Image, and if we are right that the Image and the present-day Shroud are one and the same object, where could it have been during the so-called 'missing years'? One historical fact needing to be recognized is that the Pharos Chapel, where both the Image and the *sindon* (irrespective of whether they are one and the same object) *should* have been kept at the time of the Crusader sack, was mostly spared the looting and the mindless destruction that happened elsewhere. The priceless relic collection that it housed remained intact. This was carefully catalogued by Garnier de Trainel, Bishop of Troyes – ironically a predecessor of the Pierre d'Arcis, Bishop of Troyes, who would try so hard to prove the Shroud a fake a little less than two centuries later. De Trainel's list of what he found in the chapel has survived,[10] and although other objects that had been in the imperial collection are all there, there is no image-bearing cloth. Nor, surely significantly, is there the *sindon* or shroud that had appeared so mysteriously in the twelfth-century lists.

A possible alternative scenario needing consideration, given that Robert de Clari reported the image-bearing shroud being shown at the Church of the Virgin Mary at Blachernae, is that the Image was actually being kept at Blachernae rather than in the Pharos Chapel at that time. The Blachernae area was certainly among the first to be overrun when the Crusaders poured into the city in 1204, and the Image's inevitably no-expense-spared golden casing would have made it a sure-fire target.

But why then, even if the gold case was quickly melted down for its cash value, did the thief not make public his find? Contrary to what we might expect today, there was at the time of the Crusades no reason for anyone to have kept such a relic hidden through fear of punishment for their having stolen it. Not long after Constantinople's sack the Crusader leader Count Baldwin of Flanders imposed a weak and tardy ban on the looting of relics. This was a half-hearted attempt to check abuses, but also to siphon off at least some of the spoils to pay the Venetians, to whom he and his Frankish companions-in-arms were heavily indebted for transport services. But too much loot-ing had already gone on for Baldwin's order to be treated with much respect. And there are numerous accounts of knights and high-ranking churchmen who returned home to be greeted with adulation, not recrimination.

One prime example among the churchmen was Nivelon de Cherisy, Bishop of Soissons, who had travelled to Constantinople with the Crusade. In June 1205 he proudly returned to his diocese to distribute among its larger shrines two large pieces of the True Cross, the head and forearm of John the Baptist, the finger of the blessed apostle Thomas 'which he placed in the side of the Lord', 'one thorn from the crown of the Lord', 'a portion of the towel with which the Lord girded himself at the [Last] Supper', the head of the apostle Thomas, the staff of Moses, a part of the reed with which Jesus was beaten up prior to his crucifixion, the head of the apostle Thaddaeus, and more.[11] Some of these relics had come from the very chapel of the Pharos in which the Image of Edessa was normally housed. The highly respectable monastic establishment at Cluny likewise delightedly received the head of St Clement, stolen from Constantinople's Church of St Theodosia. Relics brought trade through pilgrimages to wherever they came to rest. The income received from them helped to fund the construction and reconstruction of cathedrals and churches. Enormous sums of money were paid for them, the crown of thorns alone changing hands for the princely sum of ten thousand marks. So there was every incentive to publicize their whereabouts.

But no one brought to light the Image of Edessa or, if it really was a different object, the shroud that had been housed in Constantinople's Pharos Chapel.

If the cloth survived – and de Clari does not suggest that it was destroyed – it seems necessarily to have gone 'underground' under some rather unusual and devious circumstances, which actually helps us to develop a working 'suspect profile' of the unknown keeper or keepers. The very fact that this period of 'going underground' extends through more than a century implies a certain continuity of ownership, suggesting in its turn ownership by a group of people rather than a single individual. If this hypothetical group was not actually among the Crusaders who took Constantinople in 1204, logically they are likely to have had some reasonably close connection with them. Also they were either very wealthy – otherwise there would surely have been too great a temptation to sell such a priceless relic, either openly or on the black market – or they were possibly contemptuous of personal wealth owing to a strong and sincere religious leaning.

If the latter – that is, that the owners were strongly religious – they must have had some special reason for keeping the Image secretly for themselves rather than making its existence generally known among their fellow Christians, as might otherwise be expected. As a group they must also have had not only the self-discipline and the means to maintain such well-kept silence over their collective secret, but some suitably secure building or buildings to ensure their sacred possession's safety and concealment, moreover housing that was appropriate to its sanctity. This would have been no mean feat over five generations.

Last but not least, if the Shroud/Image of Edessa identification is valid, by the early fourteenth century these unknown owners must have had some historical link with Geoffrey de Charny, the relatively lowly French knight who was the Shroud's improbable first authenticated owner.

While nothing of this 'suspect profile' may appear to provide anything of real substance, as it happens there is one historical group that fits these requirements with quite uncanny precision.

Some eighty years before the capture of Constantinople two French knights, Hugh de Payens and Geoffrey de Saint-Omer, with seven companions, founded the Crusader Order of Knights Templar, or 'Poor Knights of Christ of the Temple of Solomon', so-called because they were given land close to the site of the ruined Temple in Jerusalem. Created with the idea of protecting pilgrims visiting Jerusalem, and with a rule composed by the contemporary Churchman St Bernard of Clairvaux, the Templars were an order of warrior-monks who made priestly vows of poverty and chastity but were expected to train to become fighting men so fearless and well disciplined that they would never flee in battle. Independent of Europe's kings, their allegiance was solely to the Pope. Well-to-do families would often provide their second or third sons to the Order, and they would bring with them all their worldly goods – lands, horses, servants, etc. Such was the general support for the idea that from its initial small beginnings in the region of Troyes, by the early thirteenth century the Templar Order had become a very powerful entity, with its own fleet, regional monastic headquarters called preceptories – each under the leadership of a preceptor, where recruits would receive their training – and a series of virtually impregnable fortresses from eastern Europe to the Near East.

In a precarious age, these fortresses were of no small importance and became recognized as useful safehouses and storehouses for treasures and valuables of all kinds. Just as people today safeguard their money in banks, so in the late twelfth and thirteenth centuries the 'poor knights' fulfilled this role, inadvertently creating an impression that they possessed enormous wealth. With an efficient support structure of clerks and servitors, the Order was able to act as guardians, traders and pawnbrokers for the flourishing trade in relics, genuine and false alike, that ensued after the Fourth Crusade. Thus the means of acquiring the Image of Edessa/Shroud were in place. Also, the well-distributed and heavily guarded Templar monastery-fortresses provided suitable means for keeping the cloth's whereabouts secret for a considerable period.

The question that therefore arises is this: is there any valid reason for believing that this actually took place?

The answer is surprisingly affirmative. The Templars' daytime business dealings with the money entrusted to them were impeccable; many modern-day banking and accountancy practices can be traced back to the Templars. But at night dark mysteries surrounded their internal doings. Chapter meetings were held at midnight behind locked doors. Initiates were sworn under pain of death never to reveal the details of the ceremony by which they were admitted to the Order. Most significant of all, at the beginning of the fourteenth century all Europe buzzed with gossip and rumours of a mysterious 'head' that they were believed to worship idolatrously at secret ceremonies. According to one account, 'it was a certain bearded head, which they adored, kissed and called their Saviour'.[12] Another described this head as an idol, which

'They venerate as God, as their Saviour. Some of them, or most of those who attend the chapters, say that this head can save them, that it has given them all the wealth of the order, that it makes the trees flourish and the earth fruitful. Also they bind or touch this idol with the cords with which they gird themselves.'[13]

The mystery cult was no mere rumour. It was of great importance for the Templars, and one of the key factors in the Order's downfall. For with the tinge of heresy it raised, it provided the excuse needed by the forceful French king Philip the Fair – who had overreached his credit limit with the Templars, frustrating his ambitions to wage war with England – with the opportunity to lay his hands on their wealth on the grounds of their corruption.

In view of the Templars' military strength, Philip could only achieve his plan by surprise. He succeeded brilliantly. At daybreak on Friday, 13 October 1307, acting on orders they had been allowed to open only the day before, seneschals throughout France seized and imprisoned without warning every Templar they could lay their hands on, from the lowliest servant to Jacques de Molay, Grand Master of the Order, along with his private bodyguard of sixty knights. In the hands of Philip's interrogators, the Templars were subjected to merciless torture and forced to confess to a series of highly perverted religious practices, the truth of which has long been debated by historians. Entangled with the knights' 'confessions' of lurid sexual aberrations are to be found other confessions of great importance to us: the Templars' own descriptions of the mysterious 'idol' or 'head'.

It is important to consider the possibility that if indeed the Templars had acquired the Shroud, then they, like the Byzantines, may have presented it mostly in the 'disembodied' form it takes when 'doubled in four'. In practical terms the cloth is much more manageable folded to this size, and the face is its most immediately meaning-ful feature. Also likely is that the Templars had some special 'holy' copies made, kept in jewelled cases similar to those popular at around this same time for other similarly sacred items, such as the Acheropita of Rome's Sancta Sanctorum Chapel. It is significant, therefore, to discover that while there is a wide variation in accounts of the 'head', for reasons to be considered shortly, the consistent picture of it was 'about the natural size of a man's head, with a very fierce-looking face and beard', and 'of a red-dish colour'.[14]

Philip the Fair actually appears to have known such information before he laid hands on a single Templar. He had informers, but he may even have gathered something of it first hand, having taken refuge in the Paris Temple during a city riot in 1307. He knew, for instance, that viewing the head was the privilege only of a special inner circle of the Order, as is clear from his instructions to his seneschals that the 'idol' was in the form of 'A man's head with a large beard, which head they kiss and worship at all their provincial chapters, but this not all the brothers know, save only the Grand Master and the old ones.'[15]

But why should there have been such a fuss about some representation of a man's head? Why should it not have been clear who the man was? Why should only select brethren have been allowed to see it? And why should it have been regarded with such fear?

Seasoned knights as they were, those Templars who spoke of it – and many refused point-blank – described themselves as having been quite literally terrified. One used this as his excuse for not being able to describe it, saying that it had filled him with such terror he hardly knew where he was.[16] Another, Raoul de Gizy, who was preceptor for the Templar 'home' province of Champagne, had this to say:

INQUISITOR: Now tell us about the head.

BROTHER RAOUL: Well, the head. I've seen it at seven chapters held by Brother Hugh de Peraud and others.

INQUISITOR: What did one do to worship it?

BROTHER RAOUL: Well, it was like this. It was presented, and everyone threw himself on the ground, pushed back his cowl, and worshipped it.

INQUISITOR: What was its face like?

BROTHER RAOUL: Terrible. It seemed to me that it was the face of a demon, of a *maufé* [evil spirit]. Every time I saw it I was filled with such terror I could scarcely look at it, trembling in all my members.[17]

Puzzling as this testimony may seem, Preceptor de Gizy's 'terror' appears to have been no act to throw his inquisitors off the scent. That there was something very potent about this object of Templar worship is readily confirmed by an English Franciscan's most graphic description of an incident that took place at a Templar preceptory in England some time before the persecution began:

He was guest of the Templars at the preceptory of Wetherby in Yorkshire, and when evening came he heard that the preceptor was not coming to supper, as he was arranging some relics he had brought with him from the Holy Land. And afterwards, at midnight, he heard a confused noise in the chapel, and getting up he looked through the keyhole, and saw a great light therein, either from a fire or from candles, and on the morrow he asked one of the brethren the name of the Saint in whose honour they had celebrated so great a festival during the night, and that brother, aghast and turning pale, thinking he had seen what had been done among them, said to him, 'Go thy way, and if you love me, or have any regard for your own life, never speak of this matter.'[18]

The crucial question is the head's identity. It is of considerable significance that it is precisely the most holy images that caused real fear at this time. Pope Alexander III (pontificate 1159 to 1181) had ordered the Sancta Sanctorum Acheropita image of Christ to be veiled because it caused a trembling dangerous to life. In the romance of the Holy Grail, which had its origins in much the same circles as the Templars' founders, it was the image of Christ that similarly caused the otherwise valiant knight Galahad to tremble. So what was the Templar head?

There has been a lively literature of suggestions regarding its identity, both by serious scholars and by 'non-fiction' writers of the more tabloid variety. Candidates have included a Gnostic or Muslim demon that the Templars had taken to worshipping during their long sojourns in the east, the head of St Euphemia[19] ('bearded', and frightening to seasoned warriors?), even the head of one of St Ursula's eleven thousand virgins. None of this ilk has been exactly convincing. Respected British historian Malcolm Barber, for his part, has suggested the head was a complete fiction on the part of Philip the Fair and his hired interrogators, along the same lines as their accusations of the knights' sexual misdemeanours.[20] But those who manufacture lies usually like to have some element of truth upon which to build their fictions.

Accordingly, a possibility at least worthy of consideration is that rather than anything heretical or satanic, the Templar head was actually the divine likeness in a form mortal men were not normally privileged to view; in other words, something not only Christian, but with much the same 'too holy for common gaze' mystique that we have seen associated with the Image of Edessa.

In this connection there are several significant clues.

Among the evidence given against the Templars was a deposition from the Abbot of Lagny to the effect that at Templar chapters the priest had nothing to do but repeat Psalm 67. Other depositions alleged that the words of consecration were omitted at Templar Masses. The words of Psalm 67 are therefore of some interest, particularly the phrases here italicized:

> God be merciful to us and bless us
> *And cause his face to shine upon us.*
> Selah
> That thy way may be known upon earth,
> Thy saving health among all nations,
> Let all the people praise thee . . .
> Selah
> Let the people praise thee, O God,
> Let all the people praise thee,
> *Then shall the earth yield her increase*
> And God, even our own God, shall bless us.
> God shall bless us,
> And all the ends of the earth shall fear him.

As was noted well over a century ago by the great Inquisition historian Sir Charles Lea, such a chant is hardly what you would expect to hear coming from the lips of men alleged to be idolaters and worshippers of demons.[21] Rather, it makes a lot more sense as having been addressed to some very special likeness of Christ, one in which Jesus was believed to be *physically* present in such a real way that no priest, as intercessor for Christ, was needed, and likewise no words of consecration were needed. The psalm also makes sense of the independent descriptions that the 'idol' 'made the trees flourish and the earth fruitful'.

In fact, for independent corroboration that the

Templars indeed had a special veneration for Christ's holy face we need look no further than the wax seals certain German Masters of the Temple used for their official documents. One such, of Master Brother Widekind, appended to a charter dated 1 April 1271, is in the Lower Saxony State Archives, Wolfenbuttel.[22] Another, of Master Frederick Wildergrave, appended to a charter dated 7 December 1289, is in the Main Bavarian State Archives, Munich, Germany (pl. 28b).[23] Both of these seals show a Shroud-like Christ face, notably without the cruciform halo normally identifying such an image.

By far the most fascinating piece of evidence, however, came to light during World War Two when some ceiling plaster fell to the floor in the outhouse of a cottage in the English village of Templecombe, Somerset. The tenant of the cottage, a Mrs Molly Drew, found a face staring down at her. The plaster's removal revealed, carefully wired into the ceiling, a panel painting some fifty-seven inches wide by thirty-three inches high that had been created on four oak planks, tongued and grooved together (pl. 28a). Because the painting appeared to be a head of Christ the cottage owner took it to the local vicar who reportedly scrubbed it in his bath, according to Mrs Drew ruining the original 'bright blues and reds' of the head's surround.

Templecombe owes its name to the fact that from 1185 through to the early fourteenth century it was the site of a Templar preceptory that managed the Order's estates in the region, and was used for training new members and their horses for service in the east. From the painting's medieval style there seems little doubt that it belonged to the Order, and a sample taken from the panel's wood in 1986 indicated that the oak tree from which it came had been felled some time around the years 1280 to 1310. It readily conforms to some of the more rational of the Templar descriptions such as 'a painting on a plaque',[24] 'a bearded male head', 'life-size',[25] 'with a grizzled beard like a Templar's', and of someone whose identity, because of the omission of any halo, goes unspecified. The face's disembodied character clearly relates it to the face on copies of the Image of Edessa and on the Turin Shroud.

Fascinatingly, when the panel was given conservation treatment in 1986, conservator Anna Hulbert noted tiny traces of blue paint in the painting's background (exactly as remembered by Molly Drew). Hulbert further observed that there were stars on this blue, which she documented on a working drawing (fig. 30a). The suggestion therefore is that the head was set against a background of the sky, consistent with its location in the ceiling having been intended as part of a vigil for the initiate Templar knight in which he was granted a celestial vision of the Holy Face (fig. 30b). This is readily consistent with what we know of St Bernard of Clairvaux's ideas for the Order.

There is certainly a strong temptation, therefore, to identify the bearded face possessed by the Templars, of which the Templecombe panel would appear to be one of the local copies, as the cloth we know today as the Turin Shroud. Ownership by the Templars would readily account for much of the time between its disappearance as the Image of Edessa and its emergence as the Shroud known to have been owned by the later French knight Geoffrey de Charny of Lirey.

The very ethos of the Templar Order certainly makes them logical candidates as the Shroud's secret owners at this time. St Bernard of Clairvaux, the creator of the Templar Rule, referred repeatedly in his writings to the contemplation of 'the glory of the unveiled countenance of the Lord Jesus' as the ultimate human experience, one that was only admissible to those who kept themselves chaste in the manner expressly required of each Templar knight: 'if any of the Brethren do not keep chastity, he cannot

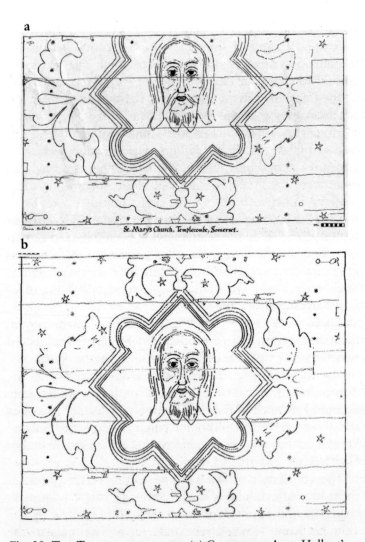

Fig. 30 The Templecombe panel. (a) Conservator Anna Hulbert's working drawing of the panel's main elements, including the today near-invisible stars. (b) Rough reconstruction by the author of the panel's original appearance, including its missing top section, showing how it may have been displayed as a 'celestial vision' based on the face on the Shroud.

come to perpetual rest nor can he see God, as witness the Apostle who says, "Search after peace with all, and keep chastity, without which nobody can see God." '[26]

Conceivably the granting of a highly privileged foretaste of this experience, in the form of sight of either the Shroud face itself or a special holy copy of this, was the culmination of the initiation ceremony each Templar had to undergo before achieving his knightly status. The 'little cord' which each Templar knight was given to wear perpetually after his initiation – a cord specifically accredited with having been placed in contact with the 'idol' – would have been touched to the Image of Edessa/Shroud to empower the knight with something of its protective properties. And the total prostration which the Templars were reported to perform before their 'idol' – an act which those not 'in the know' would very understandably perceive as the grossest idolatry – together with the fear some exhibited, makes perfect sense if the object in question was what we today know as the Turin Shroud, terrifyingly, awe-inspiringly bearing Christ's *real* body and his *real* blood.

Although many of the witness statements and confessions that were recorded during the trials of the Templar knights indicate regional copies of the Shroud face, of which the Templecombe panel would have been an example, there are others that sound strikingly like the original itself, in the form of our Shroud. The servitor Stephen of Troyes, although not himself an initiate, reported seeing at one of the Templars' Paris ceremonies the mystery object 'brought in by the priest in a procession of the brethren with lights; it was laid on the altar; it was a human head without any silver or gold, very pale and discoloured, with a grizzled beard like a Templar's'.[27] Similarly intriguing, from the *Chronicles of St Denis*, is a reference to it looking like 'an old piece of skin, as though all embalmed and like polished cloth'.[28]

Potentially most compelling of all is a document that has only recently been discovered, by Vatican archivist Dr Barbara Frale, a globally recognized specialist in the documentation on the Templars (pl. 28b). Preserved in the Paris National Archives,[29] this as yet unpublished document[30] includes the 'confession' of a French Templar, Arnaut Sabbatier. As is evident from what Sabbatier told his interrogators, his ceremony of initiation took place in the year 1287 at the Templars' commandery of Mas Deu at Roussillon in Provence, forty miles north of Marseille. In the course of the ceremony he was taken to 'a secret place to which only the brothers of the Temple had access'. There he was shown 'a long linen cloth on which was imprinted the figure of a man', and was ordered to venerate this by kissing it three times.[31] As immediately recognized by Dr Frale, this description, brief though it is, can hardly be interpreted as anything other than the cloth we know today as the Turin Shroud.[32]

Because Dr Frale's discovery is so recent, its implications have yet to be fully explored. The story of the Templars' last days at their Mas Deu commandery at Roussillon seems to have been colourful in its own right, involving families with names that will subsequently figure large in the history of the Shroud.[33] The territory of Roussillon, although today part of France, at the time of the fall of the Templars belonged to the kingdom of James II, King of Majorca. Politically, it was therefore outside Philip the Fair's jurisdiction. So if the Shroud was indeed there in 1287, and remained there during the next two decades, it would have escaped Philip the Fair's dawn arrests of 13 October 1307, which would readily explain why Philip the Fair's men never found the Templar 'idol'. But even if this 'idol' was what we now know as the Shroud of Turin, how could it have got from the Templars into the hands of the mid-fourteenth-century

French knight Geoffrey de Charny of Lirey?

As in the case of the Shroud's stay in Constantinople, the clue comes in some very 'final' moments of the Templar Order's history. Acting quite illegally, because constitutionally the Templars came under the authority of the Pope, Philip the Fair behaved with abominable cruelty to those he had captured, submitting them to extreme torture in order to obtain the confessions with which he could justify their seizure. There duly ensued a power play between King Philip and the France-based Pope Clement V, during which the Pope initially gamely insisted upon his prerogative to check on the truth or otherwise of the king's allegations. In August 1308, cardinals appointed by Pope Clement convened at Chinon, near Tours, to conduct their own questioning of leading Templars – without the duress imposed by Philip the Fair. During this and similar papal inquiries the severity of Philip the Fair's tortures became clear. One Templar, Bernard de Vado, told the papal commissioners, 'So greatly was I tortured, so long was I held before a burning fire, that the flesh of my hands was burned away, and these two bones, which I now show you, these came away from my feet. Look and see if they be not missing from my body.'[34]

Some strange and questionable elements of the Templar initiation ceremony do seem to have been admitted by the knights, and these led to a curious stalemate situation between the increasingly vacillating Pope and the forceful French king. Clement neither wanted to condemn the Templars as heretics – indeed Frale's researches have showed that Clement definitely absolved them of heresy[35] – nor to insist upon the Order being reinstated. During this period, the Templar leaders languished in prison for seven years while waiting for an outcome. Eventually, on 19 March 1314, with papal commissioners present, the four principal French masters of the Order were brought out on to a public scaffold in front of Notre Dame Cathedral in Paris. The instructions the Templars had been given were to repeat their confessions, with the promise that if they did so, although they would remain in prison, they would be spared execution.

Two of the masters accepted the 'deal' and were led off to spend the rest of their days in Philip the Fair's dungeons. But the remaining two acted differently. Pale from considerable suffering and long captivity, the Grand Master, Jacques de Molay, together with the Order's Master of Normandy, stepped forward with quite unscripted confessions. They were guilty, they said, not of the crimes imputed to them, but of uttering falsehoods about their Order under the duress of unbearable torture and the threat of execution. Instead of the scandalous iniquities of which it had been accused, the Templar Order was pure and holy. It had nobly served the cause of Christianity.

This is of course entirely consistent with the 'idol' genuinely having been the secretly owned imprint-bearing shroud of Christ. But this was not what Philip the Fair wanted to hear. While Clement V's cardinals tamely retired to deliberate, Philip ordered that the two Templars should immediately be burned at the stake as relapsed heretics. The site chosen was a small island in the Seine, the Ile des Juifs, today incorporated into the Ile de la Cité. As the sun dipped below the Paris housetops, Jacques de Molay and his companion, facing at their own request the Cathedral of Notre Dame, slowly burned to death (pl. 28c). They refused all offers of pardon for retraction, and bore their torment with a composure that won for them the reputation of martyrs in the eyes of many of the spectators.

De Molay reputedly called upon the Almighty to bring both Philip the Fair and Pope Clement V to justice, and it is one of those strange quirks of history that both would die painful deaths, Clement within a month, and Philip within the same year. But for us there is another, altogether more relevant twist to the story, concerning the name of de Molay's companion at the stake, the aforementioned Master of Normandy.

It was Geoffrey de Charny.

Chapter 15

The Knight's Tale

He was a verray, parfit gentil knyght.
Geoffrey Chaucer, Prologue to *The Canterbury Tales, c.* 1370

GEOFFREY DE CHARNY, the Templar Master of Normandy, was burned at the stake in 1314. Geoffrey I de Charny, of Lirey, was the first recorded owner of the Shroud in the west, beginning his appearance in French historical records around 1337 and living until 1356. So was there some family connection between the two by which the Shroud could have passed from the Order of Knights Templar into the de Charny family of Lirey? The short answer is that although the available genealogical records are insufficient to provide absolute proof, enough has been learned about both individuals during the last three decades to make the family connection near definite.

Thus, although some modern historians have rendered the Templar Geoffrey's surname as 'de Charnay' or 'de Charney', as if to indicate a different family, this distinction is unnecessary, and can be positively misleading. Just as English spelling was non-standardized up to the time of Dr Johnson's famous dictionary, medieval

French was the same. In the original documents, the Templar Geoffrey appears as Charny, Charneyo, Charnayo and Charniaco. In a register from the reign of the French king Philip VI we find Geoffrey the Shroud owner in the forms Charneyo, Charni, Charnyo and Charniaco.[1] The two surnames cannot be distinguished by the spellings.

Location-wise, Geoffrey de Charny the Templar is known to have come from a knightly family and to have been admitted to the Order near Etampes to the south of Paris in the year 1269. The preceptor who received him was Amaury de la Roche.[2] Amaury, who had been Grand Commander of the Templars in the East, held the key post of Preceptor of the Paris Temple between 1265 and 1271.[3] After his reception Geoffrey spent some time in Cyprus. In 1294 he was in charge of the Templar house of Villemoison in Burgundy.[4] A year later he was made Master of Normandy. His nickname was 'the berruyer' ('of the berry'),[5] associating him with the Champagne berrichonne district of France, on the borders of the territories of medieval France's great Counts of Champagne and Burgundy.

But what do we know of Geoffrey I, the first recorded owner of the Shroud in the west? And do we have any idea how he acquired the Shroud? With an older brother, Dreux, he was the younger son of the Burgundian knight Jean de Charny and his wife Margaret de Joinville (see family tree, fig. 31, pp. 278–9).[6] Important for our story is the description of his family coat of arms as 'gules [red] three silver shields'.[7] Geoffrey's paternal grandfather was a Hugh de Charny, and intriguingly a Templar of this same name is recorded in a charter of Provins, to the south of Paris, in 1241.

As a result of the Fourth Crusade, many parts of Greece had been taken over by the French, and Jean de Charny himself, together with his eldest son Dreux, served under Louis of Burgundy in an expedition to Greece in 1313. During this expedition, Dreux – the heir to his father's second Mont St Jean title – won the hand in marriage of Agnes de Charpigny, heiress of the Greek barony of Vostitza. It is possible that Geoffrey I may have been in Greece with his father and elder brother, though too young to bear arms. There he, and most certainly his family, would have mixed with people displaced by the recent suppression of the Templars, as the Order had been given extensive lands in the same region. So, just possibly, at this stage he may have been in a position to acquire the Shroud.

Again, a possible means to Geoffrey I's acquisition of the Shroud is that his second wife, Jeanne de Vergy, had members of the Besançon-based de la Roche family among her ancestors. Jeanne de Vergy's great-great-great-grandfather Otho de la Roche had been with the Fourth Crusade, and in 1205 he acquired the lordship of Athens, living on until 1224. As we saw earlier, in 1269 Geoffrey de Charny the Templar was admitted into the Order by leading Templar Amaury de la Roche. Geoffrey de Charny and Jeanne de Vergy's granddaughter Margaret's second husband Humbert de Villersexel, Count de la Roche, also came from this same family.

In a society in which, by the law of primogeniture, the eldest son inherited the knightly estates, it was common in the thirteenth century, both in France and in England, for any second and third sons to be given over either to the Church or to a knightly order such as the Templars. This certainly seems to have been the case with both the de Charny family and the de la Roche family. And the problem for establishing absolutely certain genealogical links is that because of the Order's celibacy, whenever anyone became a Templar knight they became a genealogical dead-end. Also, all the Templars' registers and books of accounts disappeared almost as mysteriously as their mystery 'head'. So in the case of several Templars with 'illustrious' family names, although their connection with that family can be considered very likely, it is often lacking the kind of documentary proof we would wish for. What can certainly be said of the family trees of the knightly families who provided sons for the Templar Order is that they frequently intermarried over several generations. This was definitely the case with the de Charnys and the de la Roches.

Noteworthy in this connection is a relatively modern transcript of a page from the lost *Chartularium Culisanense*, suggesting that the Shroud went to Athens, and was thereby under Otho de la Roche's control, immediately following Constantinople's sack. Dated 1 August 1205, the document takes the form of a letter to Pope Innocent III from Theodore Angelos-Komnenos, nephew of the former Byzantine emperor Isaac II. In it, Angelos-Komnenos complained about the sack of Constantinople by 'troops of Venice and France' and stated (my italics): 'The Venetians appropriated the treasures of gold, silver and ivory, while the French did the same with the relics of the saints and *the most sacred of all, the linen in which our Lord Jesus Christ was wrapped after his death and before his resurrection*. We know that the sacred objects are preserved by their predators, in Venice, in France and other places, *the sacred linen in Athens*.'[8]

On the basis of this letter, if it is genuine – and Dr Barbara Frale has pointed out that its diplomatic form certainly suggests that it is[9] – was this the way the Shroud passed from the de la Roche family into Templar hands along with one of their sons, the lack of any record of this owing to the loss of the Templar archive?

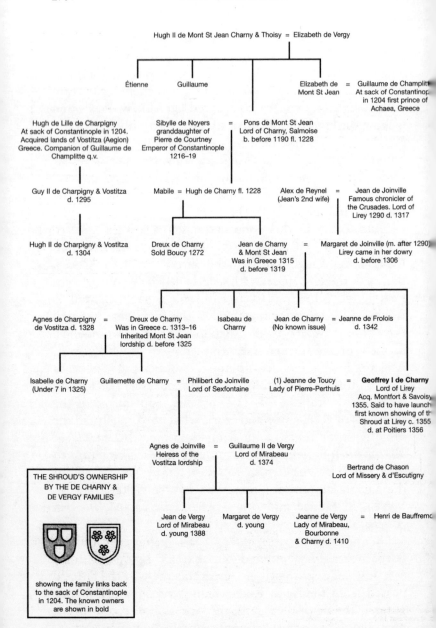

Fig. 31 **The de Charny family tree**

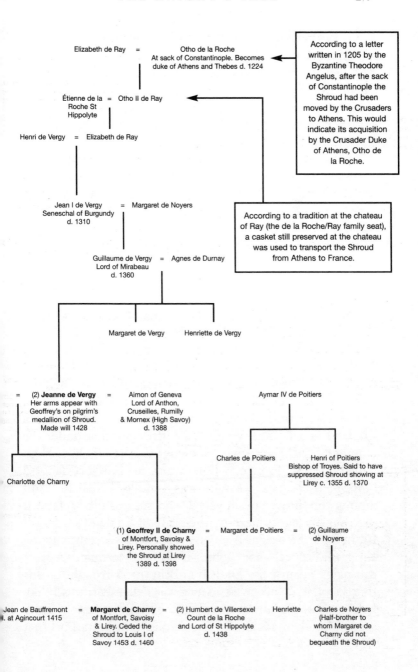

Elizabeth de Ray = Otho de la Roche
At sack of Constantinople. Becomes duke of Athens and Thebes d. 1224

According to a letter written in 1205 by the Byzantine Theodore Angelus, after the sack of Constantinople the Shroud had been moved by the Crusaders to Athens. This would indicate its acquisition by the Crusader Duke of Athens, Otho de la Roche.

Étienne de la Roche St Hippolyte = Otho II de Ray

Henri de Vergy = Elizabeth de Ray

According to a tradition at the chateau of Ray (the de la Roche/Ray family seat), a casket still preserved at the chateau was used to transport the Shroud from Athens to France.

Jean I de Vergy Seneschal of Burgundy d. 1310 = Margaret de Noyers

Guillaume de Vergy Lord of Mirabeau d. 1360 = Agnes de Durnay

Margaret de Vergy   Henriette de Vergy

= (2) Jeanne de Vergy Her arms appear with Geoffrey's on pilgrim's medallion of Shroud. Made will 1428 = Aimon of Geneva Lord of Anthon, Cruseilles, Rumilly & Mornex (High Savoy) d. 1388

Aymar IV de Poitiers

Charles de Poitiers   Henri of Poitiers Bishop of Troyes. Said to have suppressed Shroud showing at Lirey c. 1355 d. 1370

Charlotte de Charny

(1) Geoffrey II de Charny of Montfort, Savoisy & Lirey. Personally showed the Shroud at Lirey 1389 d. 1398 = Margaret de Poitiers = (2) Guillaume de Noyers

Jean de Bauffremont d. at Agincourt 1415 = Margaret de Charny of Montfort, Savoisy & Lirey. Ceded the Shroud to Louis I of Savoy 1453 d. 1460 = (2) Humbert de Villersexel Count de la Roche and Lord of St Hippolyte d. 1438

Henriette   Charles de Noyers (Half-brother to whom Margaret de Charny did not bequeath the Shroud)

Of further interest in this same connection is a wooden casket conserved to this day in the castle of Ray-sur-Saône, near Besançon, still owned by a direct descendant of Otho de la Roche, Countess Diana-Régina de Salverte. With internal dimensions just over fourteen inches long by six-and-a-half inches wide and ten inches high (which would have required the Shroud to be folded to one forty-eighth of its actual size), this casket carries the enigmatic label 'Thirteenth-century coffer in which was preserved, in the castle of Ray, the Shroud of Christ brought by Otho de Ray [i.e. Otho de la Roche][10] from Constantinople, 1206'.[11]

Supporting this same tradition, also preserved at the castle of Ray is a canvas copy of what can only be the Shroud's frontal image, the difficulty being that the date of this artwork is undetermined.[12]

Whatever the value of these suggestions of a de la Roche–Shroud–Templar connection, what is quite certain is that Geoffrey I de Charny's career was that of a very God-fearing individual who positively lived and breathed the knightly ideal in ways strongly reminiscent of the relatively recently defunct Templar Order. As early as 1337, the very start of the so-called Hundred Years War between England and France, we find Geoffrey fighting under the command of Raoul de Brienne, Count of Eu, High Constable of France,[13] in wars in Languedoc and Guyenne. The next year he was at Lille,[14] then in 1340 among the 'flower of chivalry' who successfully defended Tournai during its siege by the army of England's Edward III. In 1341 he was at Angers, preparing to accompany the King of France's son John, Duke of Normandy, on a campaign against the English in Brittany. Throughout this time Geoffrey's 'home' residence was Pierre-Perthuis,[15] like Lirey one of the tiniest of villages, with in its near vicinity several family-associated names: Toucy (the seat of his first wife, Jeanne), Noyers (the seat of his paternal great-grand-mother), and Charny itself. In exactly the manner of Geoffrey de Charny the Templar's 'border' status, some of the sources refer to Geoffrey I as Burgundian, others as *champagnois*, or from Champagne.

In September 1342, with the Brittany town of Morlaix under siege from the English, in spite of his first-class horsemanship Geoffrey was taken prisoner, and fifty of his fellow knights were killed. He himself was whisked across the Channel to spend a short time behind bars at Goodrich Castle in Herefordshire.[16]

As noted by the modern-day French historian Philippe Contamine,[17] the names of certain of Geoffrey's companions-in-arms keep recurring in the records of these actions – Beaujeu, Houdetot, Moisy, Noyers and others – and among them that of de Beaujeu deserves more attention than it has received hitherto. In the annals of the Knights Templar, William de Beaujeu was the Order's last Grand Master to be based in the Holy Land, dying a hero's death defending its key fortress of Acre in 1291. As legend has it, while de Beaujeu was battling to hold a breach the Turks had made in Acre's walls he was seen to fall back, causing one of his fellow knights to protest that he was breaking his Templar vow never to retreat. 'I am not running away. I am dead. Here is the wound!' was de Beaujeu's response, lifting up his arm to show his injury, whereupon he fell dead on the spot.

The member of the de Beaujeu family who was a companion-in-arms to Geoffrey I was Edward de Beaujeu, an individual of higher social status to Geoffrey, and just twenty-seven years old in 1343. At the beginning of that year papal envoys brokered what would turn out to be a three-year truce between the English and the French. With no 'old enemy' to fight, Edward de Beaujeu accordingly turned his energies to the Turks, who were now seriously menacing the entire eastern Mediterranean. Reportedly he told the new Avignon-based Pope Clement VI how much he wanted to go to Rhodes with a body of armed men to help fight the Turks. This was music to the ears of Clement, who was very keen to reverse the Turkish inroads upon Christian territories. As a result, on 23 September the Pope wrote a letter to Hélion de Villeneuve, Grand Master of the Malta-based (and surviving) order of Knights Hospitallers, commending Edward's fighting services. The following day Clement granted de Beaujeu the privilege of a portable altar, enabling him to have a Mass conducted by his own personal chaplain while he was overseas in places where a Mass in church might not be available to him.[18] Ten months later we find Clement granting Geoffrey de Charny exactly the same portable altar privilege.[19]

A few months after that, on 28 October 1344, a Christian fleet, inclusive of Cypriots, Rhodians, Venetians and forces supplied by the Pope, mounted a surprise attack on the then Turkish-held harbour fortress of Smyrna, then 'the finest of the Turkish ports'[20] and today the smart port of Izmir. The Christians managed to capture it, thereby gaining a tiny but vitally strategic foothold on the otherwise strongly Turkish-held Anatolian mainland. Definitely Edward de Beaujeu took part in this victory because in January 1345 Pope Clement, clearly having just heard the news, wrote to de Beaujeu's wife Marie congratulating her on the valour her husband had displayed during this action. Shortly after that came the rather more alarming news that retaliating Turks had ambushed and slaughtered a number of the holding force who had left the protection of the fortress to attend Mass at a church in the town. But de Beaujeu was not among them and was safe, arguably thanks to his portable altar. And that Geoffrey was there along with de Beaujeu, and likewise survived, can be inferred from later events.

Did Geoffrey acquire the Shroud during a visit to Smyrna as some theorists both for and against the Shroud's authenticity have suggested? Definitely he was there, confirmed by the fact that Humbert II, Dauphin of Viennois, who led a later, major expedition to reinforce the Smyrna garrison, made a payment to Geoffrey for his services to this venture.[21] There is also a reference in Geoffrey's poems to a long sea journey to the east.[22] But the puzzle has long been that Geoffrey could not actually have been with Humbert's slow-moving expedition proper because by the time Humbert's forces actually engaged the Turks at the formal battle of Smyrna on 24 June 1346, Geoffrey was back in action in south-west France, fighting the English again at the siege of Aiguillon. As proof of this there survives a receipt for Geoffrey's and his men's services in the latter action dated 2 August.[23] So Geoffrey's services to Humbert had to have occurred beforehand. And we know from a recently discovered document[24] that Geoffrey was specifically in the company of de Beaujeu and others at Smyrna, also (from separate sources) that de Beaujeu was back in France by the early summer of 1345. The likelihood is that Geoffrey returned to France with de Beaujeu, and his services to Humbert were in the form of a valuable pre-departure briefing on what opposition his expedition could expect on reaching Smyrna.

It is true that Smyrna was Geoffrey's closest encounter with the Near East, and theoretically, therefore, a likely place for him to have obtained the Shroud.[25] However, the harbour fortress area that the western forces captured was so limited in its extent – the Turks, when they eventually retook it, would completely erase it – that the idea that someone might have been there waiting with the Shroud on the harbourside, like some modern-day tourist tout, lacks all credibility.

At the battle of Aiguillon, Geoffrey was promoted to the rank of chevalier, and 'Captain of Saint-Omer', giving him the powers of France's king in that locality. From this point on – arguably due to his well-demonstrated valour both in France and overseas – he seems to have commanded the greatest respect in France's highest circles, quickly appointed a member of King Philip VI of France's secret council. His friend Edward de Beaujeu was likewise elevated, having been appointed Marshal of France after his return from Smyrna.

Early in 1347 Geoffrey helped to broker a peace treaty between the Dauphin Humbert II, with whom he had been associated concerning the Smyrna expedition, and Count Amadeus VI of Savoy. As Geoffrey could surely never have dreamed, just over a century later his granddaughter Margaret – whom he would never live to see – would bequeath the Shroud to Count Amadeus's great-grandson Louis I of Savoy.

In this same year, 1347, Geoffrey, his friend Edward de Beaujeu and other notables negotiated none too success-fully with England's King Edward III to try to lift the siege of Calais. Then in 1349, at a time when the Black Death was at its height all around Europe, Geoffrey was again in correspondence with Pope Clement VI, this time putting forward his intention to build a collegiate church at Lirey – one rather intriguingly, given the earlier-noted association of the Image of Edessa with the Annunciation, dedicated to the Blessed Virgin Mary of the Annunciation.[26]

Despite this plan, the relevant documents include not the slightest mention of any intention to install the Shroud in this church, nor historically is there any indication of Geoffrey's ownership of the Shroud at this point. But this curiosity aside, events overtook Geoffrey before he could take up Clement's readily given approval of his venture.

In order to try to recapture Calais, Geoffrey had entered into secret negotiations with a Lombard mercenary serving with the English garrison. In return for a handsome bribe the Lombard promised Geoffrey he would let him and a carefully picked fighting force sneak into the city overnight, thereby enabling them to link up with Calais's French citizens and open the gates to the full French army. Unfortunately for Geoffrey, the Lombard was a double agent who had warned Edward III of Geoffrey's intentions. Edward accordingly laid an ambush, so that when Geoffrey and his men made their dead-of-night entry, the king himself, together with his son Edward the Black Prince, was waiting to receive them (pl. 29b). Geoffrey did not give in without a struggle, receiving a head-wound in the process. The all-too-inevitable outcome – after Geoffrey had chivalrously been given a supper at which King Edward personally waited on him – was a second spell of imprisonment in England.[27]

During Geoffrey's absence, King Philip VI of France died, to be succeeded by his son John, under whom Geoffrey had served on earlier campaigns. John II 'the Good' of France appears to have personally paid the English the very substantial ransom of twelve thousand écus to secure Geoffrey's freedom. But tragedy struck only shortly after Geoffrey's return. In June 1351, reunited with his friend Edward de Beaujeu, the pair were riding together at Ardres near Calais when they came upon a seven-hundred-strong troop of English in the process of leaving the town. Although it was an all-too-rare instance of the French winning the day, it was not without the loss of Edward de Beaujeu, who was killed in the mêlée.

Later that same year King John took an initiative which seems to have been very close to Geoffrey's heart. French knightly society in Geoffrey's time had become increasingly immoral and hedonistic, more interested in fashion than in learning the craft of a soldier. This was being increasingly perceived as responsible for their repeated defeats in battles against the English, the most disastrous of which had been at Crécy just five years earlier. While Geoffrey was in prison in England he seems to have used the time to set down his thoughts on the correct behaviour for the ideal knight, for surviving among his writings is a *Book of Chivalry*[28] with this advice: 'Do not take up arms nor in any way put your life in danger without first seeking to be in . . . a good state in relation to God . . . And if you want to continue to achieve great deeds, exert yourself, take up arms, fight as you should, go everywhere across both land and sea and through many countries, without fearing any peril and without sparing your wretched body, which you should hold to be of little account, caring only for your soul and for living an honourable life.'[29]

As has been pointed out by American historian Richard Kaeuper, such ideals of 'hard service and even martyrdom in a good cause' were essentially identical to those which had been put forward by St Bernard of Clairvaux for the foundation of the Order of Knights Templar two centuries earlier.[30] In order to implement such idealism, King John the Good therefore decided to found a new chivalric order, the Ordre de l'Etoile (Order of the Star), composed of knights who, exactly like the Knights Templar, had taken a vow never to flee the battlefield. The king called five hundred knights to a special inaugural meeting for the Order held on 6 January 1352 at its intended headquarters at Saint Ouen, then just outside Paris. There Geoffrey found himself among an illustrious gathering of French high society, including his old Smyrna patron Humbert II, a man of similar chivalric idealism. But because of France's parlous political state the meeting was poorly attended, and as events would soon prove, the high-minded vow never to flee in battle would be part of its downfall.

The following year Geoffrey resumed his plans to build the Lirey church. At just fifty hearths, the village of Lirey at this time was of much the same tiny proportions as today, and the church Geoffrey built seems to have been correspondingly modest. It needed rebuilding within two hundred years, at which time a notice placed in its stone-built successor described it as 'of wood . . . very small and insubstantial, awaiting more fortunate times'.[31]

The surprising aspect is its large clerical staff. From the original 1349 intention of five canons, by January 1354 – when Geoffrey had to resubmit his intentions to the new Avignon Pope Innocent VI – this had swelled to six canons, one of whom was the ruling dean, together with three assistant clerics.[32] Their main duties were the celebration of two daily Masses, one in honour of the Virgin, the other for their founder Geoffrey. Outside visitors were expected, because indulgences were requested for those visiting the church on the Virgin's four great feast days. And a further petition, in August 1354, extended the same indulgences to Christmas, Easter, Ascension and Pentecost. Intriguingly, this August petition also requested permission for Geoffrey and his successors to be buried in a cemetery next to the Lirey church, where previously his funerary intentions had been for his bones to be divided and buried in divers places. This raises the interesting question of what had caused him to change his mind about his last resting place.

Throughout all these moves, for which the documentary sources are reasonably extensive,[33] still no mention was made of Geoffrey's ownership of the Shroud. However, in the same year of 1354 Henri of Poitiers, a colourful individual widely rumoured to keep a concubine, by whom he had conceived several children, at an abbey in the vicinity, was appointed Bishop of Troyes, Lirey's nearest large city. Henri was the bishop who was stated by his successor Pierre d'Arcis to have been responsible for the Troyes diocese, and to have conducted appropriate investigations at the time the Shroud was so controversially first exhibited at the Lirey church. Accordingly, we can date these first expositions no earlier than Henri's arrival at Troyes in 1354, and not much later either, as events will soon make clear.

By way of reminder of what Bishop Pierre d'Arcis, writing three and a half decades later, alleged happened concerning these mid-1350s expositions, these were his words:

> The Dean ... of Lirey, falsely and deceitfully, being consumed by the passion of avarice ... procured for his church a certain cloth cunningly painted, upon which by a clever sleight of hand was depicted the twofold image of one man, that is to say, the back and the front, he falsely declaring and pretending that this was the actual shroud in which our Saviour Jesus Christ was enfolded in the tomb, and upon which the whole likeness of the Saviour had remained thus impressed together with the wounds which he bore.
>
> The story was put about not only in the kingdom of France, but, so to speak, throughout the world, so that from all parts people came together to view it. And further to attract the multitude so that money might cunningly be wrung from them, pretended miracles were worked, certain men being hired to represent themselves as healed at the moment of the exhibition of the shroud, which all believed to be the shroud of our Lord. The Lord Henry of Poitiers, of pious memory, then Bishop of Troyes, becoming aware of this ... set himself earnestly to work to fathom the truth of this matter.[34]

17a Beardless face of Jesus, from a fifth-century sarcophagus in the Istanbul Museum.

17b The sixth-century rediscovery of the Image of Edessa above a gate of the city, detail from a Russian icon.

17c (*above*) Face of Jesus, from sixth-century silver vase found at Homs, Syria.

17d (*left*) Face of Jesus depicted as Christ Pantocrator, from sixth-century icon at the monastery of St Catherine, Sinai.

18a Sixth-century face of Christ as depicted on the walls of a spring at Salamis, Cyprus; an artist's copy made following its rediscovery in the 1930s.

18b (*left*) Ruined Georgian church. In the sixth century, Assyrian monks journeyed from Edessa to decorate such churches with the 'not by hand made' likeness of Jesus as found on the Edessa cloth.

18c (*below*) Mosaic face of Jesus (nineteenth-century restoration of sixth-century original), said to have appeared miraculously in the apse of the Lateran Basilica, Rome.

19a Mosaic face of Jesus, sixth century. Fragment from an unidentified location in Şanlıurfa.

19b Wall-painting of the face of Christ, from the Spring of Nicodemus, Salamis, Cyprus. This is the actual present-day appearance of the painting reproduced via a twentieth-century artist's copy in plate 18a.

19c Face of Christ, detail from icon of Saints Sergius and Bacchus, formerly at St Catherine's Monastery, Sinai, and now in the Kiev Museum of Western and Oriental Art, Ukraine.

20a Gold coin of the Byzantine emperor Justinian II, *c.* 692, featuring the face of Jesus on the obverse or 'heads' side. This is the earliest instance of Jesus's portrait being depicted on a coin.

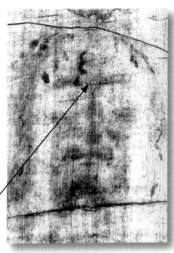

20c (*above*) and 20d (*below left*) Details of the face on the Shroud, showing the identical topless square feature in the identical location.

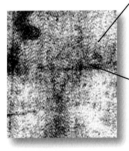

20b (*right*) Seventh-century wall-painting of Christ Pantocrator from the catacomb of St Ponziano, Rome. Note the unnaturally rectangular topless square between the eyebrows.

21a (*right*) Coin of Byzantine emperor Constantine VII Porphyrogennetos, issued shortly after the Image of Edessa's arrival in Constantinople, and featuring a particularly Shroud-like Christ Pantocrator on the obverse.

21b Hagia Sophia, Istanbul, formerly Constantinople. The Image of Edessa was brought here on its arrival in Constantinople in August 944.

21c (*right*) Detail of icon showing King Abgar with the features of Emperor Constantine Porphyrogennetos (see coin **21d** *above*) holding the Image of Edessa.

Archangel Gabriel announcing Jesus's birth

22a Typical rock church location at Göreme, Cappadocia.

22b (*above*) Image of Edessa from wall-painting in the Sakli or 'Hidden' Church, Göreme.

22c Image of Edessa from manuscript in the Greek Patriarchal Library, Alexandria.

22d Image of Edessa from the Karanlik or 'Dark' Church, Göreme.

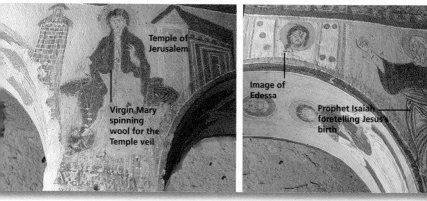

Temple of Jerusalem

Virgin Mary spinning wool for the Temple veil

Image of Edessa

Prophet Isaiah foretelling Jesus's birth

23a Sanctuary barrier in the Sakli Church at Göreme, showing the Image of Edessa intimately linked with a scene depicting the Annunciation, or announcement of the 'imprinting' of Jesus in Mary's body. The Archangel Gabriel is at left, then Mary spinning wool in front of the temple. The Image of Edessa is at the summit of the arch, and at far right the prophet Isaiah makes his prophecy of Jesus's birth.

23b (*left*) The Annunciation; twelfth-century icon from St Catherine's Monastery, Sinai.

23c (*above*) Detail of the St Catherine's icon of the Annunciation, showing the infant Jesus being imprinted in Mary's womb. Note the blood-red skein of wool, reminiscent of the later wound in Jesus's chest.

24a Seventeenth-century French artist Grelot's view of the apse of Hagia Sophia, Istanbul, showing mosaic of the Image of Edessa still extant at point L.

24b Equivalent area of Hagia Sophia today, showing a blank (arrowed) where the Image of Edessa mosaic was once located.

24c (*below*) How the Image of Edessa was conserved in eleventh-century Constantinople? Manuscript illumination of 1054 from the Alaverdi manuscript of the gospels preserved in Tbilisi, Georgia.

Now, however hostile such allegations might seem towards the Shroud's authenticity, they establish several key points. First and foremost, now firmly on the pages of history we have a reference to a cloth that seems indisputably to be the Shroud that we know today. As befitting d'Arcis's legal training, his words were admirably precise: 'a certain cloth . . . upon which . . . was depicted the twofold image of one man . . . the back and the front . . . impressed together with the wounds which he bore'. If this verbal description is not enough to show that it was our present-day Shroud which was being exhibited in Geoffrey de Charny's brand-new Lirey church back in the mid-1350s, visual corroboration exists in the form of a pilgrim's badge, made of lead, today preserved in the Paris Museum of the Middle Ages (Musée du Moyen Age) (pl. 29a). The kind of souvenir which medieval pilgrims purchased to wear on their hats after they had visited some holy shrine, this badge clearly depicts two clerics holding up for exposition a cloth with a double-figure imprint quite unmistakably corresponding to our Turin Shroud. Despite its tiny size – just under two-and-a-half inches wide by one-and-three-quarter inches high – its creator indicated the Shroud's unusual herringbone weave, and the distinctive splashes of blood from the wound in the side running transversely across the small of the back. In the badge's lower part, flanking a roundel depicting Jesus's tomb and related implements of the Passion, can be seen two shields bearing coats of arms, readily identifiable as those of Geoffrey de Charny ('gules, three inescutcheons argent [silver shields]') on the left, and those of his second wife Jeanne de Vergy on the right.

With regard to these shields, the location of Jeanne de Vergy's on the seemingly more important right-hand side has raised the question of whether, given that Bishop Henri of Poitiers held the see of Troyes until 1370, the

Shroud expositions might have been conducted after Geoffrey's death in 1356, and any time up to 1370, under Jeanne's auspices. This possibility has, however, been firmly discounted by heraldry specialist the late Noel Currer-Briggs:

> It is true that the dexter side in heraldry is the more important, but only when two or more arms are quartered together ... On the Lirey medallion the arms of Charny and Vergy are quartered, but each shown separately, with Charny on the left and Vergy on the right. Perversely in heraldry, sinister and dexter mean left and right as seen from behind the shield, not as the observer sees it, so when the shield was carried across the knight's chest to protect him, his right side was protected by the dexter side of the shield, and his left by the sinister. So far from the medallion telling us that Geoffrey was dead, it tells us that he was very much alive when the medallion was made, and that his wife was equally involved as himself.[35]

So we must envisage at least the start of the Shroud expositions, accompanied by the manufacture of marketable souvenir badges, to have taken place around 1355, in exact accord with the 'thirty-four years or thereabouts' before 1389 that Bishop d'Arcis estimated.

Plainly evident from d'Arcis's information is that back in the 1350s the Lirey/de Charny camp was insistent that the shroud they were showing was none other than 'the actual shroud in which our Saviour Jesus Christ was enfolded in the tomb'. Although later members of the de Charny family would adhere to the papal requirement to describe the Shroud as only a 'likeness or representation', back in the mid-1350s the cloth's first-known European owners were promoting it as the genuine article. And in the apparent belief, whatever d'Arcis's allegations to the contrary, that that is what it indeed was.

And with regard to that promotion, we also know from Pierre d'Arcis that news of the expositions in Lirey had spread widely, apparently causing many to flock to the tiny village even from outside France. D'Arcis's mention of people claiming to be cured by the sight of the Shroud suggests a lot of emotional fervour attached to the showings, reminiscent of scenes that had happened in Rome only five years earlier. As we know from the Italian chronicler Matteo Villani, when in 1350 the so-called Veronica cloth was exhibited, on similar feast days with similar 'indulgence' inducements and similar souvenir hat badges, 'to number the crowds was impossible, but it was estimated that from Christmas to Easter there were constantly at Rome from ten to twelve hundred thousand people, and at Ascension and Pentecost eight hundred thousand'.[36]

If we may therefore envisage that a cult of the Shroud definitely happened at Lirey something along these lines some time around 1355, and self-evidently with Geoffrey de Charny's blessing, we are nevertheless presented with many questions for which there remain no satisfactory answers. For if, as Bishop d'Arcis conveys, Bishop Henri made 'diligent' enquiries concerning where the Shroud had come from, when did these enquiries begin? And why was Geoffrey, as the Shroud's owner – a man very much in the public eye, and with the highest reputation for integrity – not providing the requisite explanations?

The answer may well lie in timing. In 1355, Geoffrey can only be construed as enjoying the highest favour in all the right circles. In February that year, as recompense for military successes he had achieved near Noyon, north of Paris, he was made lord of Savoisy and Montfort (the latter had an excellently fortified castle, not far from his old Pierre-Perthuis stamping grounds). In June, possibly for the second time, he was named *porte-oriflamme*, bearer of the sacred banner of St Denis, the royal battle standard of France, of red silk split into points like a flame, borne on a gilt-covered flagstaff. Because of the banner's sanctity, the carrying of it was an honour bestowed only on an individual considered to be of the very highest integrity, 'a knight noble in intention and deed, unwavering, virtuous, loyal, adept and chivalrous, one who fears and loves God'.[37] To receive it, Geoffrey had to take the following oath in the presence of his king, bareheaded and on his knees, in the monastery of St Denis, with the bodies of France's royal dead all around him: 'You swear and promise on the precious, sacred body of Jesus Christ present here,[38] and on the bodies of Saint Denis and his fellows which are here, that you will loyally in person hold and keep the *oriflamme* of our lord king, who is here . . . and not abandon it for fear of death or whatever else may happen.'[39]

At the end of May 1356 there then occurred, at Aix, the aforementioned formal ratification by Bishop Henri of Poitiers of Geoffrey's letters instituting the Lirey church.[40] On this occasion Henri went on record as stating, 'we praise, ratify and approve the said letters in all their parts'. He also spoke of 'ourselves wishing to develop as much as possible a cult of this nature'. And he referred warmly to Geoffrey's 'sentiments of devotion, which he has hitherto manifested for the divine cult, and which he manifests ever more daily'.

In the light of such approving statements from Henri concerning the Lirey foundation, it seems impossible that this bishop could already have made his damning enquiries into the showings of the Shroud at Lirey that were later referred to by Bishop d'Arcis. They would have to have been conducted after. Yet as events were about to prove, there was scant time left afterwards, at least for

Geoffrey himself to be asked the pertinent questions. Within a month Geoffrey was at Breteuil, playing such a commendable part in its siege that John the Good rewarded him with two houses, one in central Paris close to the royal palace, the other a suburban mansion. Shortly after that, during August, Geoffrey was with his king and his well-equipped main army as they played cat-and-mouse with the English army of Edward the Black Prince, which was in retreat to the south after having spent too long unsuccessfully trying to capture the town of Tours.

The Black Prince had reached a few miles south of Poitiers when King John the Good's army, outnumbering the English three to one, caught up with the fugitives. Coolly, the Black Prince ordered his men to take up defensive positions along similar lines to those he had successfully used at Crécy ten years earlier, when he was only sixteen. The French, for their part, thinking that they had learned their lessons from Crécy, tried different tactics. But a French cavalry charge on England's long-bowmen proved more devastating for the cavalry than for the archers. Then, when they threw three infantry divisions against the English positions, the French army's numerical superiority proved no match for English discipline and strategic ingenuity.

Geoffrey was at the side of his king, leading the third division, when John gave him the order to raise the *oriflamme*, signifying battle to the death. With both king and standard-bearer having taken their vows as members of the Order of the Star, both were committed to hold their ground, whatever the odds. As the chronicler Jean Froissart recorded of the closing moments, 'There Sir Geoffrey de Charny fought gallantly near the King. The whole press and cry of battle were upon him because he was carrying the King's sovereign banner. He also had before him on the field his own banner, gules, three inescutcheons argent. So many English and Gascons came around him from all sides that they cracked open the King's battle formation and smashed it; there were so many English and Gascons that at least five of these men-at-arms attacked one [French] gentleman.'[41]

When one of those surrounding King John seized the bridle of his horse, Geoffrey, still holding aloft the *oriflamme*, cut him down. But having no shield for his own personal protection (the big handicap for any standard-bearer), Geoffrey was then in turn quickly cut down. He fell dead at John's feet. In Froissart's words, 'Sir Geoffrey de Charny was killed with the banner of France in his hands, and the French banners fell to the earth.'[42]

So there died, pierced by the lances and battle-axes of Englishmen, the Shroud's first certain owner during the European phase of its history. Froissart, despite writing on behalf of the English side, was generous in his verdict on Geoffrey: 'The bravest and most worthy of them all.'[43] And with Geoffrey's last breath, so also there died whatever he knew about how the Shroud had come into his possession. We now need to look to those who lived on after him.

Chapter 16

Charny Bequest

. . . a cloth, on which is a figure or representation of the Shroud of our Lord Jesus Christ, which is in a chest decorated with the arms of Charny . . . we have taken and received into our care . . .

Humbert de Villersexel, Count de la Roche, 6 July 1418

WHEN THE NEWS OF Geoffrey's death eventually reached Lirey, two hundred miles to Poitiers' north-east, it can only have come very hard for his still young second wife Jeanne de Vergy. Jeanne had been left with an infant son poignantly bearing the same name as his father.[1] She could not even give her husband the decent burial in Lirey he had asked for. With so many French fatalities from the battle to cope with, including many other notables, his body had been very hastily interred at a Franciscan monastery close to Poitiers. Not least, Jeanne had ultimate responsibility for the nine clergy of the collegiate church that her husband had so recently founded.

Similar agonies to Jeanne's were being felt right across France. For King John the Good was now a captive of the English. When Geoffrey had fallen dead at his feet, someone, possibly one of Geoffrey's pages, had snatched the *oriflamme* away to safety. But King John himself, true to the Order of the Star vow he had taken along with Geoffrey never to retreat, simply surrendered, becoming a most unexpected 'trophy' – one quite literally worth a king's ransom – for Edward the Black Prince to take back to England.

And with the French throne technically vacant, even though its king was still alive, politically France was now into a situation unprecedented in its history. The French army was decimated, many of its best leaders, Geoffrey among them, having been left dead on the battlefield. Soldiers from all sides accordingly roamed the countryside, foraging and looting at will. King John's eighteen-year-old son Charles, lacking his father's physical strengths, had prudently withdrawn from the battlefield and gone back to Paris. However, he had no viable army. The heavy taxes his father had imposed left him with little popular support among France's Estates-General. And when the English made known their ransom demand of three million écus for King John's release, the chances of Charles raising that sum of money from his lack-love French populace looked bleak indeed.

Our interest, however, concerns what effect these dire circumstances may have had for the Shroud. The Lirey clergy, now with no guarantee of receiving the annual 'rents' that John the Good had granted them, can hardly have failed to perceive that continuance of their successful Shroud expositions could be a very useful means of financial support. Despite the greater dangers the country's plight posed to travellers, some people would still be prepared to face them for a soul-saving glimpse of the true body and blood of Jesus promised by the relic in Lirey.

Meanwhile, just a few miles away in Troyes, Bishop

Henri of Poitiers was facing somewhat similar financial difficulties. His cathedral, even though it had been started more than a century earlier, was still under construction, its slow rate of progress largely due to a chronic shortage of funds, now bound to get worse because of the current crisis. The situation stemmed partly from the fact that the cathedral had no major relic as a sure-fire crowd-puller. As we noted earlier, a predecessor of Henri's, Bishop Garnier de Trainel, had been responsible for controlling a lot of the major relics looted from Constantinople back in 1204. But whereas de Trainel's Soissons-based counterpart Bishop Nivelon de Cherisy had creamed off an impressive collection to distribute around the major shrines of his diocese, Troyes' Garnier de Trainel had been much less acquisitive, and the cathedral's main relic, the body of St Helen of Athyra,[2] was not in the same league as those over in Soissons.

For Henri, a highly pragmatic individual who was Troyes' military captain as well as its bishop, the spectacle of large sums of money pouring into the tiny Lirey church rather than into his own cathedral would quickly have become rather too much to bear, particularly galling because of his lack of any jurisdiction over the Lirey enterprise. As d'Arcis's memorandum makes clear, Bishop Henri's anger was directed not against any member of the de Charny family, but against Lirey's dean, whose name we know to have been Robert de Caillac. That the transparency of the money-making aspects of de Caillac's showings of the Shroud was very much in Bishop Henri's mind is evident from d'Arcis's repeated insistence on the dean's motives being 'consumed with the passion of avarice', cunningly 'wringing' money from 'the multitude'. We may also infer that when Bishop Henri launched his 'diligent' enquiries into the Shroud's origins, Geoffrey was no longer in the land of the living to answer any questions about where the Shroud had come from. It was de Caillac who was in charge.

According to the d'Arcis memorandum, the moment Bishop Henri began to institute formal proceedings against Lirey's dean and his fellow clergy, 'They, seeing their wickedness discovered, hid away the said cloth so that the Ordinary [Bishop Henri] could not find it, and they kept it hidden afterwards for thirty-four years or thereabouts down to the present year [1389].' As noted earlier, the vagueness of d'Arcis's words 'thirty-four years or thereabouts' tells us a great deal about how little hard information d'Arcis had from his predecessor's time on which to base his allegations. Had his estimate been accurate, not only would the first certain European show-ings of the Shroud have been conducted in 1355, so also would Bishop Henri's inquiry, all therefore being over and done with before Geoffrey had set foot on the battlefield of Poitiers. But as we have already seen, Bishop Henri was still on very good terms with Geoffrey, and apparently with the whole Lirey venture, even as late as the end of May 1356. This necessitates that at least the Bishop Henri inquiry part of the events must have happened after Geoffrey's death, though arguably not a long time after.

Bishop d'Arcis, writing in 1389, seems to have had no documentation to support what he said had happened in his predecessor's time. Not only did he fail to quote from a single official record of Bishop Henri's alleged proceed-ings against Lirey, he could not even provide an accurate date for these proceedings. Despite the many problems the whole Troyes region suffered during the intervening years, the city of Troyes itself was never captured, nor ravaged by any fire throughout this time, so diocesan records from Bishop Henri's time should have remained intact.

The likely inference, therefore, is that the moment Jeanne de Vergy learned of Bishop Henri's hostile designs on the Shroud, possibly even before a single document had been filed, she decided to whisk the Shroud, herself and her young son out of any harm's way. Other events that were happening in the region would have reinforced this decision. In 1358, marauding bands of English reached the Troyes region, seizing Bishop Henri's chateau at Aix-en-Othe and attempting to capture Troyes itself. During the same year, fuelled by all the unrest over taxation, the peasant uprising known as the Jacquerie broke out, during which bands of French brigands roamed the countryside at large, terrorizing anyone not behind the safety of a city's walls. Tiny Lirey would have been a totally unsafe place to be. Whether it was through local violence or through natural causes, Dean Robert de Caillac died in 1358, but more than likely Jeanne, her young son Geoffrey II and the Shroud were already long gone.

So where did they go? The one definite fact is that within a few years Jeanne remarried. Her new husband was Aimon of Geneva, a wealthy and influential nobleman whose estates of Anthon, Cruseilles, Rumilly and Mornex[3] were in Alpine High Savoy, safe from the marauding bands ravaging neighbouring France. In 1366, Jeanne and Aimon together completed a census return on behalf of the young Geoffrey II de Charny because he was still below his age of majority.[4] So Jeanne and Aimon's marriage must have occurred some time before then, and young Geoffrey's childhood would have been spent in the Alps. A somewhat neglected figure in investigations into the Shroud's history, Aimon of Geneva was cousin to Aimon III of Geneva, and in 1367 accompanied Amadeus VI of Savoy – with whom Geoffrey I de Charny had had dealings – on an expedition to Constantinople.[5]

By 1370, John the Good's son Charles V had won back some much-needed stability for France, and in that year Geoffrey II was no doubt in Paris, at his mother's side, when his father's remains, exhumed from Poitiers, were given a solemn reburial fitting for a hero, at royal expense, in the recently founded and richly endowed Abbey of the Célestins.

Predictably, Geoffrey II followed in his father's footsteps and became a well-respected royal official and professional soldier, even assuming responsibility for a similar region. As early as 1375, when only in his early twenties, he was appointed *bailli*, the northern version of seneschal (an official responsible generally for a region of which they were not native), for the region of Caux in northern France. Geoffrey's duties were something of the order of mayor, chief of police, judge, tax collector and army mobilization officer all rolled into one. Two years later he was recorded as taking part in a spectacular tournament held in St Omer, where his father had been captain. In 1382 we hear of him being praised by France's new king Charles VI, then only fourteen years old, for the role he played in a victory over the Flemish at Roosebeke. By one of the many ironies of the Shroud's story, Geoffrey II married Margaret de Poitiers, daughter of Charles de Poitiers, the brother of the Bishop Henri who had created such a fuss over the Shroud back in the 1350s. And throughout all this time, true to Bishop d'Arcis's words that the Shroud had been 'kept hidden ... for thirty-four years or thereabouts', there is absolutely no word anywhere of any showing of the Shroud.

There can be little doubt that throughout the thirty-four years that Lirey's dean and canons were supposedly 'hiding' the Shroud it was actually being quietly kept with Jeanne de Vergy somewhere in her husband's domains in High Savoy. This is because it was only upon Aimon of Geneva's death in 1388 that, the very next year, we hear of the Shroud being brought out again – and back to Lirey. The first document to indicate this, a file copy preserved in the Vatican archives, is a letter dated 28 July 1389 from the new Avignon Pope Clement VII to Geoffrey II de Charny.[6] This formally ratified the permission that Clement's legate Cardinal Pierre de Thury had apparently already granted Geoffrey II to re-exhibit what Clement carefully called the 'image or representation of the Shroud of our Lord Jesus Christ'.[7]

But once again a Bishop of Troyes perceived himself to have been sidelined, and the current incumbent, Pierre d'Arcis, a former lawyer, was a man of too formidable a mettle not to make a strong protest. Indicative of the speed with which Bishop d'Arcis acted, taking his complaint straight to Geoffrey II de Charny's superiors at the royal court, is a letter of 4 August 1389, signed by King Charles VI in Paris, ordering his *bailli* in Troyes, Jean de Venderesse, to seize the Shroud from the Lirey church:

> The Bishop of Troyes has asserted before our Curia that 'in the collegial church of Blessed Mary in Lirey, a certain hand-made and artificially depicted cloth was kept, bearing the figure or likeness of, and in commemoration of, the holy *sudarium* in which the most precious body of our Lord Jesus Christ the Saviour was wrapped after his holy Passion'. And although the faithful are in danger of idolatry, the knight Geoffrey de Charny has himself displayed or caused to be displayed the cloth with full ceremony as if it was the true *sudarium Christi* and he has not ceased, though we have tried to impede this practice. And so we command you, *bailli*, to get the cloth and bring it to me, so that I might relocate it in another church in Troyes and place it under honest custody.[8]

From this and from the later memorandum d'Arcis sent to the Pope it is apparent that when d'Arcis tried to impose his ecclesiastical authority over Lirey's clergy, Geoffrey II de Charny, via a royal warrant, took personal charge of the Shroud. This made any attempt by d'Arcis to stop the showings a civil rather than an ecclesiastical matter. But if Geoffrey, as a royal official, had obtained King Charles VI's approval for the showing – as d'Arcis himself acknowledged – how was it that the same king was ordering the Shroud to be seized from Geoffrey by a fellow *bailli*? As there is no evidence of any ill will between Charles and Geoffrey II de Charny – rather the reverse – surely the logical procedure would have been for Charles to take up the matter with Geoffrey direct?

Troyes' *bailli* Jean de Venderesse must have been aware that Geoffrey II de Charny was a fellow *bailli*, and there is a distinct air of half-heartedness in the way he tried to carry out his orders, as evident from the report he sent back after going to the Lirey church on 15 August, the Feast of the Assumption, when he would have expected the Shroud definitely to be exhibited:

We went to the church at Lirey and by virtue of the Royal papers asked that the cloth be delivered to us by command of the King. The dean [his name was Nicole Martin] responded that he could not give it to us because it was sealed in a treasury where vestments, relics, precious books [records] were kept, and locked with several keys. He had only one key. My procurer was for breaking in, but the dean opposed this saying the cloth was not there. We placed our seal on the treasury door, left a guard, and went to dinner. That evening the dean again said the cloth was not there and requested that we remove our seal. Then he said that the other key resided with the people of the lord of Lirey. We said we would continue to keep a seal on the treasury until the other key should arrive. The dean replied that he did not know when the keeper of the lord's key might come. We said we would wait until the next day;

but when the dean with all his canons filed an official appeal we did not proceed further in the matter.[9]

So why was de Venderesse so easily deflected by the official appeal? Why, on a major feast day of the Virgin, was the Shroud not being shown at all, let alone in the manner that had been causing Bishop d'Arcis such offence? And where was Geoffrey II de Charny, who, having taken all his previous personal initiatives over such showings, was apparently nowhere to be found in Lirey on 15 August 1389?

The answer again lies in timing. Over in Paris, 20 August was a day that had long been chalked into the French social calendar as the start of five days of celebrations to mark the official arrival in the capital of King Charles VI's queen, Isabel of Bavaria, preliminary to her formal coronation as Queen of France. Froissart devoted several pages of his *Chronicle* to a description of the proceedings.[10] Everyone who was anyone had already arrived in Paris for the festivities, Geoffrey II de Charny quite definitely among them, because he went on record as a participant in colourful joustings by a group of nobility, including Charles VI himself. Geoffrey would almost inevitably have gone to Paris several days beforehand to prepare for his part in this entertainment programme. Hence he was absent from Lirey at the time of Jean de Venderesse's visit, but present in person in Paris to answer any *genuine* concerns his monarch might have had over the showings of the Shroud at Lirey.

On 5 September, a royal serjeant arrived rather tardily from Paris to inform the Troyes *bailli* and Lirey's dean and his canons that the Shroud was now '*verbally* put into the hands of the lord our king'.[11] 'Verbally' is an interesting turn of phrase, because physically the Shroud most certainly remained under lock and key with Nicole Martin and his canons – and it would stay that way. The main intention of this gesture seems to have been to stall any further activity by Bishop d'Arcis – otherwise he would immediately have the king to reckon with. As later events make clear, not least the absence of any royal seizure of the Shroud, there was positively no animosity between Geoffrey II de Charny and his king over the Shroud, or any other matter. And almost certainly the seizure letter from the king sent to de Venderesse on 4 August had been prepared for him by persons unknown and slipped into a pile described 'unimportant' which the king had signed without reading it.

From 20 August onwards, King Charles VI, and very likely Geoffrey II de Charny along with him, were happily preoccupied with a 'royal tour'. All the main participants in the festivities for Queen Isabel's coronation went 'on the road', first to Burgundy, then south to Avignon, where the party were suitably entertained by Pope Clement VII. Which makes it all the more interesting that it is precisely around this time that there would have arrived in Avignon Bishop d'Arcis's impassioned and now so famous memorandum[12] to Pope Clement VII arguing for the Shroud's fraudulence, setting out all his frustrations over the sidelining to which he had been subjected, and pleading for Clement to use his ultimate authority to get the showings stopped. Unfortunately for d'Arcis, what he almost certainly did not know was that Clement and Geoffrey II de Charny were rather better acquainted than anyone might have expected. Before Clement became Pope he had been known as Robert of Geneva. His father was the cousin of Aimon of Geneva, Geoffrey II de Charny's stepfather, who had died the year before. Geoffrey II de Charny was thereby 'family' for Clement. And although we cannot be certain that he and Geoffrey were actually together at the time Bishop d'Arcis's letter arrived in

Avignon, the family tie certainly goes some way to explaining what happened next.

The first event, from an authority higher even than Pope Clement, actually makes you feel a little sorry for Bishop d'Arcis. It was another of those 'acts of God' that have occurred every now and then in the course of our reconstruction of the Shroud's history with a wry sense of timing. During the Christmas period in 1389 the entire nave of d'Arcis's Troyes Cathedral collapsed, apparently because of the failure of one of the arches supporting the upper tier, or clerestory. Shortly afterwards the rose window of the north transept fell out.[13] Not only were decades of fund-raising and construction work ruined, d'Arcis now had a roofless cathedral filled with rubble and broken stained glass. He had to hire thirty labourers just to clear the mess.

Next, sent from Avignon on 5 January 1390, arrived a stern letter from Pope Clement VII, ordering d'Arcis to keep silent about the Shroud, under threat of excommunication. On the same date Clement sent a letter to Geoffrey stating that he could continue to hold the Shroud expositions, though he should limit the lavishness of the accompanying ceremonial. There was not even any mention that Geoffrey was required to describe the Shroud as a 'figure or representation' of Jesus's shroud.

In the event, probably because Clement had used this same ambivalent formula of words in his letter to Geoffrey back in July 1389, Geoffrey and his daughter Margaret would continue to use it for a few more decades, despite what we know from Bishop d'Arcis's memorandum of what was being said in private in and around Lirey: 'although it is not publicly stated to be the true shroud of Christ, nevertheless it is given out and noised abroad in private, and so it is believed by many, the more so because, as stated above, it was on the previous occasion [i.e. back in the 1350s] declared to be the true shroud of Christ'.

This passage is extremely important because it conveys in a nutshell the delicate tight-rope Geoffrey was treading. A cautious king's official, trained in patient diplomacy by the court of Charles V, Geoffrey had very likely prearranged the form of words with Pope Clement in order to re-establish the foothold of respectability for the Shroud, lost when Bishop Henri first tried to discredit it back in the 1350s.

By being so lavish with the ceremonial, Geoffrey was probably also conveying, rather too easily to those of his time, what he believed in his heart: that the Shroud was actually genuine. For us it is this, what he genuinely believed, which is of the utmost importance. It is ironic that Bishop d'Arcis's insistence, without any document-ation to support the allegation, that Bishop Henri had found the Shroud to be the work of an artist actually con-firms for us the belief held in the de Charny camp that the Shroud's imprint was *not* the work of human hands.

Why, given this belief, Geoffrey could not have explained to all comers whatever he knew about how and where his father had obtained the cloth, and thereby satisfied con-temporary doubters as well as those of the present day, is one of those mysteries that simply has to remain unsolved. Given the unexpectedness of his father's death, and his infancy at the time, it is possible that Geoffrey II may have known little, if any, of the facts himself. Albeit highly speculative on the evidence available thus far, it is also possible that on reach-ing his majority Geoffrey II learned a great deal more about his Shroud inheritance than he ever revealed, thereby per-haps deliberately maintaining a chivalric secret stretching back to the Templars, and ultimately to Byzantium. Whatever the case, the continued upholding of the de Charny belief in the Shroud's authenticity into the next generation is certainly indicated by subsequent events.

Again following in his father's footsteps, in 1396 Geoffrey II was one of thousands of French knights and soldiers who set out from the Burgundian capital Dijon to aid the Hungarians against the Ottoman Turks. The Turkish sultan Bajazet had been boasting that once he had rolled back Hungary he would continue on into Italy to make St Peter's in Rome a stable for his horses. Geoffrey formed part of a multinational expedition that reached Turkish-held Nicopolis in what is today Bulgaria. Although they laid siege to it, they were too inexperienced to dislodge the Turks and suffered some heavy losses.

Geoffrey managed to get back home to France but he may well have been wounded in the heavy fighting, or have caught a fatal infection, for soon after he died, still only in his mid-forties, on 22 May 1398. He was buried in the church of the Cistercian abbey of Froidmont, near Beauvais. There his tombstone, with an elaborately carved effigy of him as a knight in armour, survived until the First World War when, caught up in a rain of heavy shelling from both sides, the abbey and all its buildings were reduced to rubble. Thankfully, the tombstone's appearance was preserved in a drawing collected by the seventeenth-century antiquary François Roger de Gaignières (fig. 32) – the only surviving record of the facial features of any of our three Geoffrey de Charnys.

Geoffrey had left behind a daughter, Margaret, a pious and strong-willed woman who had her own quite remarkable part to play in determining the future course of the Shroud's history. Margaret would have been little more than a girl at the time of her father's death, and because her mother swiftly remarried, it was she who inherited the paternal seigneuries of Lirey, together with Montfort and Savoisy which her grandfather had acquired during the year before his death.

In the year 1400 Margaret married Jean de

Bauffremont, a union that would be marked by double sadness. First, it was childless, which in the light of subsequent events seems to have been due to infertility on Margaret's part. Second, in 1415 husband Jean suffered the same fate as Margaret's grandfather Geoffrey, death at the hands of the English, this time at the battle of Agincourt. Following her mother's example, Margaret de Charny lost little time in remarrying. Her new husband was Humbert de Villersexel, who was a member of the earlier-mentioned de la Roche family, and who bore the titles Count de la Roche and Lord of St Hippolyte sur Doubs.

The marriage was opportune because once again, exactly as had happened after Poitiers, the French countryside was ravaged by marauding bands, and Lirey was very much in the danger zone. All parties – that is, Margaret and Humbert and the Lirey canons – agreed that it would be best for the Shroud's safety for it to be taken to the well-fortified de Charny castle of Montfort, near Montbard, well to the east. On 6 July 1418, Humbert duly issued this receipt to the canons:

> 'During this period of war, and mindful of ill-disposed persons, we have received from our kind chaplains, the dean and chapter of Our Lady of Lirey, the jewels and relics of the aforesaid church, namely the things which follow: first, a cloth, on which is a figure or representation of the Shroud of our Lord Jesus Christ, which is in a chest decorated with the arms of Charny ... The aforesaid ... we have taken and received into our care ... to be well and securely guarded in our castle of Montfort.'[14]

Humbert's receipt is an important and interesting document because it carries the only surviving description of how the Shroud was stored while in Lirey, in 'a chest

Fig. 32 TOMBSTONE OF GEOFFREY II DE CHARNY AT FROIDMONT, from a seventeenth-century ink drawing of its appearance in the Roger de Gaignières Collection, Bibliothèque Nationale, Paris.

decorated with the arms of Charny'. Although it uses the same problematic 'agreed' formulaof 'figure or represent-ation' to describe the Shroud, it notably acknowledges the Shroud as a relic. Equally interestingly, it lists the other relics held in the Lirey church,[15] among them a vial of miraculous oil of the icon of Our Lady of Sardonnay. The cult of the last item was specifically associated with the Templars.[16] Also intriguing is 'an angel of gilded silver holding between his hands a vase in which is a hair of Our Lady. And the angel is seated on a tower with three supporting columns and in this tower is a pedestal on which is a knight armed with the de Charny arms.'

Humbert's receipt ended with a promise that when France's troubles were over, all would be returned to the Lirey canons, including the Shroud. Most likely this was the genuine intention at the time, probably because up to then Margaret had seen little of the strange cloth that was part of her inheritance.

But over the next few years Margaret changed her mind. The Shroud was soon moved from Montfort to Humbert's domain at St Hippolyte sur Doubs, a tiny village on the river Doubs east of Besançon, close to today's Swiss border. There it was stored in a chapel today known as the Chapel of Our Lady of Pity,[17] which has a modern-day plaque commemorating its one-time resident. Here in St Hippolyte, Margaret and her husband began the custom of holding annual showings of the Shroud every Easter Sunday on the banks of the Doubs in a meadow known as the Paschal Field.[18] In the St Hippolyte church there is a modern-day stained-glass window show-ing, according to its inscription, 'Count Humbert de la Roche' in armour holding the Shroud, a depiction which may well have been based on an older version since destroyed.

Altogether more interesting for this particular phase of

the Shroud's history is a discovery that was made only recently in a neighbouring church at Terres de Chaux, four miles to St Hippolyte's east.[19] In 1997, fragments of decaying nineteenth-century plasterwork fell from the ribbed vaulting in this church and disclosed paintwork hidden beneath. Removal of the plaster revealed an extensive scheme of frescoes depicting instruments of the Passion, datable to the early fifteenth century, that had clearly been plastered over when their condition had deteriorated over time. On the arch to the choir, in very poor condition, came to light a depiction of a knight holding a casket between his hands – almost certainly again our Count Humbert de la Roche with the casket of the Shroud.

By far the most interesting find, however, was at the topmost point of the arch separating the choir from the nave: a depiction of a large piece of only partly unrolled cloth bearing the face of Christ, with angels supporting it either side (pl. 29c). Had such a fresco been found anywhere in the Byzantine East it would unhesitatingly have been identified as the Image of Edessa, there being numerous examples with exactly this iconography. Anywhere else in western Europe it would have been taken to be Rome's Veronica cloth, except that the Veronica was rarely if ever depicted as of such a width. But here in Terres de Chaux, created at this particular time, in a region with such close associations to showings of the Shroud, it can hardly be other than our Shroud, depicted, whether intentionally or otherwise, in a form most eerily harking back to its days as the Image of Edessa. And on the other side, the side facing the congregation, is the Annunciation.

Humbert de la Roche's death in August 1438 left Margaret widowed for a second time. Even at the end of this second, twenty-year marriage she remained childless, and was therefore going to die without direct heirs. This raised the problem of who should inherit the Shroud, and one can detect from subsequent events the extent to which this occupied her thoughts during her remaining years. She appears to have been determined that despite the promise Humbert had made to the Lirey canons some two decades earlier, the Shroud should not be returned to them.

There were sound reasons for this. The church founded by her grandfather had been a modest enough wooden structure, and after ninety very turbulent years of French history it was in a poor state of repair. Margaret appears to have regarded it as far too inadequate a home for a relic she clearly now cared for very deeply. Also, she appears not to have been prepared to trust her own relatives. The nearest relation, Francis de la Palud, her nephew by marriage, was a most colourful character who had lost his nose at the battle of Anthon and wore a false one made of silver. His exploits had been the cause of no little international friction, and although he inherited all Margaret's husband's estates, she seems to have regarded him as quite unsuitable as an heir to the Shroud. Neither did she choose to bequeath the relic to her half-brother Charles de Noyers, whom she at least trusted to negotiate on her behalf on several occasions. Nor did she give it to her cousin Antoine-Guerry des Essars, to whom, at her death, she left the lands of Lirey.

As she was under considerable pressure, beginning shortly after her husband's death, to return the Shroud to the dean and canons of Lirey,[20] Margaret's recalcitrance was remarkable. In May 1443 she was summoned before the parlement of Dôle, and agreed to hand over all the jewels and other relics that Humbert had taken into safekeeping in 1418 – except for the Shroud. In return for various payments towards the upkeep of the Lirey church she was allowed to keep the Shroud for three more years. At the end of this period she continued to show no sign of being prepared to surrender the Shroud, and was summoned this time before the court of Besançon. She was allowed to keep the Shroud for another two years, on payment of the Lirey canons' legal costs and more church upkeep. Two years later this agreement was renewed for a further three years, this latter extension negotiated on Margaret's behalf by Charles de Noyers.

One senses in these negotiations a tacit recognition that this intrepid woman was unlikely to hand over her beloved Shroud in her lifetime. Both parties seemed to believe that the problem would be solved by Margaret's death, which, as she was now in her sixties, could not be far off. But the canons reckoned without Margaret's determination and farsightedness. Although of advanced years, she had already set out on a deliberate hunt for suitable heirs.

Perhaps in a move to check out Philip the Good, the then Duke of Burgundy and a great patron of the arts, in 1448 she travelled north to Hainaut, then part of Philip's dominions, today part of Belgium. In the 1448 archives of Hainaut's capital, Mons, Margaret, as 'Mme de la Roche', is recorded as arriving in the city with 'in her care what is called the Holy Shroud of Our Lord'. Apparently she ordered some wine. Not long after that she was at Chimay, in the diocese of Liège, where the Benedictine chronicler Cornelius Zantiflet recorded her exhibiting 'a certain sheet on which the shape of the body of our Lord Jesus Christ has been skilfully painted, with astonishing artistry, showing the outlines of all the limbs, and with feet, hands and side stained with blood-red, as if they had recently suffered stigmata and wounds'.[21] Just like the bishops of Troyes, the Hainaut clergy were suspicious of the Shroud's credentials, and the Bishop of Liège asked Margaret to produce some certification. All she could offer were the bulls of Clement VII describing the cloth as only a 'figure or representation' of Jesus's shroud.

In the light of Zantiflet's clearly sceptical remarks concerning the 'astonishing artistry' of the cloth's image, Margaret evidently failed to find among the hard-headed northerners the interest she was seeking, so she returned to the south. And an exposition held in 1452 at the castle of Germolles near Chalon-sur-Saône, not far from France's border with Savoy, may have been rather more forthcoming.

For early the very next year the search was over. On 22 March 1453, Margaret was in Geneva, and a curious transaction took place between her and pious forty-year-old Louis, Duke of Savoy, the current Count of Geneva in succession to the cousin of Margaret's stepfather Aimon. According   to the still extant document drawn up between Margaret and Louis,[22] she received from Louis the small, picturesquely sited chateau of Varambon – a strange acquisition, because for umpteen generations this had been the residence of the Palud family to which her silver-nosed nephew Francis belonged. She also received the revenues from the estate of Miribel near Lyon. In return for this real estate, Louis acknowledged having received from Margaret 'valuable services'.

But what 'valuable services'? Without anyone stating it as such – this was an age in which trading in religious relics was considered unseemly – Margaret had just ceded to Louis the Shroud. And a new, five-hundred-year episode in the Shroud's history was about to begin.

Chapter 17

The Wise Mule

The said Shroud of the Lord the perfidious woman handed over . . .

16th-century French comment on Margaret de Charny

ACCORDING TO A Savoy legend, Margaret de Charny's decision to cede the Shroud to Duke Louis of Savoy was made by a mule. When Margaret first showed the Shroud to Louis and his wife Anne, the couple expressed their wish to look after it, but still Margaret could not bear to part with it. However, when she was leaving with the casket containing the Shroud strapped on to the back of a mule, the animal refused to go through the gate. Margaret interpreted this as an expression of God's will, so the Shroud passed from the knightly family of de Charny to the ducal family of Savoy.

Although it is likely this legend really is fictional, if it was a mule that prompted Margaret's decision it was a wise one, aided in part by some sensible family ties. The line of the counts of Geneva, to which Margaret's step-father was related, had become extinct in 1394. The closely related Savoy family had thereafter taken over their territory and title, creating sufficiently substantial a domain that from 1416 onwards they became styled as dukes. Both Margaret's father and her second husband Humbert had been conferred with the coveted and exclusive Collar of Savoy,[1] the Savoy equivalent of England's Order of the Garter, hence there was a strong and affectionate bond.

The other aspect that no doubt impressed Margaret about Duke Louis and his wife Anne was their piety – a trait that ran deep through their lineage, and would continue to do so. Louis was a descendant of St Louis, the thirteenth-century French king who had purchased the Holy Crown of Thorns looted from Constantinople in 1204. St Louis specially built the beautiful Sainte Chapelle in Paris to house this relic. In the same century Boniface of Savoy had become England's forty-seventh Archbishop of Canterbury. Duke Louis's father Amadeus VIII of Savoy, the first of the family to bear the ducal title, had retired to become a monk – only to find himself elected as antipope Felix V. Duke Louis himself cultivated a constant retinue of Franciscan friars who served as his confessors, and his eldest son Amadeus, renowned for his concern for the poor, would be beatified.

Duke Louis's wife Anne de Lusignan, to whom he had been married for nearly twenty years, was the daughter of Janus I, King of Cyprus, Armenia and Jerusalem. Back in 1186, Anne's Crusader ancestor Guy de Lusignan had crowned himself King of Jerusalem at none other than Jerusalem's Church of the Holy Sepulchre. Although the title became a hollow one following Jerusalem's final fall to the Turks in 1244, the Lusignans had doggedly retained it, even through their own island of Cyprus's effective fall to the Turks in 1426. Soon, as we will see, they would pass it on to the Savoys.

Before her marriage to Louis, Anne had spent the first

fourteen years of her life on Cyprus. Most likely she would have become aware how every 16 August those of the Orthodox faith on the island celebrated the feast of the bringing of the Image of Edessa to Constantinople, instituted by Constantine Porphyrogennetos five hundred years earlier. Although as a Roman Catholic Anne would not have taken part in these Orthodox celebrations, her parents' palace was not far from the island's Church of the Acheiropoietos, with its own special cult of the Image.[2] Any stories she heard about the Image while she was on Cyprus may well therefore have heightened her intrigue concerning the similarly Christ-imprinted cloth that was now in her and her husband's immediate possession.

But Margaret de Charny was not yet dead, and in the interim there were some unresolved legal hurdles to be overcome. In Lirey, the canons appear not to have been told what had happened. In May 1457 they again petitioned Margaret for the Shroud's return to them, threatening her with excommunication, and actually carrying this out when she failed to comply with their demand. Two years later Margaret's half-brother Charles de Noyers seems to have persuaded them that they would never now recover the Shroud, and that it was best for them to negotiate for compensation. Having become aware that the Shroud was now in Savoy hands, they appear grudgingly to have recognized the sense of this. A sixteenth-century manuscript in the Bibliothèque Nationale's collection for Champagne conveys their feelings: 'The said Shroud of the Lord the perfidious woman handed over, and, it is said, sold it to Duke Louis of Savoy.'[3] In any event, Margaret's excommunication was lifted, and on 7 October 1460 she was at last able to go to her rest content that her duty to the Shroud had been done.

And she had done well. Even during her lifetime the

always unsettling 'figure or representation' formula for referring to the Shroud had ever so subtly begun to be dropped from formal documentation. 'Precious jewel of the Holy Shroud' was how it was referred to during the legal manoeuvres of 1443; 'The Holy Shroud of Our Lord Jesus Christ' was its description during Charles de Noyers' dealings with the Lirey canons in 1449, and again in 1457; 'The Most Holy Shroud representing the Image of Our Lord and Saviour Jesus Christ' was how it was described in an accord[4] drawn up in Paris on 6 February 1464, by which Duke Louis agreed to pay the Lirey canons, as compensation for their loss, an annual rent to be drawn from the revenues of the castle of Gaillard, near Geneva. This same accord specifically noted that the Shroud had been given to the church of Lirey by Margaret's grandfather Geoffrey de Charny, and that Margaret had transferred it to Duke Louis, thereby providing substance and clarification to the otherwise cryptic real-estate transaction that had taken place between Margaret and Louis in March 1453. Leonardo da Vinci theorists should note that this accord of 1464 establishes beyond reasonable doubt that the Shroud subsequently owned by the Savoys can be traced *at least* as far back as the 1350s – and Leonardo was not born until 1452.

As mentioned earlier, Duke Louis was continually in the company of Franciscan friars. While considering a more permanent home for the Shroud, he lodged it in their church in Geneva. In 1464, the Franciscan order elected as its head the humbly born but highly respected theologian Francesco della Rovere, who was at that time involved in an intense theological controversy concerning the blood of Christ. Rovere, if he was not actually a member of Duke Louis's retinue at some stage, certainly seems to have seen the Shroud at first hand, because in this very same year he alluded to it in these words: 'the Shroud in which the body of Christ was wrapped when he was taken down from the cross. This is now preserved with great devotion by the Dukes of Savoy, and it is coloured with the blood of Christ . . .'[5]

Della Rovere's phrase 'when he was taken down from the cross' indicates the view, to become widely accepted, that the Shroud may have been the cloth in which Jesus was wrapped when he was taken down still bloody from the cross, rather than the cloth in which he was actually buried. This concept allowed for the widely held belief that Jesus's body was washed before burial, there being a stone in Jerusalem's Church of the Holy Sepulchre on which this is supposed to have happened. It also usefully allowed for other shroud relics without figurative imprints not to have their authenticity challenged by this 'new-comer' relic owned by the Savoy family.

The deeper and subtler shift was that a senior church-man and leading Renaissance academic was now speaking of the Shroud as if it was a fully authentic relic of Jesus's Passion, rather than a mere 'representation' of one. And that particular Churchman was about to become even more senior. Three years after his headship of the Franciscans, della Rovere was made a cardinal. Two years after that he became Pope Sixtus IV, to whom we owe the Sistine Chapel. By the magic of the rapidly developing new printing technology, della Rovere's book *The Blood of Christ*, written in 1464, was published in 1473, the second year of his pontificate. This gave the Shroud for the first time in its history, and less than a century after the adverse allegations by Bishop d'Arcis, some proper recognition from the Holy See.

By this time Duke Louis had died, likewise Anne de Lusignan. Their pious son Amadeus, now thirty, had become his father's successor, as Duke Amadeus IX. Though Amadeus was showing no aptitude for ruling, and managed to neglect the agreed payments to the Lirey canons, he and his wife Yolande of France upheld their predecessors' devotion for the Shroud. They also set in train the completion of the already partly built chapel at Chambéry with a view to this eventually becoming the Shroud's permanent home.

Meanwhile, Amadeus's brother Louis had mimicked his namesake father by becoming betrothed to his generation's equivalent Cypriot princess, Charlotte de Lusignan. On the death of her father King Jean II in 1458, Charlotte had become the full-blooded Queen of Cyprus, Armenia and Jerusalem. Despite her consort Louis of Savoy dying before her, in a special ceremony held in Rome's eleven-hundred-year-old St Peter's basilica she bequeathed all her royal titles to the House of Savoy. And empty though these titles were, their acquisition – that of Jerusalem par-ticularly apposite for the owners of the Shroud – was one that later generations of the Savoys would very proudly perpetuate in their coat of arms.

For a cloth so long regarded as imparting divine pro-tection to its owners – and later we will see several signs that the Savoy family actually believed this of the Shroud, irrespective of any awareness of its Edessan origins – such protection was none too evident during the first few gen-erations of Savoy ownership (see family tree, fig. 33). When Duke Amadeus IX of Savoy died in 1472 he was succeeded by his six-year-old son Philibert, who died only ten years later in a hunting accident, to be succeeded by his fourteen-year-old Paris-educated brother Charles I. In 1485, Charles married Blanche de Montferrat, daughter of the Marquis de Montferrat, whose forebears had married into Byzantium's Palaeologos dynasty. A commemorative page painted into Blanche's heirloom prayer book depicts the couple reverently kneeling either side of a figure of Christ bearing wounds clearly inspired by those on the

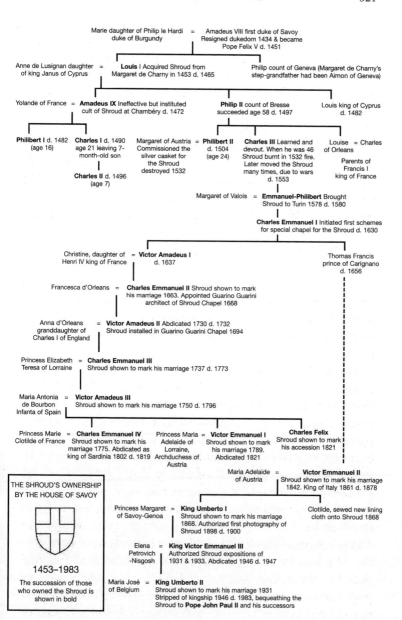

Fig. 33 **The Savoy family tree**

Shroud. But Charles lasted only to the age of twenty-one, leaving behind a seven-month-old son, Charles II, who would live only to the age of seven. Charles II's fifty-eight-year-old grand-uncle Philip then took over, as Philip II. He lasted but a year before being succeeded, in 1497, by his seventeen-year-old son Philibert II the Handsome. Philibert in his turn would last for a mere seven years.

For both the cynical and the superstitious alike, these Savoy family misfortunes might be interpreted as expressions of divine displeasure for the amount of moving around the Shroud was being subjected to throughout much of this time. Of the travels we happen to know about, simply because the documentation[6] survives, on 20 September 1471 the Shroud was moved from Chambéry over the Alps to Vercelli. Then on 2 July 1473 it went from Vercelli to Turin. Three months later it was transferred to Ivrea. Nine months after that it went to Moncalieri, followed in another month by a return to Ivrea, then back over the Alps again to Chambéry. In 1483, Savoy family chaplain Jean Renguis was paid two écus 'in recompense for two journeys which he made from Turin to Savigliano [approximately thirty miles] carrying the Shroud'.[7] All these journeys were no doubt completed on the back of a well-trusted mule, and they formed part of the peripatetic moving of themselves and their worldly goods between their castles that was a normal way of life for Europe's bluebloods during the Renaissance era.

And how was the Shroud stored? Early on in their ownership of the Shroud, the Savoys had, of course, needed to provide a replacement casket for the wooden one bearing the de Charny arms that we heard about back in 1418. An inventory of 1483 describes its conservation in that year as enveloped in a red silk drape and kept in a case covered with crimson velours, decorated with silver-gilt nails, and locked with a golden key.[8] A second inventory in 1498 makes clear that the case was of silver.

Despite the alarmingly high mortality rate among the young dukes of this time, there were some public showings of the Shroud, clearly following on from the Easter Sunday showings Margaret de Charny and her husband Humbert had held annually at St Hippolyte sur Doubs. One such showing was recorded on Easter Sunday 1488 at Savigliano, forty miles to the south of Turin. Another, at Vercelli in 1494, conducted by the now Dowager Duchess Blanche, and held this time on Good Friday, was in the presence of Rupis, secretary to the Duke of Mantua. Of this occasion Rupis reported back to his employer, 'A *sudarium* was exhibited, that is, a sheet in which the body of our Lord was wrapped before being laid in the tomb, and on which his image – both the back and the front – can be seen outlined in blood. And it looks as though blood is still issuing.'[9]

Eight years later such proceedings were in the hands of a new duchess, twenty-year-old Margaret of Austria, daughter of the Holy Roman Emperor Maximilian, who had married the young and dashing Duke Philibert II the Handsome. Margaret ordered the Shroud to be transferred from Chambéry's Franciscan church to the same town's newly constructed Sainte Chapelle, with a view to the latter now definitely becoming the cloth's permanent home. In a special ceremony held in Chambéry on 11 June 1502, the Bishop of Grenoble, in his finest vestments, solemnly made the transfer, in an elaborate procession that included the new duke and duchess, several visiting dignitaries, and all Chambéry's clergy. First the Shroud was displayed on the chapel's high altar for the benefit of the large crowds (the ordinary populace) that had followed the procession. After its return to its case, it was then taken behind the high altar to be deposited in a cavity that had been specially hollowed out of the wall for its safekeeping. The cavity was secured by a heavy iron grille with no fewer than four locks, each opened by a separate key, two of them held personally by the duke.

Despite these good intentions, the very next year the Shroud was on the move again, this time travelling the fifty-five miles to Bourg-en-Bresse, to take part in the celebrations for the return from Spain of Archduke Philip the Handsome, Duchess Margaret's brother. Again a Good Friday was chosen for the main showing, an occasion for which history again has an eyewitness report, in this instance from Savoy courtier Antoine de Lalaing: 'On Good Friday the Passion was preached in Monsignor's [i.e. Duke Philibert II's] chapel by his confessor, in the presence of the Duke and Duchess attending. Then they went with great devotion to the town's market-place, where a great number of people heard the Passion preached by a Cordelier [Franciscan]. After that three bishops showed to the populace the Holy Shroud of Our Lord Jesus Christ, and after the service it was shown in Monsignor's chapel.'[10] Lalaing also recorded his own personal impressions of the Shroud: '[The Shroud] is, I believe, the most devotional and contemplative thing on earth. It is the rich *sydoine* and noble Shroud purchased by Joseph of Arimathea. It is clearly seen to be bloody with the most precious blood of Jesus, our Saviour. It can be seen to be imprinted with all of his sacred body.'[11]

Self-evidently, now that the Shroud was in the hands of such blueblood owners, it was being taken altogether more seriously than it had been during its ownership by the humbler de Charnys.

The very next year Duchess Margaret's beloved husband Duke Philibert II died, aged only twenty-four, apparently from drinking too much iced wine after becoming overheated while hunting. Margaret was devastated, and just six days after his death she installed herself at Bourg-en-Bresse to supervise the building of a magnificent monastery as a shrine to his memory.

The following year, 1505, Margaret formally ceded the Shroud to her mother-in-law, Duke Philip II's widow Claudine, who temporarily took it with her to her castle of Bylliat, near Nantua, on the banks of the Rhône. It was a time of plague, ravaging northern Italy, and in October Claudine wrote to Margaret inviting her to 'come to see the blessed Holy Shroud that you commended [to us], that it may protect you and all your house from the epidemic'.[12] Here we can clearly see the idea of the Shroud's protective properties already being circulated among these high-born women – even despite the fact that all except Claudine had lost their husbands at an untimely age.

In 1506, Pope Julius II, the great patron of Michelangelo's Sistine Chapel ceiling and nephew of Francesco della Rovere, gave his formal assent for Chambéry's Sainte Chapelle to be known henceforth as the Sainte Chapelle of the Holy Shroud. On 21 April Julius authorized that a Feast of the Holy Shroud, initially just for the Chambéry diocese, should be assigned to 4 May, the day after the feast celebrating the finding of the True Cross. Two weeks later Julius issued a follow-up papal bull formally approving a Mass of the Shroud. This included a prayer with the words 'Almighty, eternal God, in memory of the Passion of your only begotten Son, you have left us the Holy Shroud on which his image is imprinted.' If Pierre d'Arcis had still been alive, he would not have been amused.

Margaret of Austria's recent loss of her husband was followed in 1507 by that of her no less beloved brother Philip the Handsome. Because much of the Hapsburg Empire was now being ruled by the infant Charles V, this prompted Margaret's father Maximilian to ask her to

move to the Netherlands as its regent, a duty she performed conscientiously and effectively for the next
twenty-three years. But this did not cause her to forget the
Shroud. As is known from an inventory, she kept a copy of
it with her, which may well be the one-third scale version
that is today kept in Lierre, Belgium (pl. 3a). And in 1509
she sent all the way from Flanders a beautiful new silver
casket for the Shroud that she had specially commissioned
from the local artist Lieven van Latham. The governor of
Bourg-en-Bresse personally transported this casket to the
Sainte Chapelle in Chambéry where it was installed with
great ceremony.

Two years later none other than Anne of Brittany,

Fig. 34. RECONSTRUCTION OF A SHOWING OF THE SHROUD from the
parapet to the rear of the Sainte Chapelle, Chambéry, overlooking the
walls.

Queen of France, visited Chambéry for a private viewing of the Shroud. She was followed in 1516 by her son-in-law and new King of France the twenty-two-year-old Francis I, who was said to have walked all the way from Lyon to Chambéry dressed in white linen like a monk to give thanks to the Shroud after winning the battle of Marignac. Such illustrious visitors inevitably brought gifts, as a result of which, along with the embellishments made by the Savoys themselves, the Chambéry Sainte Chapelle became a jewel of stained glass, Flemish sculpture, ornamentation from Cyprus, rich draperies and much more.

With high-born foreign visitors receiving such viewing privileges, large numbers of the ordinary populace inevitably wanted a taste too. Accordingly, when the Cardinal of Aragon visited Chambéry in 1518, the people were given a showing of the Shroud from high up on a balcony of the chapel jutting out over the castle walls (fig. 34). This venture was clearly perceived as a success because the same location was used for a showing in 1521, when Duke Philibert II's brother and successor Charles III, now thirty-five, married seventeen-year-old Princess Beatrice, second daughter of King Manuel I of Portugal. And this in turn set a precedent for Savoy family matrimonial occasions that would become a dynastic tradition for centuries to come.

Recently there came to light in the tiny Spanish village of Noalejo two life-size copies each bearing the inscription in Latin 'Here you can see the image of the Holy Shroud which is kept in the chapel at Chambéry, in its true dimensions, by Gir, 1527'. As the village of Noalejo was effectively founded by Doña Mencia de Salcedo, a former laundry mistress to the new duchess's sister Isabel, wife of Emperor Charles V, the copies were very likely sent as a gift from Beatrice to Isabel during the late 1520s. Empress

Isabel then in turn made a gift of them to Doña Mencia when she retired from royal service in 1548. At least sixty similar 'facsimiles' would be created during the next two centuries.

Also from this period is a stained-glass window in the chapel of Pérolles, at the chateau of the Deisbach family near Fribourg, depicting four mitred bishops holding out the Shroud (pl. 30a). Featuring castle walls in its background, this stained glass undoubtedly depicts one of the castle-wall balcony showings at Chambéry. It is also the earliest known depiction of a showing of the Shroud since the Lirey pilgrims' badge of two centuries earlier. Made at much the same time, for one of the main windows of the Sainte Chapelle, was a stained-glass depiction of the Shroud by the artist Jean del'Arpe.

But whenever the Shroud begins to look well established usually something happens to reverse matters, and the first half of the sixteenth century was no exception. On the night of 4 December 1532, the now infamous fire broke out in the Sainte Chapelle, the flames quickly consuming most of the furniture and fittings, including del'Arpe's recently completed stained glass. Because of the four-lock security system to the iron grille guarding the Shroud's storage cavity there was insufficient time to summon all the key-holders, most notable among them Duke Charles III. The Shroud's very survival depended on the heroism of a local blacksmith, Guillaume Pussod, braving the flames. With the help of the chapel's canon, Philip Lambert, and two other canons, Pussod managed to prise open the grille, remove the casket and whisk it away to safety. The empty cavity can still be seen in the chapel's apse wall, complete with scars where the grille had been secured to the stonework. As we learned earlier, it was only in the nick of time, and not without suffering some serious damage, that the Shroud was rescued that night.

In the aftermath of the fire, rumours began to spread that the Shroud had been completely destroyed, particularly when the next year there was no showing on the Shroud's now appointed feast day, 4 May. The French satirist Rabelais, writing his *Gargantua* at this time, included a scene in which some attacked soldiers invoked the intercession of various saints and relics: 'Some made a vow to St. James, others to the Holy Shroud of Chambéry, but it caught fire three months later so that not a single scrap could be saved.'[13] Clearly measures were needed to quell such gossip and restore public confidence, also some cosmetic and structural repairs if the Shroud was going to be suitable for any future public exposition.

At eight a.m. on 16 April 1534, a solemn procession, including the now forty-two-year-old Duke Charles III, the local cardinal Louis de Gorrevod, two bishops, an ecclesiastical notary and numerous other clergy, assembled to carry the Shroud to Chambéry's convent of the Poor Clares, where a table had been specially prepared. Once the Shroud was laid out on this table, Cardinal de Gorrevod formally asked those present to testify that the cloth before them was the same as that which they knew before the fire. As the cardinal himself solemnly affirmed, 'It is the same sheet as we ourselves before the fire have many times held in our hands, seen, touched and shown to the people.' (Gorrevod's words 'many times' interestingly suggest there to have been rather more showings from the chapel balcony than appear on the historical record.) But what truly arrested those present was the way that the fire damage the Shroud had sustained had, most eerily, missed the all-important image. As remarked at the time by an eyewitness, Baron Pingone, 'Indeed, we all clearly saw it (for I was then present), and were amazed.'

Four Poor Clare nuns, including the Mother Superior, the Abbess Louise de Vargin, had been appointed to carry

out the repair work. After various ceremonial prelim-
inaries, and the nuns being given absolution and a meal,
an embroiderer arrived with the wooden frame for stretching the Holland cloth on to which the Shroud would be
sewn to strengthen it for display purposes. Then, according to Louise de Vargin's surviving account of her nuns'
work, 'After two hours of fixing the cloth to the frame and
to the open-work we laid out the precious Holy Shroud on
top of it, and we sewed it turn by turn using tacking
stitches.'[14]

The four nuns were taking on a lengthy and painstaking
task, and although an iron grille held back large crowds of
daily onlookers, their work was constantly interrupted by
special visitors. In Louise de Vargin's words:

> When His Highness [Duke Charles III] learned that the
> crowds were so great . . . he was obliged to take charge of
> the key to the grille. Nevertheless he often gave it back to
> his chief steward in order to satisfy the pious wishes of a
> large number of pilgrims who had come from Rome, and
> Jerusalem and several distant countries. They were shown
> the Holy Shroud in the presence of lighted candles, while
> we knelt down and chanted . . . They went away consoled,
> satisfied that it was the same Shroud that they had viewed
> on earlier occasions.

De Vargin's detailed description of what she saw of the
Shroud's image at such privileged close quarters is an
intriguing mixture of piety and cool-headed observation:

> All our thoughts were with God . . . We saw on this rich
> tableau sufferings which could never have been imagined.
> We could see the marks of a face completely battered and
> bruised from being struck. His divine head was pierced
> with large thorns from which trickles of blood ran down

his forehead and separated into different branches, coating it with the world's most precious purple. We noticed on the left side of the forehead a drop that was thicker and longer than the others, undulating from one side to the other . . .

This was clearly the Shroud's reverse '3'-shaped forehead stain from the crown of thorns, similarly noticed by coin-engravers and others.

The eyebrows seem to be well-formed and the eyes a little less so. The nose, being the most prominent part of the face, has left a clear imprint. The mouth is very well-formed and fairly small. The cheeks are swollen and disfigured, revealing that they have been cruelly beaten, particularly the one on the right. The beard is neither too long nor too short . . .

Made without the benefit of our post-Secondo Pia knowledge of the Shroud face when it is seen in negative, de Vargin's description readily cross-checks with how we have seen Byzantine artists interpret the same stains on the cloth they knew as the Image of Edessa, notably including 'eyes'.

Similarly impressive is de Vargin's assessment of the evidence for crucifixion and for the fatal lance-thrust:

A large cloth of blood marks the holes in the feet. The left hand, which is very well defined, crosses over the right hand, covering its [nail] wound . . . From them [the hands] meanders a stream of blood extending up to the shoulders . . . The wound in the divine side appears large enough to accommodate three fingers. It is surrounded by a blood-stain the size of four fingers, narrowing at the bottom . . . On looking at the Shroud from underneath, while it was stretched out onto the holland cloth or canvas, we could see the wounds as though looking at them through glass.

The nuns' work took them fifteen days – with a deadline to be met because the duke and his clergy did not wish to let another 4 May feast day go by without a showing. When the delegation of clergy arrived to collect the Shroud, now firmly stitched on to the Holland cloth and patched with around two dozen specially shaped pieces of altar cloth to hide the holes from the fire, they held it up so that the nuns could take one last look at their handiwork.

Then, as reported by de Vargin, 'they rolled it on the roller and covered it with a sheet of red silk'. From this information we are able to glean that those with overall responsibility for the Shroud in 1534 had made a very far-sighted decision concerning the Shroud's future conservation. Rather than the multi-fold arrangement we know (from the damage) the Shroud was kept in at the time of the 1532 fire – an arrangement which could cause crease lines to form – henceforth it would be stored rolled around a roller. Although this in its turn would have some disadvantages, as noted earlier, this arrangement would prevail right up to the 1990s.

Hardly had the status quo been restored, however, than a fresh hazard loomed. France's King Francis I, who had appeared so peaceable when as a twenty-two-year-old he visited Chambéry back in 1516, was now at odds with the Hapsburg emperor Charles V, with whom the Savoys were closely allied via Charles's aunt Margaret of Austria. Being a relatively small duchy, and right on his eastern border, Savoy was an easy target for some French bullying to aggravate the emperor. French invasions accordingly began in 1537, causing Duke Charles III and his family to abandon Chambéry, taking the Shroud with them on some frenetic travels, this time for its own safety.

Once again it went over the Alps, its route trackable as passing through the Lanzo valley via Bessans, Averoles,

Ceres and Lanzo itself, culminating in Turin where the annual feast day showing was held on 4 May. The following year it was in Milan. Then, because of fresh invasions by the French, it was transferred first to Vercelli, then to Nice, still at that time within Savoy territory.[15] In 1543, Duke Charles III – clearly rather more of a survivor than his predecessors – brought it back to Vercelli to be kept in its St Eusebius Cathedral.

Ten years later, with the Shroud still in Vercelli, the troops of Francis I's vigorous son King Henri II sacked the town, and six of the French soldiers strode into the cathedral specifically in search of the Shroud. According to one legend, the canon, Antoine-Claude Costa, invited the French soldiers for a meal, in the course of which the wine flowed so freely that they fell asleep, forgetting what they had come for, during which time the canon had the Shroud hidden under his cloak. When they woke up they learned that Hapsburg troops were in the vicinity and had to make a hasty escape. Whatever the exact details of the story, the Shroud survived yet another hazard.[16]

In the event, only a few months earlier the now elderly Duke Charles III had died, at which time virtually all of what had been the duchy of Savoy lay in French hands. The only one of Charles's nine children to have survived to adulthood, twenty-five-year-old Emmanuel-Philibert, now succeeded to what might have seemed a very empty title indeed. However, before he became duke, Emmanuel-Philibert had distinguished himself as a military commander, serving his dynasty's long-term ally, the Hapsburg empire of Charles V. As a result, he was appointed Governor to the Netherlands, personally led the Spanish attack on northern France, and won a brilliant victory at St Quentin in August 1557. On Charles V's death two years later, bringing in a new mood of diplomacy rather than conflict between France and

Spain, Emmanuel-Philibert had virtually the entire former domains of Savoy returned to him by the terms of the treaty of Cateau-Cambrésis, plus the hand in marriage of Margaret de Valois, King Henri II's own sister.

Emmanuel-Philibert and Margaret's wedding was held in Paris – surely one of history's most low-key because in a nearby room Margaret's brother King Henri lay dying from a terrible jousting injury he had received during a tournament that was one of the preliminary entertainments for their nuptials. Then an outbreak of plague delayed the couple's journey back to Savoy. Eventually they reached Chambéry, and at the beginning of June 1561 they had the Shroud brought there from Vercelli. After being temporarily housed in the church of the Franciscans, on 4 June it was returned to the partly restored Sainte Chapelle in a now expected elaborate procession accompanied by trumpets and torches. On 15 August, the Feast of the Assumption, and again two days later, the Shroud was first displayed on the chapel's high altar then taken out on to its castle-wall balcony for the once again hugely popular showings to the general populace massed below.

With stability restored, there was a fresh wave of making 'facsimile' copies of the Shroud. One still extant specimen bears the inscription 'At the request of Signor Francesco Ibarra this picture was made as closely as possible to the precious relic which reposes in the Holy Chapel of the Castle of Chambéry and was laid upon it in June 1568'.[17]

Three years later, the very saintly Pope Pius V asked Duke Emmanuel-Philibert of Savoy to join a Holy League alliance of western powers, of which the largest were Spain, Venice and Naples, aimed at trying to strike a decisive blow against the continuing build-up of Turkish naval might in the eastern Mediterranean. Emmanuel-Philibert provided three ships for what became a combined Christian fleet of two hundred galleys. On 7 October 1571, this force engaged the numerically superior and highly impressive main fleet of Turkish war galleys in a set-piece naval battle known to history as Lepanto, off the shores of western Greece.

Savoy's flagship fought with the Centre division under the overall command of the dashing Don Juan of Austria, illegitimate son of the now deceased emperor Charles V. At the height of the action, Savoy's admiral, on Duke Emmanuel-Philibert's instructions, hoisted a banner embroidered with an image of the Shroud held out by the Virgin Mary and two angels, accompanied by the inscription 'Behold O God our protector, and look on the face of thy Christ'.[18] It is difficult not to recall the protective powers that were attributed to the Image of Edessa, arguably the very same cloth, a thousand years earlier.

Whatever efficacy we may impart to Emmanuel-Philibert's Shroud-decorated banner, it is a fact of history that the Christian fleet soundly defeated the Turks that day. As a souvenir gift for the victorious Don Juan, Pope Pius V duly ordered from Savoy a facsimile copy of the Shroud. Both this copy and the original banner survive, the former in Alcoy, Spain,[19] the latter in Turin's Church of San Domenico.[20]

By one of those quirks of history, less than twenty years before Duke Emmanuel-Philibert so providentially used the Shroud as a battle standard, two thousand miles away the notorious Tsar Ivan the Terrible of Russia had deployed a flag embroidered with the face on the Image of Edessa for much the same purpose, and with much the same result. With the flag held aloft, bearing the inscription 'May God protect his followers from the cunning of the enemy', in 1552 Tsar Ivan scored a decisive victory against the Turkic Tatars at the battle of Kazan. In gratitude he dedicated Kazan's first-ever Christian church to the Image. The original flag survives to this day in the Museum of Arts in Moscow (fig. 35).

Geoffrey de Charny would have been proud.

Fig. 35 BATTLE STANDARD OF IVAN THE TERRIBLE depicting the protective Image of Edessa which was deployed at the battle of Kazan in 1552.

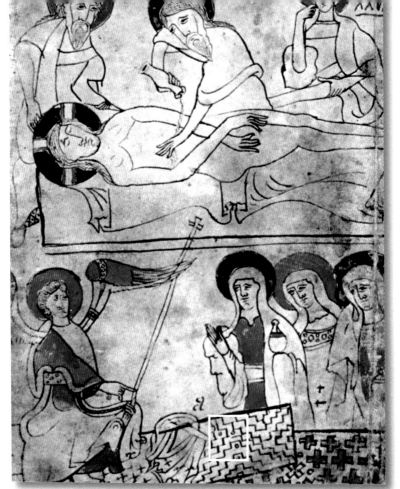

25a Entombment of Christ and Visit of the Holy Women, scenes from Hungary's Pray manuscript, *c.* 1192. Note the Shroud-like figure in the upper scene and the herringbone in the lower one, together with a possible indication of the Shroud's triple burn-holes (highlighted).

25b Wall-painting depicting the Image of Edessa, *c.* 1250, Church of Hagia Sophia, Trebizond, today Trabzon, north Turkey.

25c The Church of Hagia Sophia at Trabzon, today a museum.

26a (*above*) Wall-painting depicting the Image of Edessa, late thirteenth century, in the monastery of Sopocani, Serbia.

26b (*left*) The equivalent face area of the Shroud, with the burn damage from the 1532 fire digitally removed.

26c (*left*) Fourteenth-century wall-painting from the church of Matejce, Macedonia, showing the Image of Edessa as a large-size piece of cloth. The accompanying description specifically refers to it as a sindon. Albanian terrorists destroyed this and other paintings in the Matejce church in 2002.

27a (*right*) Eastern Orthodox liturgical cloth known as an epitaphios, specifically symbolizing Jesus's burial shroud. Example of the early fourteenth century from Serbia, preserved in the Museum of the Serbian Orthodox Church, Belgrade.

27b (*above*) Processional carrying the epitaphios – reminiscent of how the Image of Edessa was brought to King Abgar?

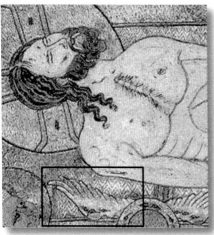

27c (*above*) Detail of another epitaphios, from Thessaloniki, Greece, showing Shroud-like herringbone pattern (highlighted).

28a (*above*) Panel painting with Shroud-like face, found hidden in the ceiling of a cottage 'outhouse' in the village of Templecombe, Somerset, England, the site of a former Templar preceptory.

28b (*above*) Archivist Barbara Frale of the Vatican Secret Archives, Rome, with the seal of German Templar master Frederick Wildergrave, featuring a similar face. In 2008, Frale, who has long argued for the Templars' innocence of heresy, discovered a hitherto unknown document directly linking the Templars and the Shroud.

28c (*left*) Geoffrey de Charny the Templar, and the Templar Order's last Grand Master, Jacques de Molay, being burned at the stake as heretics in Paris in 1314.

29a (*above*) Pilgrim's souvenir badge, a unique relic of the mid-fourteenth-century showings of the Shroud in Lirey, as found in the mud of the river Seine, Paris, in the mid-nineteenth century. The coat of arms on the left is that of Geoffrey I de Charny of Lirey, that on the right, his second wife Jeanne de Vergy.

29b (*right*) Geoffrey I de Charny ambushed by King Edward III at Calais, from a contemporary manuscript.

29c (*left*) Wall-painting of the Shroud discovered in 1997 in the church of Terres de Chaux, near St Hippolyte sur Doubs where the Shroud was mostly kept between 1418 and 1453.

30a A showing of the Shroud from the castle walls parapet of the Sainte Chapelle, Chambéry, *c.* 1520–3, as depicted on a stained-glass window in the Chapel of Pérolles, Fribourg, Switzerland.

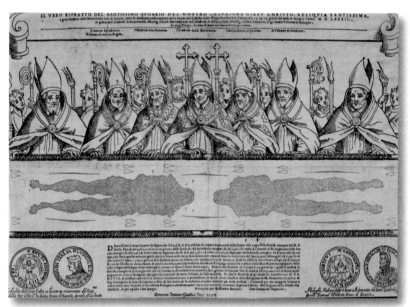

30b Showing of the Shroud (now bearing the scars of the 1532 fire) in Turin in 1582. This was the occasion of the second pilgrimage of Cardinal Charles Borromeo, seen at the centre of the line of bishops. Below the Shroud are depictions of ducal commemorative medals.

31a Showing of the Shroud in the Piazza Castello, Turin, 1684; detail of painting by Pieter Bolckmann in the Castello di Racconigi, Turin.

31b (*left*) Altar of the Shroud, designed by Bertola, in which the Shroud reposed between 1694 and 1993.

31c (*below*) Showing of the Shroud on the cathedral steps, 1933, at the end of the expositions of that year. Crown Prince Umberto, later King Umberto II, is seen in uniform at the left.

32a (*left*) Meeting of Pope John Paul II and King Umberto II of Savoy in Portugal, 14 May 1982. Umberto bequeathed the Shroud to the Pope and his successors.

32b (*below left*) Fire engulfs the Shroud Chapel in Turin Cathedral on the night of 11 April 1997.

32c (*below right*) At the height of the fire, Turin fireman Mario Trematore smashes open the Shroud's bullet-proof display case.

32d (*right*) Pope John Paul II in prayer before the Shroud, 24 May 1998.

Chapter 18

Backbone of a Dynasty

It is an amazing thing to see how that Most Holy Linen, made of one-doesn't-know-what material, but one opines that it is of linen, so fine that it seems to be silk – is found to be whole and undamaged after a thousand and five hundred years.

Future cardinal Agostino Cusano in 1578

As the 1570s wore on, Duke Emmanuel-Philibert, now safely back in Savoy, became increasingly minded that the days of the Shroud's stay in Chambéry were numbered. The duke was too good a strategist not to be keenly aware that Chambéry's Alpine location, pleasant though it is, was rather too high, too close to France, and not central enough within Savoy, to remain as his capital. Rapidly growing, and much more centrally located within Savoy's boundaries, Turin was the logical alternative. Accordingly, the duke began to develop Turin as his main base, and because of the Shroud's close association with his dynasty he wanted it to be with him there, rather than remain in Chambéry. A good excuse was therefore needed for the transfer, in order not to offend Chambéry's populace too greatly. That excuse was not too long in coming.

In 1578, Cardinal Charles Borromeo, the very pious forty-year-old Archbishop of Milan, made it known that he and twelve companions were planning to walk all the way from Milan to revere the Shroud in Chambéry, fulfilling a vow he had made when Milan was gripped by a plague epidemic two years earlier. Emmanuel-Philibert volunteered to have the Shroud brought to Turin to save the cardinal the ordeal of crossing the Alps. The ruse worked.

On 14 September, the Shroud, having been secreted out of Chambéry and across the Alps, arrived at the castle of Lucento just outside Turin. There Duke Emmanuel-Philibert met it and accompanied it to the city gate, where a procession of the local clergy then escorted it into the city, accompanied by volleys of cannon-fire from the ducal artillery. It was temporarily lodged in Turin's Chapel of San Lorenzo, tiny, but conveniently on the ramparts adjacent to the palace gardens.

Three weeks later, Borromeo, having been informed that the Shroud now awaited him, set out on his long walk. It took him and his companions four days, and when they arrived, it was to an emotional and similarly noisy welcome. The cardinal went directly to the cathedral, then to the San Lorenzo chapel. However, the main private viewing of the Shroud was arranged for the next day, when it was displayed in the cathedral, on a silk-covered table.

With the intention of putting the Shroud on public view on 12 October, a specially erected platform was built in the cathedral. However, so many people were seen to be streaming into the city, in such a confined space public safety could seriously be jeopardized. It was therefore decided to stage the event outdoors in the public square closest to the cathedral, the Piazza Castello. A member of Borromeo's party, Agostino Cusano, described what he witnessed:

The Most Holy Linen was carried in procession from the Cathedral to Piazza del Castello where a multitude of people, practically innumerable from every region around about, had come together, filling all that huge piazza so thickly that one saw nothing but heads, so that it looked like the Last Judgment; it was estimated that there were forty thousand people, and there upon a big platform all richly adorned, the most holy relic was displayed, while cries of mercy rose even to the sky.[1]

The specially made platform in the cathedral was still used. After the Shroud was returned to its casket – the cedar chest in which it was brought over the Alps is preserved in Turin's Museum of the Shroud – it was taken in procession to the cathedral. There the closed casket was set on the platform, guarded by civic dignitaries dressed in their finest robes, as the centre point of Forty Hour devotion, with sermons on every hour by relays of churchmen starting with Borromeo himself.

For the late-comers there followed a second public showing, together with a second private showing for Borromeo and his companions. Agostino Cusano, who would become a cardinal ten years later, was almost breathless in the way he recorded his personal impressions of this second viewing:

It is an amazing thing to see how that Most Holy Linen, made of one-doesn't-know-what material, but one opines that it is of linen, so fine that it seems to be silk – is found to be whole and undamaged after a thousand and five hundred years. But it is even more amazing to see here impressed the image and effigy of the true and natural body of the Lord with all his form and with the signs and vestiges of the scars and wounds that he sustained for our sins; impressed, I say, not with the human art of a painter

nor with a variety of colours, but miraculously stamped
and portrayed by his own body. The whole figure is rather
obscure, like a dark shadow, or like the first sketch of a
painting that now you see it, now you don't, and that
arouses greater desire and diligence to see it again better;
now it is seen better up close, now farther back . . .[2]

Clearly aware of the still awkward questions surrounding
the Shroud's provenance, Cusano also made the following
commendably modern-minded observation: 'For certification this one [the Shroud] does not need other
approbation since it carries on it the testimony of its
veracity, with its miraculously imprinted image.'[3]

Politically, the Borromeo event was an outstanding
success both for Duke Emmanuel-Philibert and for Turin.
The duke had licensed Turin engraver Giovanni Testa to
create a twenty-inch-wide copper engraving depicting the
eleven principal attending ecclesiastics, with Borromeo at
their centre, holding up the Shroud.[4] This carried the headline in Italian 'The Most True Image of the Most Holy
Shroud of our Saviour Jesus Christ'. Testa probably prepared this in advance, because it shows no evidence of
eyewitness reporting. Even so, the souvenir prints he made
from it evidently sold out very quickly, because for every
subsequent showing of the Shroud more engravings of this
kind would follow.

Indeed, when four years later Cardinal Borromeo
returned to Turin, by which time Duke Emmanuel-
Philibert had been succeeded by his now twenty-two-
year-old son Charles Emmanuel, not only was Borromeo's
visit marked by fresh showings of the Shroud, on 13, 14
and 15 June, there was also a new, improved souvenir
print (pl. 30b), with the Shroud itself, by a second impression, reproduced in an ochre colour. In this the bearded
cardinal at Borromeo's right hand is Gabriele Paleotto,

Cardinal-Archbishop of Bologna, who had come accompanied by his young cousin Alfonso Paleotto.

The young Paleotto was so deeply affected by what he observed at first hand that he went on to write a full book on the Shroud,[5] published in 1598. It is evident from this, the first such book to be created within the era of the printing press, that Paleotto, benefiting from his era's rapid advances in anatomical knowledge, actually recognized the Shroud's indication that the crucifixion nails pierced through the wrists rather than the palms. In his own words, 'It appears on the Holy Shroud that the [nail] wound is seen at the joint between the arm and the hand, the part anatomists call the carpus, leaving the backs of the hands without wounds.'[6] As Paleotto further deduced, pre-empting the French surgeon Dr Pierre Barbet's insights by three and a half centuries, the Romans must have driven in the crucifixion nail 'so that it passed through the hand . . . towards the arm, where the hand gets thicker and the bone stronger. Thus the point of the nail comes out at that part of the back of the hand in the middle of the joint, and that is because the nail [in the palms] would not have supported the body. Instead the weight of the body would have torn the hand, according to the experiments made by master painters and sculptors with dead bodies that they intended as models for their artworks.'[7]

One of the main reasons for Borromeo's fresh visit to Turin was to advise the young Duke Charles Emmanuel on how best to permanently house the Shroud, now that it had been removed from the purpose-built Sainte Chapelle in Chambéry. For temporary purposes the duke had had a small chapel built for it within his palace, but something more fitting and lasting was needed. Another concern was how and where the Shroud might be most safely and reverently exhibited to the general populace in the light of the huge crowds such showings evidently attracted. As some of Borromeo's surviving letters indicate, in the months that followed he consulted an architect, Pellegrino Tibaldi, on how to locate the Shroud in Turin Cathedral in such a manner that it could be 'preserved and displayed without the danger of folding and unfolding it'.[8]

From a still extant drawing[9] made by Tibaldi, it seems that he and Borromeo envisaged creating a shrine within a niche at the back of the cathedral choir. This was immediately rejected by the duke. For the duke, if the Shroud were to be kept permanently in the cathedral, this would effectively mean its passing to ecclesiastical control. His alternative idea was for a new church personally endowed by him to be built close to his palace, whereby he could maintain both the appearance and the actuality of his family's all-important personal possession of the Shroud.[10]

Nevertheless, the duke acknowledged that the cathedral was the most fitting place for the Shroud during the interim, and by 1587 a shrine on four wooden columns had been erected over the cathedral's high altar, its form resembling that of the shrine used for housing and showing the Veronica cloth in old St Peter's, Rome.[11] Around 1604, the Shroud was also provided with a new four-foot-long purpose-built casket in which it could be kept rolled up around a velvet-covered staff to eliminate the need for any further folding or unfolding. This casket, of silver-gilt wood decorated with relief plaques of the instruments of the Passion, would serve as the Shroud's container until the 1990s.

During the decades that followed, the public showings of the Shroud every 4 May quickly developed into a well-refined and practised tradition. First there would be an elaborate preliminary court ceremony in which the Shroud was brought down from its shrine, taken out of its casket and laid out fully extended on a specially prepared table near the altar. The protocol required that the duke and hereditary prince should reverently kiss the Shroud's major bloodstains[12] – intriguingly reminiscent of the kissing associated with the secret Templar rituals. A procession would then be formed and the cloth carried out to the huge crowd gathered in the adjacent Piazza Castello, today relatively little changed from its early seventeenth-century appearance.

In 1613, the engraver Antonio Tempesta produced an ambitious twenty-five-inch-by-eighteen-inch souvenir engraving that was the first properly to depict the sheer spectacle of these occasions (fig. 36). In the engraving, a sea of people can be seen filling every possible vantage point. Servants perch precariously on rooftops. Every balcony is filled to capacity, the one on the Castello, at the top centre of the picture, brimming with the leading ladies of the Savoy court. At the sides of the square, temporary 'corporate boxes' provide a high vantage point for those willing to pay for this privilege, while at ground level thousands of the humbler folk fill the square, surrounded by the ducal cavalry. In the foreground, in what seems to be the first of two separate moments that the artist has conflated into one, we see musketeers and halberdiers struggling to open a path for the procession of torch-bearers who accompany the high square canopy beneath which the Shroud is being carried by mitred Church dignitaries. This procession leads our eye to the second moment the print encapsulates: the showing of the Shroud from a high platform that has been erected in the middle of the square. At a height that is comfortably beyond the crowd's reach, eight mitred bishops and archbishops hold out the Shroud to the populace. Below them, members of the crowd throw up *corone di Cristo*, rosary-like strings of beads, for the bishops to press against the Shroud then return duly sanctified to their owners. Behind the bishops

Fig. 36 SHOWING OF THE SHROUD IN THE PIAZZA CASTELLO, TURIN, 1613, from an engraving by Antonio Tempesta.

can just be glimpsed the faces of Savoy's duke and duchess. In the sky above, a banner reads 'Happy House of Savoy, which, endowed by so great a pledge [i.e. to keep and protect the Shroud] is glorified by this sacred gift'.

For us in the twenty-first century, used to garments that fall apart after only a few years of wear, one of the daunting facts the Shroud presents is that it has come down to us in such remarkably good condition despite the amount of handling it received at this time. A popular story concerning the 1613 showing relates how the future saint Francis de Sales was among the clergy holding up the Shroud before the people. It was a hot day, and despite the shade provided by the large overhead canopy, perspiration from the saint's brow accidentally fell on the Shroud. Standing next to him was the then Cardinal of Turin, Prince Maurice, who chided him for his carelessness, causing St Francis to reflect that Jesus had not held back from mixing his blood and sweat with ours.[13] The story underlines just how much human handling and contamination the Shroud was subjected to during this particular phase of its history.

The ensuing decades of the seventeenth century were characterized by the upholding of this tradition of 4 May showings, interspersed with the staging of others to celebrate Savoy family weddings, such as that of the future Duke Victor Amadeus to Princess Christine of France in 1620, and Duke Charles Emmanuel II to Francesca d'Orleans in 1663. But deeply engrained in the psyche of Savoy's dukes remained the need for the Shroud to be given a proper purpose-built chapel of its own that would be commensurate with its standing as a sacred relic, yet which would also glorify the Savoy dynasty.

Accordingly, in 1668, the era of Duke Charles Emmanuel II, a particularly talented architect, Guarino Guarini, was recruited from France, where he had been working for Louis XIV. Guarini was an extraordinary combination of theologian, philosopher, mathematician, physicist, astronomer, engineer and architect. Some of the groundwork had already been done by a previous architect, but Guarini took it to a whole new dimension.

The basic plan was for the Shroud chapel to be sited between the royal palace and the cathedral, and at first-floor level, so that members of the family could access it easily from the palace apartments. Externally, the chapel's dome was to rise higher than that of the cathedral. The shrine containing the Shroud itself should form the centre-piece of the chapel, and be viewable from the main body of the cathedral, though elevated from it. The chapel should also be able to accommodate statues and possible tombs for members of the Savoy dynasty.

Although Guarini did not live to see his design to completion, he achieved a positive tour de force (fig. 37). The cupola was a marvel of geometry and optical effects, its drum and dome perforated to the maximum extent possible to reduce the structure's weight and allow in the maximum amount of light. To link the first-floor chapel with the cathedral, he removed the cathedral's apse wall. A first-floor gallery connected it to the palace's *piano nobile*. So that the chapel could be reached from the main body of the cathedral, Guarini built behind the high altar two black marble staircases, their steps cleverly made convex to slow any ascent. The chapel itself he created circular, in black marble, with the Shroud's altar – an extraordinary baroque confection designed by the local architect and engineer Antonio Bertola (pl. 31b) – in its centre.

When in 1684 Charles Emmanuel II's son Victor Amadeus II married his first wife, Anna d'Orleans, a particularly lavish public showing of the Shroud was staged in the Piazza Castello. To supplement the usual souvenir

Fig. 37 GUARINO GUARINI'S CHAPEL OF THE HOLY SHROUD – cut-away view showing its appearance before the fire of 1997. The two flights of stairs (**a**) lead up from the main body of the cathedral (**b**). The Bertola altar is seen at (**c**), the traditional repository of the Shroud having been in the upper tier (**d**). The connecting Royal Palace is at (**e**).

prints, Dutch artist Pieter Bolckmann was commissioned to create a huge commemorative oil painting, which today hangs in Turin's Castello di Racconigi (pl. 31a). In the distance, beneath a large red canopy, the usual line of bishops and archbishops can be seen unfurling the Shroud's red silk cover to display the cloth to the crowds, which are even more extensive than in the Tempesta engraving of seven decades earlier, with every rooftop filled. And just behind the Royal Palace in the background can be seen the spire of Guarino Guarini's chapel showing that externally it was complete.

By the beginning of June 1694 the chapel's interior was finished and Bertola's central altar installed, complete with a safe-like top storey that was to be the Shroud's new, 'permanent' home. As recorded in Savoy's log of court ceremonies for that month:

His Royal Highness [Duke Victor Amadeus II] took himself with the court to S. Giovanni [Turin Cathedral] . . . and having taken the Most Holy Shroud carried it into the upper chapel, passing by way of the stairs next to the sacristy that lead to the same . . . The Most Holy Shroud was spread out on a table placed between the chapel altar and the balustrade facing S. Giovanni, then from that same balustrade it was shown to the people who were in the church, and from that same table it was held up also to display it to those who were in the chapel, then it was put back in its casket and placed in that assigned location.[14]

The balustrade that separated the newly built first-floor chapel from the main body of the cathedral had been planned by Guarino Guarini as a natural location for any showings of the Shroud to the congregation of the cathedral one floor below.

At the end of that month, the Savoy family's spiritual

director, the Blessed Sebastian Valfré, carried out the first main conservation work on the Shroud since the 1532 fire, reinforcing and adding to the patches sewn on by the Poor Clare nuns, also supplying the Shroud with a new black silk lining cloth which would last until 1868.

But again, it seems every time something 'permanent' was arranged for the Shroud, external events would conspire against this. In 1701, France's King Louis XIV inherited the title of King of Spain, and when he tried to implement this, the threat this posed to Europe's balance of power set off the War of the Spanish Succession. Savoy, taking the Hapsburg side as usual, was the equally usual easy target for a French invasion, and in 1706 a French army was besieging Turin itself. The Shroud, however, had already been quietly moved south to Genoa for its safety. In the event, largely thanks to the duke's cousin Prince Eugene of Savoy, who brilliantly led the relieving Hapsburg forces, the city was saved from capture. The House of Savoy even benefited from this episode, because by the treaty that ended the war Victor Amadeus was able to add to his titles King of Sicily, one he exchanged in 1720 for the more geographically practical King of Sardinia.

From this time on, although the annual 4 May events seem quietly to drop from the calendar, whenever there was a major Savoy dynasty wedding the same crowd-pulling Shroud showings would take place. On 1 April 1737, when Charles Emmanuel III married his third wife Princess Elizabeth Teresa of Lorraine, the Shroud was shown on the subsequent traditional feast day, and on 5 October 1775 it was again shown when Piedmont Prince Charles Emmanuel married Princess Marie Clotilde of France, sister of the King Louis XVI who within two decades would lose his head during the French Revolution.

The end of the eighteenth century saw a new threat, this time to all Europe: the meteoric rise to power and insatiable territorial ambitions of Napoleon Bonaparte. Charles Emmanuel, now Savoy's Duke Charles Emmanuel IV, saw no option but to quit Turin and withdraw to his kingdom of Sardinia. Before he and his family departed, he arranged for them all to have one last private viewing of the Shroud, but he left the relic behind. Two years later Pope Pius VII, having been summoned to Paris to crown Napoleon, made a special stop-over in Turin on his journey north, and at his request the Shroud was specially brought out for him. The sixty-four-year-old pontiff spent a long while examining and venerating it, 'kissing it with tender devotion'.[15]

Twelve years later, when a new duke, Victor Emmanuel I, was able to take up residence again in Turin, the Shroud was brought out for him as part of the celebration. The following year, 1815, Pope Pius VII, now well into his seventies, made a second stop-over in Turin, this time on his way back to Rome after having all the while been kept a semi-captive by the now defeated Napoleon. On 21 May, Pius stood at the centre of the usual line of clergy showing the Shroud from the balcony of Turin's Madama Palace. Though he is the only Pope known to have done so, sadly the event seems to have gone unrecorded by any souvenir print.

Victor Emmanuel's four children were all daughters, and since Salic law demanded that the title pass to the next male in line, in the late 1820s Victor Emmanuel's brother, Charles Felix, became the new duke. Apparently to eliminate draughts in the chapel, Charles Felix ordered the installation of a huge glass partition in the location of the balustrade separating the chapel from the cathedral below. An unsightly ruination of Guarino Guarini's intentions for how the Holy Shroud Chapel should look from the cathedral nave, this also put an end to any future showings of the Shroud from the chapel's balustrade.

In 1842, by which time there had been a further shift in the dynastic succession, the Shroud was once again shown to mark a ducal marriage, that of Victor Emmanuel II to Maria Adelaide of Lorraine, Archduchess of Austria. This time a commemorative print was issued (fig. 38). It shows five mitred bishops and archbishops holding up the Shroud from the same Madama Palace balcony used by Pope Pius VII in 1815. The newly married couple can be seen behind them, together with members of their family.

This would turn out to be the very last occasion on which the Shroud was formally held up by clergy outdoors, and without any kind of frame. In April 1868, when Victor Emmanuel II's son Umberto married Princess Margaret of Savoy-Genoa, the cloth was exhibited statically for four days on the cathedral's high altar, simply mounted on a board. The day after these showings Umberto's sister Clotilde, working on her knees, unstitched the former lining cloth of black silk that had been sewn on to the Shroud by Sebastian Valfré back in 1694, replacing it with a new one of red silk.[16]

During Victor Emmanuel II's lifetime, the several Italian-speaking states, of which Savoy was merely one, were unified and became the single country of Italy. Accordingly, when Umberto succeeded his father in 1878 he assumed the title King of Italy. It was during the autocratic Umberto's reign, in 1898, that the Shroud was brought out again to celebrate the fiftieth anniversary of Italy's constitution. The plan was to display it statically in the cathedral, this time properly framed. However, because a senior cleric back in 1868 had measured the Shroud inaccurately, the frame was made too wide, and not long enough. This meant that both foot ends of the cloth had to be partly folded underneath, thereby causing one of the several imperfections of Secondo Pia's famous first ever official photographs of the Shroud, as taken on this same occasion.

Fig. 38 SHOWING OF THE SHROUD FROM THE BALCONY OF TURIN'S ROYAL PALACE, 1842, to celebrate the marriage of Victor Emmanuel II to Maria Adelaide of Lorraine. The newly married couple can be seen either side of the middle of the five bishops. This was the last occasion on which the Shroud was displayed in public loosely, without any frame.

The first quarter of the twentieth century saw the accession of Umberto's son King Victor Emmanuel III, and the outbreak of the First World War. When Italy became caught up in the conflict and Austrian-German air-raids on Turin became a serious possibility, the king ordered that somewhere safe from such dangers should be found for the Shroud. He also insisted that the cloth should not leave the territory of Turin's now royal palace, so in the

strictest secrecy – not even the contractors were told of the
room's intended purpose – an underground chamber was
constructed two floors below ground level on the palace's
south-eastern side. A large strongbox with a complex
combination lock was then installed. On 6 May 1918, the
casket containing the Shroud was taken down from its
Bertola altar home, wrapped in a thick blanket of
asbestos, laid in a chest made of tin plate, hermetically
sealed with cold solder, then solemnly carried down to the
secret chamber and placed inside the strongbox.[17] Not
even Constantinople's Pharos Chapel could have been
more secure.

When the war ended, the Shroud was returned to the
Bertola altar. It was brought out again in 1931 to mark
the marriage of King Victor Emmanuel III's son Umberto to
Princess Maria José of Belgium. This time it was displayed
for three weeks in the cathedral, and in a new frame made to
the correct dimensions. This was the occasion on which
Giuseppe Enrie took his definitive black-and-white photographs. It was also the last time the Shroud would be shown
to mark a Savoy marriage. And when two years later the
Shroud was put on display again, at the request of Pope Pius
XI, in celebration of the Holy Year of 1933, this was the last
sight any reigning member of the Savoy family would have
of the Shroud (pl. 31c).

For in September 1939, with the outbreak of World
War Two, King Victor Emmanuel III ordered the Shroud
to be moved out of Turin altogether, south to the
Benedictine Abbey of Montevergine in the province of
Avellino, north-east of Naples. When it arrived, only the
prior, the vicar-general and two of the monks were
entrusted with the knowledge of what they were protecting. And during these seven 'monastic' years that it lay out
of sight, but safe, the events going on in the outside world
proved catastrophic for the Savoy dynasty.

As early as the 1920s King Victor Emmanuel III had put Fascist dictator Benito Mussolini in charge of Italian politics in order to counter Communism. Under Mussolini, Italy invaded Ethiopia and Albania, which gave the king the fleeting satisfaction of adding these countries to his long list of royal titles. But under Mussolini Italy also entered into an alliance with Adolf Hitler, one with which Victor Emmanuel, for all his avowed enthusiasm for constitutional monarchy, certainly seemed to comply. Italy thereby took what would turn out to be the losing side in World War Two.

At the eleventh hour, acutely conscious of his un-popularity, Victor Emmanuel abdicated in favour of his son, Crown Prince Umberto, at whose marriage the Shroud had been exhibited fifteen years earlier. But the action was too little and too late. When in June 1946 the Italian people were given the opportunity to vote on whether they wanted to retain their monarchy or become a republic, they chose the latter. Victor Emmanuel died a year later in Alexandria, Egypt. Umberto, who had reigned for just thirty-three days as King Umberto II, quietly left Italy to go into a strictly enforced exile in the Villa Italia near the fishing village of Cascais, Portugal. He would never see Turin, Italy or the Shroud again.

Not that Umberto forgot the Shroud. Far from it. When the cloth was brought back from Montevergine at the end of October 1946, for all practical purposes it, along with Guarino Guarini's Chapel of the Shroud, which gener-ations of the Savoys had insisted should be kept independent of ecclesiastical control, became the property of the Italian state, along with all other Savoy possessions within Italy. But in exile, Umberto – undoubtedly hopeful that Italian public opinion might one day turn back in his favour – punctiliously upheld the principle of his con-tinued ownership of the Shroud, along with all other formalities associated with his blueblood birth. And in Turin, those most closely associated with the Shroud's care felt duty-bound quietly to consult him, albeit through intermediaries, on matters pertaining to the Shroud. This certainly happened prior to the exposition of 1978 and the major scientific examination that was carried out in that year.[18]

Even so, the effects of the events of 1946 went beyond robbing Umberto of his kingdom. Directly or indirectly, they also broke up his family. His wife Maria José left him in 1947, taking their only son Prince Victor Emmanuel and his three sisters with her to Switzerland, while Umberto remained behind in his Cascais villa. In 1970, the now thirty-three-year-old prince married in Las Vegas in a civil ceremony, having consulted neither of his parents, let alone invited them to attend. He thereby broke an inviolable Savoy family law stretching back forty-four generations, requiring the crown prince always to consult his father concerning his choice of marriage partner. This left Umberto with no option but to formally disinherit his son.

Then, in August 1978, Victor Emmanuel was involved in a bizarre overnight shooting 'accident' on a yacht moored off the island of Cavallo, south of Corsica, in which a bullet from a gun fired by him killed nineteen-year-old Dirk Hamer, who was asleep on a neighbouring yacht. Although Victor Emmanuel was cleared of murder, this along with other wayward behaviour on his part would seem to have minded Umberto that he had no suitable male heir to whom to bequeath the Shroud. He therefore saw it as his duty to the Shroud at long last to pass it to the Roman Catholic Church, in the person of the promising new Pope John Paul II, whose elevation to the papacy had occurred during the weeks when his son was embroiled in the murder inquiry.

On 27 March 1981, at Geneva – the very location where his ancestor Louis I of Savoy had received the Shroud from Margaret de Charny over five hundred years earlier – the now seventy-seven-year-old Umberto signed the formal decree that 'after my death the entire property of the Holy Shroud shall be donated to the Holy See'. He instructed his nephews King Simon of Bulgaria and Prince Maurice of Hesse to 'effectuate all those issues necessary to the notification of these, my wishes, to the Pontiff'.[19]

Fourteen months later, Pope John Paul II, well recovered after the assassination attempt by Turkish gunman Mehmet Ali Ağca, made a very special pilgrimage to Fatima in Portugal.[20] From Cascais, Umberto let it be known that he wished for a private audience, which was granted on 14 May. A photograph taken on this occasion[21] shows a clearly very frail Umberto being warmly embraced by the pontiff (pl. 32a), and there can be little doubt that during this audience Umberto discussed his intentions for the Shroud.

Ten months later, Umberto died peacefully at his villa in Cascais. For the first time ever in its arguably near two-thousand-year history, the Shroud became the formal property of the Roman Catholic Church.

Chapter 19

A Challenge to our Intelligence

Since it is not a matter of faith, the Church has no specific competence to pronounce on these questions. She entrusts to scientists the task of continuing to investigate, so that satisfactory answers may be found to the questions connected with this Sheet which, according to tradition, wrapped the body of our Redeemer after he had been taken down from the Cross.

Pope John Paul II speaking in Turin, 24 May 1998

JUST AS MARGARET DE CHARNY chose very carefully when she ceded the Shroud to Duke Louis I of Savoy, so Umberto II of Savoy did the same when he bequeathed the Shroud to Pope John Paul II. For Pope John Paul II had his own very genuine personal interest in the Savoys' remarkable family heirloom.

On 1 September 1978, while he was still Karol Wojtyła, Archbishop of Krakow, he had quietly knelt before the

Shroud in Turin Cathedral, hardly noticed among the thousands who daily came to do the same during the expositions held between 26 August and 8 September of that year. Only five days earlier he had attended the conclave during which, in the very same hour that those Shroud expositions began, the genial sixty-six-year-old Albino Luciani was elected Pope in Rome, taking the name John Paul I. Twenty-seven days later, while the Shroud was still on display in Turin, came the news that Luciani was dead, necessitating the summoning of another conclave. Over the course of the following ten days the Shroud underwent its most intensive ever scientific examination, at the hands of the American STURP team. Eight days after the Americans had completed their examination, Wojtyła found himself elected Pope. He duly took the name John Paul II in honour of his short-lived predecessor.

On 13 April 1980, John Paul II made his first visit to Turin as Pope, during which he was now accorded his own special private showing of the Shroud. Reportedly he examined the cloth intently, and kissed its hem. In a special pastoral address to Turin's citizens, he told them, 'At the beginning of September 1978 I came to Turin eager to venerate the Shroud, extraordinary relic related to the mystery of our Redemption ... unique witness, if we accept the arguments of so many scientists, of Easter, of the Passion, the Death and the Resurrection. Silent witness, but at the same time, surprisingly eloquent . . .'[1]

When he referred to 'many scientists', the Pope clearly had in mind the state of scientific knowledge that existed at that time – the many medical insights, the pollen findings, the VP-8 3D image, also the early data coming in from the STURP team's recent examination. On 13 May 1981, the team's leader, Dr John Jackson, together with some of his associates, was actually in St Peter's Square awaiting a prearranged papal audience to brief him on the team's findings when Turkish gunman Mehmet Ali Ağca opened fire and the Pope was rushed away for emergency surgery. At the beginning of the 1980s the new advances in radiocarbon dating were very quickly being perceived as the next logical step in scientific approaches to the Shroud. So when, on Umberto II's death in 1983, Pope John Paul found himself history's first papal 'owner' of the Shroud, his progressive mind welcomed such an initiative for his unexpected new possession.

Nevertheless, because of some highly convoluted politics, ownership of the Shroud, even by a Pope, carried some serious limitations. One of the conditions of Umberto's bequest seems to have been that the Shroud should not be removed from Turin.[2] Also, while it remained housed in Guarino Guarini's Shroud Chapel, it was technically on Italian state property; any initiative therefore required delicate diplomatic negotiations with the Italian government. Furthermore, Turin's ecclesiastical establishment, having come to enjoy the 'hands-on' power over the Shroud they had been able to exercise during King Umberto's exile, had adopted some very proprietorial attitudes towards it.

Thus, during the events leading up to the 1988 carbon dating, there occurred the earlier-described jostling for power between John Paul II's chief scientific adviser, Brazilian-born Dr Carlos Chagas, and Cardinal Ballestrero of Turin's chief scientific adviser Professor Luigi Gonella. On any basis of scientific clout, it should have been no contest. Chagas, the long-established president of the Pontifical Academy of Sciences, had international recognition for his work in the field of neuroscience and was the son of a Brazilian scientist of the same name twice nominated for a Nobel Prize for his discovery of Chagas' disease. Gonella was merely a lecturer in physics at Turin's polytechnic. But the Shroud was housed on Gonella's home turf. And when Chagas, with a command of English far superior to Gonella's, began initiatives to hold a high-powered international scientific workshop to discuss the carbon dating, Gonella vehemently resented his personal authority being thus undermined.

Gentle, diplomatic and in his mid-seventies, Chagas was simply no match for the Gonella brand of street-fighting. To document even a fraction of the behind-the-scenes manoeuvres would be wearisome and distasteful. The upshot was that Chagas's well-considered protocol, based on his workshop's recommendations, for how the radio-carbon dating *should* be conducted was virtually scrapped, much to the consternation of the seven laboratories he had invited to the workshop and whose participation in the project they thought to be assured. In a letter of 1 July 1987 to Cardinal Ballestrero, they warned,

> We urge Your Eminence before making a final decision on this question to reconvene a meeting of the seven carbon dating laboratories and the British Museum with your science adviser, Professor Gonella, to more fully apprise him of the dangers of modifying the Turin Workshop Protocol in this fundamental way. The protocol was care-fully crafted to meet your charge that the results of the measurements be credible to the general public and to knowledgeable scientists alike. As participants in the Workshop who devoted considerable effort to achieve our goal we would be irresponsible if we were not to advise you that this fundamental modification in the proposed procedures may lead to failure.[3]

On 10 October, Ballestrero responded to the laboratories, telling them very firmly that there would be no reconvened workshop, that their number had now been reduced to three, and that the Pontifical Academy of Sciences would henceforth take no further part in the exercise, merely being invited as observers.[4]

Thus Gonella and his colleague Giovanni Riggi peremptorily snatched executive command of what subsequently transpired during the carbon dating of 1988 – with results that are now part of history. With both men now dead, their mistakes may be enumerated:

1. the choosing only of laboratories using the AMS radio-carbon dating method, all three of them clones of one another;
2. no involvement of a professional textile conservator for the process of taking the sample;
3. the taking of just one single sample, from one single area of the Shroud;
4. the choice of site for that one sample being well documented as having been subjected to prolonged, repeated handling throughout centuries;
5. no provision for any chemical analysis of the sample;
6. the unofficial purloining, on the part of both Gonella and Riggi, of unused portions of that sample for their own personal research purposes;
7. the denial of any other, synchronous scientific approaches to the Shroud.

Had someone of Chagas's scientific standing remained in overall charge they might have been able to ensure that the laboratories' results were presented in a way that at least made clear the limitations of radiocarbon dating. In particular, they might have recommended that media commentators take serious note of the fact that radio-carbon dating is not infallible, and that contamination from the Shroud's historical adventures and misadventures could have significantly affected the result. In the event, Gonella and his cardinal effectively let the laboratories make the announcement themselves – unheard of in the case of any normal archaeological artefact. With the British Museum's participation little better than window-dressing, the laboratories all too predictably turned it into a PR exercise for their AMS radiocarbon dating method – and an unmitigated assassination for the Shroud.

Within months of the carbon-dating result being announced, the British Museum announced plans to stage a major public exhibition entitled 'Fake: The Art of Deception'. The exhibition was duly staged between 9 March and 2 September 1990, complete with a life-size colour photograph of the Shroud. It seems not to have been a coincidence that Oxford radiocarbon laboratory founder-director Professor Edward Hall was a British Museum trustee.

In any event, there occurred during precisely the time of this exhibition another of those 'act of God' happenings that suggest some wry comment from on high. On 4 May, during Turin's annual celebration of the Feast of the Shroud in Guarino Guarini's Chapel of the Shroud, several large chunks of stone came crashing to the floor from the dome a hundred feet above. Shortly afterwards the chapel had to be closed for major repairs to its fabric. A bullet-proof plate-glass display case was quickly constructed as a temporary new home for the Shroud, immediately behind the cathedral's high altar. The Shroud was duly installed in this, still in its seventeenth-century casket, on Ash Wednesday 1993.

Four years later, with the repair work on the Guarini chapel nearing completion, on the evening of 11 April everyone who was anyone in Turin was gathered at a gala dinner held in honour of special guest UN General Secretary Kofi Annan at the Savoys' former Royal Palace, adjoining Turin Cathedral. Among the guests was Cardinal Ballestrero's successor as Archbishop of Turin, Cardinal Giovanni Saldarini. Around eleven p.m., the Turin police received a cryptic warning that a fire might break out in the cathedral. This was followed thirty-five minutes later by a frantic phone-call, over the din of fire alarms, from parish priest Fr Francesco Barbero. Guarino Guarini's Shroud Chapel, which was still filled floor-to-ceiling with scaffolding, was uncontrollably ablaze.

Although Turin's fire brigade was quickly upon the scene, confronting them as they burst into the cathedral was a mass of flames consuming the first-floor chapel, with fiery debris from this conflagration raining down at the high altar end – the very area housing the recently installed bullet-proof display case containing the Shroud. Now part of Turin folklore is the story of fireman Mario Trematore's response to the danger. Exactly as had happened in Chambéry back in 1532, no one sufficiently senior and 'in the know' was on hand to go through the lengthy security procedure for opening up the case, nor was there the time for any such person to be summoned. As Trematore recognized, if the dome overhead gave way, the Shroud's display case and its precious contents would be wrecked and incinerated amid the falling rubble. Resolutely, Trematore swung his fireman's axe with all his might against the case's two-inch-thick toughened glass (pl. 32c). True to its specifications, it went a milky colour but remained intact. Undeterred, Trematore swung the axe again and again. Gradually he opened up a hole in the display case sufficiently large to enable him to reach in, pull out the Shroud's casket and rush it to safety.

Three days later, with Cardinal Saldarini's own residence the Shroud's temporary place of refuge, those principally responsible for the Shroud's care, together with a small group of invited specialists, gathered to conduct a formal check on how well the cloth had survived its most recent ordeal. Much as expected, it was found to be unscathed, and photographs circulated around the world to confirm this.

Such had been the scare generated by the fire that the Shroud's location was kept a well-guarded secret throughout the rest of 1997 and well into 1998. But as it was the hundredth anniversary of Secondo Pia's photographic discovery, the year 1998 had long been scheduled for fresh showings of the Shroud, the first in twenty years, to be held over two months, and fully approved by Pope John Paul II. So the activity was frenetic. To hide the massive fire damage to the Shroud Chapel a huge screen was erected behind the high altar. Set designer Fulvio Lanza from the La Scala opera house in Milan was commissioned to paint this with a *trompe l'oeil* depiction of the chapel as it would have looked from the cathedral before the installation of the unsightly glass partitioning (thankfully destroyed by the fire) back in the 1820s. Because the fire had also destroyed the cathedral's former sacristy, where the sample for the carbon dating of the Shroud had been cut away just nine years earlier, a new Tardis-like prefabricated sacristy was installed on a convenient platform on the cathedral's south side, with high windows to ward off prying eyes. On 15 April 1998, the Shroud was secretly brought into this new sacristy for textile conservator Mechthild Flury-Lemberg to prepare it for the expositions scheduled to commence just three days later.

One month in, these expositions received their most important visitor, in the person of Pope John Paul II. He was now an altogether frailer figure than the vigorous fifty-eight-year-old who had first visited it twenty years earlier. Despite the carbon dating, he had personally authorized these showings, the first ever to be held under his ownership, and he had a steely determination to be present to make a point. The many photos of the Pope, in simple white, praying on his knees before the Shroud (pl. 32c) speak louder than any words that he was far from persuaded by the carbon-dating findings. And the words of his homily reinforced this:

> The Shroud is a challenge to our intelligence. It first of all requires of every person, particularly the researcher, that he humbly grasp the profound message it sends to his reason and his life. The mysterious fascination of the Shroud forces questions to be raised about the sacred Linen and the historical life of Jesus. Since it is not a matter of faith, the Church has no specific competence to pronounce on these questions. She entrusts to scientists the task of continuing to investigate, so that satisfactory answers may be found to the questions connected with this Sheet which, according to tradition, wrapped the body of our Redeemer after he had been taken down from the Cross. The Church urges that the Shroud be studied without pre-established positions that take for granted results that are not such; she invites them to act with interior freedom and attentive respect for both scientific methodology and the sensibilities of believers.[5]

Likewise in defiance of the carbon-dating result, Turin Cathedral received more than two million visitors[6] during the two months that the Shroud was on display. Although this was less than back in 1978, it was still impressive for a cloth that had been so publicly branded a fake just ten years earlier. And to underline the point that the Shroud's branding as a fake was in no way proven, the Pope asked for it to be exhibited again as part of the celebrations for the year 2000 Holy Year. These expositions, presided over by Cardinal Saldarini's successor as Archbishop of Turin, Cardinal Severino Poletto, were staged between 12 August and 29 October, and attracted a further million and a half visitors.

Ironically, only shortly after the carbon-dating result, and well before the 1997 chapel fire, a Turin-based committee had been formed to consider the best possible measures for the Shroud's long-term conservation. Following such a scare for the Shroud's preservation so soon after the cloth's passing into Church ownership, measures for its ongoing conservation were bound to assume a greater urgency.

The many creases the Shroud accumulated each time it was rolled and unrolled meant that storing it flat ought to be the preferred choice for the future. However, a laid-out-flat mode of storage raised its own issues, because the greater exposure increased the oxygen, and thereby oxidation, to which the cloth was subjected. Too much oxygen would yellow the cloth and lessen the degree to which the image stood out from its background. The proposal was therefore to conserve the Shroud in an air-less atmosphere; argon was suggested because it is readily available, chemically inert and does not support the development of micro-organisms. In the event a gas system was chosen containing oxygen in such a low concentration that it would constitute little or no risk of oxidation and the development of aerobic organisms, yet which would be sufficient to poison anaerobic micro-organisms.[7]

A fifteen-foot-long conservation case was duly con-structed in which the Shroud could be permanently stored flat under these conditions, but tilted ninety degrees when-ever there was a call for it to go on public display. And with no definite date predicted for when repairs might be complete within Guarino Guarini's fire-ravaged Shroud Chapel, a new semi-permanent home for this case was created beneath the 'royal box' in the left transept of Turin Cathedral. Back in the heyday of the Savoy dynasty, members of Turin's first family used to attend Mass in the 'royal box', secure and private from the rest of the congregation below them.

As an extension of the same conservation measures, during the summer of 2002 the supportive backing cloth and cosmetic patches sewn on by the Poor Clare nuns after the fire at Chambéry in 1532 were removed. This intervention was very necessary, because abrasion from repeated rolling and unrolling of the Shroud over the last four centuries had loosened thousands of sooty particles from the edges of the charred areas beneath the patches. Beneath each patch a 'bomb' of soot had formed. Had the Shroud been doused with water during the 1997 fire, it would have been disfigured with soot stains. Sooner or later, therefore, such an operation had to be carried out.

As described earlier, the work was carried out by professional textile conservator Dr Mechthild Flury-Lemberg, assisted by Irene Tomedi, with the full approval of Pope John Paul II as the Shroud's formal owner. From a personal examination of it in September 2002, it seemed to me to have been carried out with extreme delicacy and professionalism, the aesthetic effect being exemplary. Furthermore, it opened up every square inch of the Shroud, back and front, for photographic documentation. With the great advances in digital photography in recent years, this has provided many opportunities for further research on the Shroud yet to be realized.

The death of Pope John Paul II in April 2005 brought about a new owner for the Shroud in the person of the present Pope, Benedict XVI, who clearly holds the same progressive attitudes towards the Shroud as his predecessor. It was Benedict XVI who approved the ultra-close-up digital photography and high-definition filming of the Shroud that was carried out in January 2008. From both of these projects has come a wealth of visual data which, when it becomes more freely available, will keep researchers busy for decades to come.

Much to general surprise, not least in Turin, the Pope also specially called for the year 2010 expositions of the Shroud which brought into being the writing of this book.

Chapter 20

On the Third Day

We heard him say, 'I am going to destroy this Temple made by human hands and in three days build another, not made by human hands.'

Jerusalem's chief priests' accusations against Jesus, Mark 14:58

THROUGHOUT THE greater part of this book we have put together a reconstruction of two thousand years of the Shroud's history that is more comprehensive and more complete than anything done before. Nevertheless, it is just that, a reconstruction, based on one simple premise: that the cloth we know today as the Shroud of Turin, which surfaced so mysteriously in France in the mid four-teenth century, is one and the same as the Image of Edessa, which began its life in Jerusalem in AD 30 and disappeared from Constantinople in 1204.

That premise is either broadly right, solving the two-thousand-year mystery of the Shroud, or it is fundamentally, humiliatingly wrong. There is simply no comfortable middle ground in which to hide. As pointed out by distinguished German art historian Hans Belting at a colloquium in Rome in 1996, 'Art historians dislike the

Shroud, as the latter either is an original (thus antedating Christian imagery), or it is a late medieval fake (thus post-dating the history of intelligent and beautiful images).'[1]

King Umberto of Savoy certainly seems to have favoured the Edessa explanation for the Shroud's early history. His daughter Princess Maria Gabriella of Savoy recalled fondly of him in 1998, 'My father enjoyed setting the dynasty of the Christian king Abgar of Edessa ... alongside ours, in that both had for centuries been jealous guardians of Christ's winding sheet. Both were rulers of small mountain states which did everything possible to remain neutral in spite of the threatening presences of bordering powers.'[2]

And if the Shroud was indeed the Image of Edessa during the early centuries, the future looks reassuring for further supportive findings. The research expedition to Cappadocia, the Şanliurfa region and north-east Turkey that was undertaken for this book, partly in partnership with Mark Guscin, yielded much new data yet to be properly evaluated, including hitherto unknown depictions of the Image of Edessa from during the period of Iconoclasm. Mark Guscin is following up his pioneering academic book on the Image of Edessa with a Ph.D. thesis for the University of London. Also in train is a compilation of all known pre-fifteenth-century depictions of the Image of Edessa, embracing not only Turkey but also Serbia, Bulgaria, Greece, Cyprus, Russia, Georgia and Egypt. This is a hitherto neglected academic exercise with the potential to reveal a number of important new insights, particularly in Georgia, where several early examples are as yet known, certainly to western scholarship, only by name[3] rather than from proper photographic documentation.

Likewise, fresh academic studies continue to add to the picture. In the United States, Dr Sharon Gerstel, Associate

Professor of Mediaeval Art at the University of California, Los Angeles, has shown that depictions of the Image of Edessa in Byzantine church iconography are often symbolically linked to the Eucharist – that is, to Jesus's sacred body and blood, the very elements that we see so graphically imprinted on the Shroud.[4] In the former Soviet republic of Georgia, Dr Irma Karaulashvili has been making ground-breaking studies of the manuscript background to the Abgar story[5] that supplement the work of Mark Guscin at the monasteries of Mount Athos.

Despite these promising developments, the ongoing historical difficulties are not to be minimized. The once remote rock churches of Cappadocia, one of the few places in Turkey where depictions of the Image of Edessa have survived deliberate Islamic destruction of Christian imagery, have now become the focus of mass tourism, with all its concomitant dangers for unprotected artworks. Elsewhere much lies long ago vandalized, and neglected, sometimes covered by centuries of soot.

In Şanliurfa, the discovery of the Haleplibahce mosaics has brought to the city competent archaeologists of the calibre of Mehmet Önal and Selçuk Şener. If their work continues, this could in time lead to the discovery of foundations of some of the lost Christian churches of Edessa, also an unearthing of remains from the time of the kings, of the city's as yet unexcavated citadel. In a city which has seen centuries of wilful destruction of anything Christian, the recent discovery of the sixth-century Christ face mosaic has been a most welcome ray of hope, like finding one child alive amid the carnage of a crashed jumbo jet. But there is room only for very guarded optimism. All too recently a mosaic depicting an Abgar, possibly one of the kings, simply disappeared from beneath museum officials' noses as part of the looting of antiquities that continues to be rife in Turkey.[6] Whole mosaics are not exempt, simply hacked away, base and all. And in Turkish-held northern Cyprus far too many churches with priceless ancient depictions of the Image of Edessa continue to lie abandoned and neglected, with access either prohibited or hampered by excessive bureaucracy.

Epitomizing the dangers, in the course of writing this book I learned of a hitherto unknown cycle of wall-paintings depicting the story of King Abgar and the Image of Edessa that throughout the last six centuries have been displayed on the walls of the church at the monastery of Matejce in Macedonia.[7] The monastery was founded by the Serbian king Stefan Dusan, a direct contemporary of Geoffrey I de Charny, and in one of the paintings the Image of Edessa featured as a significantly large piece of cloth specifically described as a *sindon* (pl. 26c).[8] But then came the downside. As recently as 2001, a gang of Albanian Muslims took over the church and desecrated its frescoes with graffiti, among them the double-headed eagle symbol of Albanian nationalism, and also daubed beards, cigarettes and sunglasses on to the images of the saints. Although the church was theoretically then placed under UNESCO protection, far worse followed. In April 2002, all the paintings were found to have been stripped off the walls and wantonly destroyed with pickaxes.[9]

On the more positive side, prime among the opportunities for fresh understandings of the Shroud's history must rank further insights from Dr Barbara Frale's findings concerning the Knights Templar and the Shroud, which I learned of only when this book was already at an advanced stage. Although the theory of Templar ownership was first advanced by me thirty years ago, this was essentially as a stop-gap explanation for these 'missing years', and in the interim a Serbian connection had begun to look promising. Accordingly, it was no less a surprise to me than to anyone else to learn from *The Times* that

Barbara Frale was telling journalists 'her discovery vindicated a theory first put forward by Ian Wilson, a British writer, in 1978'.[10] Dr Frale's further work on the key document is awaited with interest.

History aside, although every attempt has been made to make this book as comprehensive an approach to the subject as possible, there are inevitable omissions, some worthy of at least a passing mention. A cloth sometimes associated with the Shroud is the so-called Sudarium of Oviedo, the stains on which are supposed to match those of the Shroud.[11] This has suggested to some that it may have been used as a cover for Jesus's face while he was being transported to the tomb. The science and history of the Shroud seemed difficult enough to handle without complicating matters by introducing another cloth with similar issues, so non-inclusion of the Oviedo Sudarium should not be interpreted as indicating my rejection of that cloth's authenticity.

For much the same reasons, similarly omitted here have been studies by American historian Dr Daniel C. Scavone associating the legend of the Holy Grail with the early history of the Shroud. The Grail is another difficult and extremely tangled topic in its own right, sadly adulterated by many tabloid-level absurdities written on the theme in recent years. This book's neglect of the issue should not be regarded as any kind of judgement on its seriousness and relevance. In the course of the Shroud's transmission from the Templars to the de Charnys to the Savoys, there are some tantalizing hints of the Shroud's guardianship having had an almost Dan Brown-like 'secret chivalric brother-hood' element to it, possibly requiring a rather more searching investigation of France's short-lived Order of the Star and the Savoys' Order of the Collar/the Annunciation than either has yet received.

Los Angeles artist Isabel Piczek has done some

extremely valuable work with life models to reconstruct the exact 'pose' of the Shroud man's body as he lay in death. Her findings are highly intriguing, revealing the knees to have been markedly bent, the right shoulder dropped quite significantly relative to the left, and the elbows elevated – all indicative of a body still partly in a rigor mortis that had set in while hanging dead on the cross. It proved difficult to find a proper context for Piczek's findings in the writing of the present work, but they have been explored in more detail elsewhere.[12]

A frequently asked question about the Shroud is whether DNA might be extracted from the bloodstains and analysed. The answer is that in the early 1990s the unofficial samples taken from the Shroud by Giovanni Riggi, the Turin microanalyst who cut the sample for radiocarbon dating, were analysed for DNA at the University of Texas Center for Advanced DNA Technologies in San Antonio. Three gene segments were found, indicative that the DNA was that of a human male. No more meaningful data were forthcoming from them. However, even if there had been, the segments effectively prove nothing. Not only were Riggi's samples unofficially obtained, and therefore not recognized by Turin's archbishop at the time, Cardinal Saldarini, the big problem with DNA, just as in the case of radiocarbon dating, is the high risk of contamination. As we have seen, it was a frequent practice of the clergy and dukes of Savoy, during times of the Shroud's known history, to kiss the major wounds on the Shroud. We have also learned how it was not uncommon for sweat and tears to fall on to the cloth in the course of the public showings. Any of these known activities could and would have transferred more recent DNA on to the bloodstains, rendering any scientific findings from them meaningless.

Certain other omissions from this book have been more

intentional. Internet sites often abound more with mis-information than reliable information on the Shroud, one favourite of this genre being the 'seeing' of other items in the Shroud's stains besides the universally agreed double imprint of a male figure. Among these there are just a few, most notably the occasional plant specimen, which might just be plausible. When none other than Avinoam Danin, Professor of Botany at the Hebrew University, Jerusalem, very forcefully insists on seeing the images of certain plants,[13] it deserves to be taken seriously – particularly given that he can in no way be accused of any pro-Christian bias. Most other such claims, however, are extremely doubtful, particularly the Latin, Greek and Hebrew inscriptions which have been 'seen' by two French researchers, André Marion and Anne Laure Courage.[14] Also the claims of a French 'scientist', Thierry Castex, that he has found the Aramaic word 'found' written in Hebrew letters. In any event, all claims of this nature have been omitted on the grounds of their flimsiness and their im-pedance of the main line of argument.

Such distractions and controversies aside, where does the future lie for the Shroud? All too often both outsiders and insiders have tended to opt for 'another radiocarbon dating test' as the most logical way forward. This is totally the reverse of my stance, on the grounds that radiocarbon dating has had far too stifling an influence on the subject for far too long already. Until we can be sure how the carbon dating of 1988 was skewed, if it was skewed (and microbiological contamination is merely one possibility among several), there is simply no point in destroying yet more pieces of the cloth for the sake of blindly 'having another try'. As the Harwell scientist P. J. Anderson sagely warned back in the 1960s, it may well be that the Shroud, with all the vicissitudes of history it has been through, is quite simply unsuitable for the procedure. End of story.

Potentially far more productive for the future is the totally non-destructive high-resolution HAL9000 digital photography and the David Rolfe/David Crute high-definition filming that was carried out in January 2008. In time, this data should become available to anyone, at the most for the price of a DVD, which represents a wonderful research opportunity. When that happens it should enable anyone, anywhere in the world, using their home computer, to study any area of the Shroud to a degree of detail as if they were holding a high-powered magnifying glass over the cloth directly in front of them. As David Crute has pointed out, via the professional TV industry's image 'grading' equipment we already have the means to analyse the individual spectra of the image. This enables the isolation of certain details, such as the bloodstains, bringing them to the fore for special scrutiny. Without the need for any further 'hands on' examination it should be possible, in Crute's words, 'to get to the bottom of how the image was formed', something that even as an acknowledged non-Christian he admits to finding hugely exciting.

But why the excitement, even on the part of an agnostic? Although many wonder why anyone should find a few stains on an old piece of linen so fascinating, it is the character of those stains, unquestionably at least six hundred years old (radiocarbon dating has at least confirmed that!), which is so compelling.

The plain fact is that no normal human body leaves behind an image of itself, certainly not one with the extraordinarily photographic character of that on the Shroud. Can it be by accident, therefore, that this phenomenon has happened uniquely in the case of Jesus Christ, the one man in all human history who is accredited with having broken the bounds of death? If the Shroud really is two thousand years old, could whatever happened at that moment in time quite literally have flashed itself on to the cloth that we have today, a now permanent time-capsule of how Jesus's body looked at the very moment of his resurrection?

If that thought is not daunting enough in itself, little less so (certainly for a historian) is what we have been able to reconstruct of the Shroud's history. In its history as the Image of Edessa we have seen the violent pagan backlash against Edessa's first Christian community; the two floods while the cloth's whereabouts remained unknown above the city gate; the devastating earthquakes of 679 and 717; the 'trial by fire'; the inception of Iconoclasm; the running fight at Edessa during the handover to John Curcuas's Byzantine army; and the wanton destruction that accompanied the Crusader sack of Constantinople in 1204. In the cloth's history as the Shroud we have learned of the near-fatal 1532 fire in Chambéry, the hazards of the French invasions, and the dangers of the fire of 1997, not to mention the rigours of the repeated journeys in some mule's pannier on rough tracks back and forth across the French Alps. For a simple sheet of linen to have survived so much when so many solid buildings have been reduced to rubble ranks as nearly as extraordinary as the imprint itself. Somehow this ostensibly frail piece of linen has survived down the years to our own age when, uniquely in all history, we have the technology infinitely to duplicate its appearance for all posterity.

In a very real sense, therefore, the image on the Shroud has become immortal, whatever else history may throw at it. Whereas little more than a century ago it lay accessible only to the privileged few and was seen in public only on rare occasions, and from a discreet distance, now anyone anywhere in the world should soon be able to make their own personal study of it, and far more closely than anyone could once have considered possible. As I have mentioned before, it is as an aid to that study that this book has been written.

The Shroud: fake or fact? I have done my best to provide you with the most serious factual evidence on which to base that judgement. The rest is up to you.

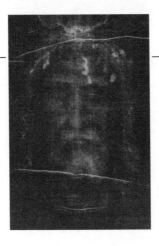

The face on the Turin Shroud

This was the look of him? This down-to-earth man?
This convinces me. None of the flimsy faces
The painters put on him. This man never arrived
At resurrection without a hard-won fight,
Nor was half air before he achieved ascension.
With him he took a look of the earth he lay in –
Rock, and a little soil, and old olive roots –
A sturdy, serene man, common sense in a riddle.
He looks like his talk, before it was pared by parsons,
Spun into sermons, and so on, transtabulated
Into theology. This man is marvellous –
Death instinct with life, life at peace,
This is man.
They say he will judge me. I'm convinced.
I am judged already. I stand before him, knowing
That like each man I am my own disaster.
He knows I know. He will be merciful.
This man looks like all that I know of God –
I can call him both me and master.

JAMES BRABAZON

Postscript

Since the publication of the hardback edition of this book, the Shroud was publicly exhibited in Turin Cathedral between 10 April and 23 May 2010. The number of registered visitors exceeded two million, thereby attracting much the same attendance as on previous occasions. On 2 May, Pope Benedict XVI paid an official visit during which he prayed before the Shroud, and his accompanying homily came the closest of any Pope to openly expressing his personal belief in the Shroud's authenticity. He stressed the Shroud's essential paradox that although its imprinted image 'is that of a dead man . . . the blood speaks of his life' – strikingly reminiscent of the Eastern Orthodox Church's long-standing attitude towards the Image of Edessa as being of Jesus while he was alive.

Shortly prior to the Turin expositions the US History Channel put out its first broadcast of *The Real Face of Jesus*, a ground-breaking TV documentary that attempted to reconstruct the face of the man of the Shroud using the latest digital imaging techniques. Although this was far-and-away one of the better documentaries on the subject, graphically demonstrating the Shroud's fidelity to a true crucified human body that other documentaries have sought to deny, New York-based producer Ray Downing had available to him only conventional photographs of the Shroud that had been taken back in the late 1970s. The hope is that within the next few years, Downing may be able to further his work using the much more advanced and detailed digital imagery now becoming available.

Appendix: Shroud Chronology
AD 30 to 2010

based on the theory that the Jesus-imprinted cloth of
Edessa and the present-day Shroud of Turin
were one and the same object

30   Likeliest date of death of Jesus by crucifixion in
Jerusalem, and his burial by wrapping in a *sindon* or
shroud, possibly one and the same as the *soudarion*
described by the gospel writer John. According to the
Christian gospels, some thirty-six hours later this
shroud is found abandoned and the tomb empty. Later
sources describe how a cloth imprinted with Jesus's likeness is taken from Jerusalem to Edessa (the present-day
Şanliurfa in eastern Turkey) during the reign of Edessa's
king Abgar V (AD 13–50). Edessa's citizens speak much
the same Syriac language as that of Jesus's disciples, and
partly as a result of King Abgar's 'exclusive' viewing of
the Jesus-imprinted cloth, he and other leading
members of his court are reputedly converted to
Christianity.

50   Death of Edessa's king Abgar V. According to later traditions, one of his successors reverts to the city's former
pagan religion, those who have converted to
Christianity are persecuted, and the whereabouts of the
Jesus-imprinted cloth become lost.

177   Accession of Edessa's king Abgar VIII, the Great.
       Christianity has apparently become re-established in
       Edessa by his time, for a chronicle of Edessa reports a
       flood in the city in 201 which damages a 'church of the
       Christians'. Also some of Abgar VIII's coins issued
       during the reign of the Roman emperor Commodus (AD
       180–192) feature a Christian cross decorating the king's
       tiara. But there is no apparent knowledge of the Jesus-
       imprinted cloth's whereabouts, or even of its existence.

212   Death of Abgar VIII the Great. Three years later the
       Romans seize his successor and transform Edessa into a
       military colony.

303   Another serious flood at Edessa.

384   Approximate date of visit to Edessa by the highly obser-
       vant lady pilgrim Egeria, who has travelled from as far
       as Spain as part of a pilgrimage to the Holy Land. In
       what survives of Egeria's travel diary she describes
       several churches and other landmarks of Edessa; how-
       ever, she makes no mention of any Jesus-imprinted
       cloth. Instead much of Edessa's fame as an early
       Christian city rests on a letter which Jesus is said to have
       written to King Abgar V, the text of which is displayed
       on Edessa's gate. This letter includes a line in which
       Jesus promises to protect the city from any foreign
       invader.

430   Death of St Augustine of Hippo. In his writings St
       Augustine mentions a general ignorance in his time con-
       cerning what Jesus had looked like. In the Christian art
       of around this time Jesus is likewise commonly depicted
       beardless, and the rare bearded examples are very
       different from the later very definitive likenesses.

503   Persian king Kavadh arrives before Edessa's walls with
       a formidable army, but fails to capture the city – reput-
       edly due to his soldiers' superstitious fears concerning
       Jesus's 'protective' letter to Abgar.

525   Edessa suffers its most serious flood yet, with thirty
       thousand citizens estimated killed in the disaster, and
       several major churches destroyed. From Constantinople
       the very able emperor-to-be Justinian initiates a major
       reconstruction programme. This involves re-routing
       Edessa's river Daisan, reconstructing the city's walls,
       and completely rebuilding several churches, including
       the cathedral, subsequently to be known as Hagia
       Sophia, like its Constantinople counterpart. These are
       the likely circumstances in which the Jesus-imprinted
       cloth became rediscovered.

544   Persian king Kavadh's son and successor Khosraw
       arrives before Edessa's walls with a yet more formidable
       army. The Persians are again repulsed – but this time
       reputedly thanks to the protective powers of the Jesus-
       imprinted cloth now confidently referred to as having
       been brought to Edessa during the reign of Abgar V five
       centuries earlier. According to later Byzantine sources
       this cloth had recently been found sealed in a niche
       above one of the city's gates. Its immediately famed
       facial imprint of Jesus is unequivocally described as 'not
       by hand made'. Some near contemporary accounts refer
       to it as on a *sindon*, also as *tetradiplon*, doubled in four,
       suggesting that it was on a large cloth folded consider-
       ably smaller than its full size.

540s   At much this same time there appear in Christian art the
       very distinctive depictions of Jesus as long-haired and
       bearded that are essentially universally accepted as his
       likeness to this day. According to manuscripts recently
       found at St Catherine's monastery, Sinai, Syriac-speaking
       monks of this time travelled from Edessa and its sur-
       rounds carrying with them depictions of this likeness.
       Directly drawn from the Jesus-imprinted cloth of Edessa,
       these depictions were used to decorate newly founded
       churches in Georgia and elsewhere. Meanwhile the origi-
       nal cloth itself is housed in Edessa's Hagia Sophia
       cathedral, reputedly one of the most beautiful shrines in
       all Christendom.

639   Never happy with their rule from Greek-speaking Constantinople, Edessa's citizens accept the suzerainty of the Arab general Iyad, one of the pioneers of the new religion of Islam. The city's Arab Muslim overlords mostly express tolerant attitudes towards Christianity.

679   An earthquake damages several of Edessa's Christian churches, including the Hagia Sophia cathedral. The Islamic caliph Mu'awiyah quickly orders the cathedral to be rebuilt. While this rebuilding is in progress, the Jesus-imprinted cloth, which had evidently survived the destruction, may well have been transferred to Jerusalem for its safety. There within the next three or four years a visiting French bishop, Arculf, reports seeing a 'trial by fire' performed on the *sudarium* which was over Jesus's head in the tomb.

692   Depictions of Jesus of the new, definitive long-haired and bearded variety now flourish throughout the Christian world. In Constantinople the Byzantine emperor Justinian II issues coins with this likeness on the obverse or 'heads' side, with his own likeness as emperor on the 'tails', or reverse.

723   Outbreak of iconoclasm or image-smashing. As part of a vogue for religious purity which will continue for the next 120 years, man-made depictions of Jesus are destroyed throughout the Byzantine and Muslim empires, and the Edessa-inspired official likeness dropped from the Byzantine coinage. But in Muslim-held Edessa, the Jesus-imprinted cloth itself apparently survives unscathed.

843   With iconoclasm having fallen out of favour, the depiction of man-made images of Christ once again becomes officially approved throughout the Byzantine empire. The Edessa-inspired official likeness of Jesus is restored on the coinage.

943   A large army sent by the Byzantine emperor Romanus arrives at the walls of Edessa, then still under Arab Muslim control. The Byzantine general promises to leave Edessa untouched, to pay a large sum of money, and to release 200 high-ranking Muslim prisoners, all in return for surrender of the Jesus-imprinted cloth. After Edessa's emir consults with the Muslim leadership in Baghdad a deal is struck, much against the wishes of Edessa's citizens, and the cloth is taken off to Constantinople.

944   15 August. After a long land journey across the breadth of what is today Turkey, the Jesus-imprinted cloth of Edessa is received in Constantinople amid great celebrations. It is accorded its own feast day, 16 August. Because of the awe in which the cloth is regarded in Eastern Orthodox thought, there is no public showing, only privileged private showings. The cloth is installed in the Pharos Chapel of Constantinople's Imperial Palace, the repository of other most sacred relics of Jesus.

945   Accession of the artistically inclined Constantine VII Porphyrogennetos as sole emperor of Byzantium. The coins that he issues from this year exhibit particularly Shroud-like features on the face of Jesus.

1050s   The earliest-known depictions of the Edessa cloth in the form of a face on cloth appear in art, one of these, now lost, in the form of a mosaic set above the arch to the apse in the Hagia Sophia cathedral, Constantinople.

1054   Although the Edessa cloth continues never to be publicly displayed at this time, a depiction of it in a Georgian manuscript dated to this year appears to show how it was stored in its casket. From the size and proportions of this the Shroud would seem to have been folded in half beneath the gold cover so that if the latter were removed the Shroud figure would look as if Jesus was standing upright.

1130   Around this time high-ranking western visitors to
        Constantinople are occasionally shown the treasures of
        the Byzantine emperors' relic collection. Reports
        percolate to Western Europe, retold by the Normandy-
        based monk Orderic Vitalis and others, that the
        Jesus-imprinted Edessa cloth, besides Jesus's facial
        imprint, displays 'the form and size of the Lord's body
        to all who look upon it'.

1146   Edessa falls to invading Turkish Muslims. Amid wide-
        spread carnage in which more than 30,000 citizens are
        killed, and 16,000 women and children enslaved, the
        city is systematically looted, and its once famed
        churches are reduced to ruin from which they will never
        recover.

1192   (approx) Hungary's 'Pray Manuscript', created about
        this year by an artist strongly influenced by
        Constantinople, shows the body of Jesus laid out totally
        naked in the identical 'crossed hands' style to the
        Shroud. The shroud depicted in the same drawings
        exhibits what appear to be the mysterious poker holes
        still visible on the Shroud to this day.

1203   The French Crusader Robert de Clari, viewing
        Constantinople at a time when he and his fellow-
        Crusaders are guests in the city, reports seeing in the
        Church of St Mary at Blachernae, Constantinople's
        rallying place in times of crisis, 'the shroud in which our
        Lord had been wrapped'. He adds: 'every Friday this
        stood upright so that one could see the figure of our
        Lord on it.'

1204   The French-led Fourth Crusade captures and sacks
        Constantinople. Many of the city's treasures are looted.
        Whatever the identity of the cloth of Edessa, it dis-
        appears amid the confusion. Robert de Clari writes of
        the 'shroud' he saw that: 'neither Greek nor Frenchman
        knew what became of this shroud when the city was
        taken'.

1205   A letter sent in this year to Pope Innocent III, reporting what happened to Constantinople's sacred treasures, refers to 'the linen in which our Lord Jesus Christ was wrapped after his death'. If the letter is genuine, the Shroud was apparently being kept at this time in Athens, which was then under the command of the Crusader leader Otho de la Roche.

1287   According to the French Templar knight Arnaut Sabbatier, on his reception into the Order of Knights Templar in this year, at Roussillon in southern France, he was taken to 'a secret place to which only the brothers of the Temple had access' and shown 'a long linen cloth on which was imprinted the figure of a man'. This recently discovered report has been interpreted as evidence of the Shroud's ownership by the Knights Templar at this time.

1307   13 October. Following rumours that the Templars have been worshipping a mysterious bearded male head at secret chapter meetings, the French king Philip the Fair orders every Templar throughout France to be arrested at dawn on a series of heresy charges. These charges included worship of the 'head', an object which, despite the surprise nature of the arrests, is never found.

1314   19 March. The Templar Order's highest dignitaries, Jacques de Molay and Geoffrey de Charny, are burned at the stake in Paris, protesting their order's innocence.

1344   Geoffrey I de Charny, of the tiny village of Lirey, France, the first known European owner of the Shroud, journeys to Smyrna, in what is today Turkey, to fight the Turks, who are then in control of the Holy Land. At Smyrna Geoffrey helps capture the port's harbour fortress, winning a hero's reputation.

1349   Back in France Geoffrey I de Charny seeks permission from Pope Clement VI to build a collegiate church in his home village of Lirey. This is where the Shroud will later be exhibited.

1350   1 January. Geoffrey I de Charny is captured during an attempt to regain Calais from the English. His imprisonment in England temporarily halts his plans to build the Lirey church.

1354   Having been freed from captivity, Geoffrey I de Charny renews his plans to build the Lirey church, this time directing his petition to the new pope at Avignon, Innocent VI.

1355   According to the later 'd'Arcis Memorandum' (see 1389), western Europe's first known public showings of the Shroud as we know it today are held around this year at the Lirey church founded by Geoffrey I de Charny. On the occasion of these showings the cloth is described as the true shroud of Jesus, and souvenir badges are sold to visiting pilgrims. A single surviving example of one of these badges features the Shroud in its full-length, double-imprint form, the earliest surviving work of art to show it in this way.

1356   28 May. Henri of Poitiers, bishop of the local diocese of Troyes, goes on record as praising Geoffrey I de Charny for his founding of the Lirey church. 19 September. Geoffrey I de Charny, fulfilling the coveted role of carrying France's sacred *oriflamme* battle-standard, is killed defending his country from the English at the battle of Poitiers. He leaves a widow, Jeanne de Vergy, a young son, Geoffrey II de Charny, and the Shroud – but no known information concerning how he might have obtained this latter.

1389   15 August. Royal officers arrive in Lirey, where reportedly Geoffrey I de Charny's son Geoffrey II has been displaying the Shroud as if the true shroud of Jesus. Apparently acting on complaints from the present bishop of Troyes, Pierre d'Arcis, the royal officials are foiled from seizing the Shroud on the grounds that the key to its repository is not available. Later in this same year Bishop Pierre d'Arcis writes a forceful memorandum to the French anti-pope, Clement VII, claiming the Shroud to have been 'cunningly painted' and asking for showings of it to be stopped, a request which is declined.

1398   22 May. Death of Geoffrey II de Charny. Ownership of the Shroud thereupon passes to his daughter Margaret, who despite two marriages appears to have been unable to bear children.

1418   6 July. Because of security risks arising from the wars with England, the Shroud is moved from Lirey to the castle of Montfort, owned by Margaret de Charny's second husband, Humbert, Count de la Roche. A few years later it is moved to St Hippolyte sur Doubs, in Alsace-Lorraine, close to Switzerland. While at St Hippolyte sur Doubs public showings of the Shroud are staged each Easter Sunday in a meadow on the banks of the river Doubs.

1453   22 March. Elderly and still childless, Margaret de Charny receives a castle and an estate from Duke Louis of Savoy, in return for 'valuable services'. The 'valuable services' are understood to have been the handover of the Shroud to the Savoy family, who will be the Shroud's owners for the next five centuries.

1460   Death of Margaret de Charny, the last of her line to own the Shroud.

1464   6 February. Duke Louis of Savoy agrees to pay the Lirey clergy a rent as compensation for their loss of the Shroud, clearly indicating the Shroud's derivation from the de Charnys of Lirey. In this same year theologian Francesco della Rovere, soon to become Pope Sixtus IV, writes a book in which he mentions 'the Shroud in which the body of Christ was wrapped when he was taken down from the cross. This is now preserved with great devotion by the Dukes of Savoy, and it is coloured with the blood of Christ.'

1502   11 June. Inauguration of a new 'permanent' home for the Shroud, the Sainte Chapelle, specially built for it at Chambéry.

1506   21 April. Pope Julius II accords the Shroud its own feast day, 4 May. This will be the day of the year on which many subsequent public showings of the Shroud will take place.

1509   10 August. The Shroud is installed in a magnificent silver casket specially commissioned for it by the Dowager Duchess of Savoy Margaret of Austria, regent of the Netherlands.

1516   King Francis I of France journeys from Lyon to Chambéry to venerate the Shroud following his victory at the battle of Marignac.

1518   On the occasion of a visit to Chambéry by the Cardinal of Aragon, the Shroud is publicly displayed from a balcony of the Sainte Chapelle that juts out over the castle walls.

1521   The Shroud is publicly displayed from the balcony of the Sainte Chapelle at Chambéry to mark the marriage of Duke Charles III of Savoy to seventeen-year-old Princess Beatrice of Portugal.

1532   4 December. A major fire breaks out at the Sainte Chapelle, Chambéry. With no time to summon the key-holders to unlock the iron-grilled repository, the Shroud is rescued chiefly by the skill and heroism of a local blacksmith, and not without serious damage. Molten silver from Margaret of Austria's ruined casket is found to have seared through one corner of the multi-folded cloth, scarring it with a patchwork of burns. Yet to general amazement the Shroud's all-important image has hardly been touched.

1534   16 April–2 May. Four Poor Clare nuns carry out repairs
        to the Shroud, stitching it onto a strong backing cloth,
        and sewing triangular-shaped patches over the more dis-
        figuring burn-marks.

1535   An invasion of Savoy by French troops necessitates the
        Shroud being moved from Chambéry for its safety. Its
        subsequent travels during the next few years include
        Turin, Milan, Vercelli and Nice.

1552   Russian tsar Ivan the Terrible, fighting Mongol hordes
        at the battle of Kazan, uses a flag bearing an image of
        the cloth of Edessa as his battle standard, even though
        the original has long been lost to the Eastern Orthodox
        world. Ivan wins the battle.

1559   Peace treaty of Cateau-Cambrésis, by which Savoy has
        most of its former domains returned.

1561   The Shroud is returned to Chambéry.

1571   7 October. At the battle of Lepanto, in which warships
        supplied by a coalition of European powers engage the
        Turkish navy, Duke Emmanuel-Philibert of Savoy's flag-
        ship hoists a battle standard decorated with an image of
        the Shroud. The Turks are decisively defeated.

1578   The Shroud is moved to Turin, ostensibly to save saintly
        Cardinal Charles Borromeo the rigours of crossing the
        Alps to Chambéry, after he had vowed to venerate it
        following Milan being spared a terrible plague.
        Following Borromeo's arrival in Turin, on 12 October
        the Shroud is publicly shown in the broad Piazza
        Castello to crowds estimated to number 40,000.

1587   A special shrine for housing the Shroud is built within
        Turin Cathedral, set on four columns over the high
        altar.

1613   A print issued in this year depicts the huge crowds that
        would gather in Turin's Piazza Castello for the annual

showings of the Shroud each 4 May feast day.

1668   Savoy Duke Charles Emmanuel appoints the talented architect Guarino Guarini to design a chapel for the Shroud. This is to be sited between the royal palace and cathedral, at first-floor level.

1684   Public showing of the Shroud in the Piazza Castello to celebrate the marriage of Duke Victor Amadeus II of Savoy to his first wife Anna d'Orleans. A painting of this occasion by Dutch artist Pieter Bolckmann shows the dome of Guarino Guarino's Shroud Chapel already complete.

1694   After conservation attention by the Blessed Sebastian Valfré, including sewing on a new black lining cloth, the Shroud is transferred to the deceased Guarino Guarini's now fully completed Shroud Chapel. It is deposited in a specially constructed repository, above the chapel altar (designed by Antonio Bertola) that will be its home for most of the next three centuries.

1706   Due to impending French invasions, the Shroud is temporarily moved to Genoa.

1737   4 May. Before huge crowds, the Shroud is displayed from a balcony of Turin's Royal Palace to mark the marriage of Duke Charles Emmanuel III of Savoy to Princess Elizabeth Teresa of Lorraine.

1750   Public showing of the Shroud to celebrate the marriage of Prince Victor Amadeus of Savoy (the later Duke Victor Amadeus III), to Maria Antonia of Bourbon, Infanta of Spain.

1775   5 October. Public showing of the Shroud to celebrate the marriage of Prince Charles Emmanuel of Savoy (the later Duke Charles Emmanuel IV), to Princess Marie Clotilde of France.

1804   13 November. Private showing of the Shroud to Pope

Pius VII, who is on his way from Rome to Paris for the coronation of Napoleon Bonaparte.

1815   Pope Pius VII, returning to Rome after years of virtual imprisonment by Napoleon, stops off in Turin and helps conduct a public showing of the Shroud from a palace balcony.

1842   Public showing of the Shroud from the palace balcony to mark the marriage of Crown Prince Victor Emmanuel of Savoy (the later Duke Victor Emmanuel II), to Maria Adelaide, Archduchess of Austria.

1868   24–27 April. Public showing of the Shroud within Turin Cathedral, the first ever in which it is displayed statically in a frame, as distinct from being held up by clergy. 28 April. Princess Clotilde of Savoy personally replaces Sebastian Valfré's seventeenth-century black silk lining cloth with a red one.

1898   25 May–2 June. Public showing of the Shroud in Turin Cathedral to mark the Savoy family's fiftieth anniversary as rulers of all Italy. 28 May. The Shroud is photographed by local councillor and proficient amateur photographer Secondo Pia, revealing for the first time the extraordinary, life-like 'photograph' when it is viewed in negative.

1902   21 April. In Paris anatomy professor Yves Delage, an agnostic, having carefully studied Pia's photographs, presents a paper to the French Academy of Sciences. He claims the Shroud to be so medically accurate that he is convinced it wrapped a genuine, crucified human body, also that the body was that of Jesus.

1931   3–24 May. Public showing of the Shroud in Turin Cathedral as part of the marriage celebrations for Prince Umberto of Savoy (later to become King Umberto II) to Princess Maria José of Belgium. 23 May. Turin photographer Giuseppe Enrie takes a series of black-

and-white photographs of the Shroud, including the face life-sized. These will be the definitive photographs of the Shroud for the next half-century.

1933   24 September–15 October. Public showing of the Shroud within Turin Cathedral as part of the celebrations for the Holy Year. The Paris surgeon Dr Pierre Barbet is among those who view the cloth at close quarters on this occasion.

1939   September. For safety, following the outbreak of World War Two, the Shroud is secretly moved from Turin to the Benedictine Abbey of Montevergine in southern Italy. Even at the Abbey only four of the monks know what they are protecting.

1946   In the aftermath of World War Two the Italian people vote for a republic, ending the very brief rule of King Umberto II of Savoy. Umberto goes into exile in Cascais, Portugal, never to return to Italy, or ever to see the Shroud again. However, although the Shroud Chapel, to which the Shroud is returned in this same year, has become a property of the Italian state, Umberto tacitly remains the cloth's legal owner.

1955   At the instigation of Group Captain Leonard Cheshire, VC, the Shroud is brought out of its casket to be briefly held by a crippled ten-year-old English girl, Josephine Woollam. However, it is not unrolled on this occasion.

1969   June. On the orders of Turin's Cardinal Michele Pellegrino, the Shroud is secretly taken out of its casket and studied by a small team of specialists. The first ever colour photographs of the cloth are also taken on this occasion.

1972   1 October. An unknown individual, after climbing over the Turin Palace roof, breaks into the Shroud Chapel and unsuccessfully attempts to set fire to the Shroud.

1973   22 November. In a specially prepared room in Turin's

Royal Palace, the Shroud is exhibited before a limited gathering of interested specialists and journalists as a preliminary to being filmed for colour television the next day. This is the only known occasion of its being displayed vertically rather than horizontally. 23 November. The Shroud is filmed for the first time for colour television. 24 November. A commission of experts secretly examines the Shroud, and a sample of its linen is snipped from one corner for study by Belgian textile expert Professor Gilbert Raes.

1978 26 August–8 October. A public showing of the Shroud in Turin Cathedral, the first since 1933. This is attended by some 3 million pilgrims. 8–13 October. Intensive scientific examination of the Shroud in a specially prepared suite in the Royal Palace. Some twenty-four American scientists and specialists, known as the Shroud of Turin Research Project (STURP), participate and take samples.

1979 24–25 March. The STURP team announce their preliminary findings that the Shroud's image was not created by an artist.

1983 18 March. Umberto II of Savoy dies at his home in Portugal. It is quickly learned that he has bequeathed the Shroud to the Pope and his successors. For the first time in its known history, the Shroud thereupon becomes formally owned by the Catholic Church.

1988 21 April. In the presence of the heads of the Arizona, Oxford and Zurich radiocarbon-dating laboratories, Turin scientist Professor Giovanni Riggi snips a sliver from the top left-hand corner of the Shroud. This is then carefully divided up so that each laboratory receives the agreed amount of Shroud sample. 6 May. The Arizona laboratory's Shroud sample produces a date of 1350. In the ensuing months the Zurich and Oxford laboratories arrive at similar dates. 13 October. At press conferences held near simultaneously in London and Turin the

Shroud is declared to date to between 1260 and 1390, according to the findings of the three radiocarbon-dating laboratories.

1990   May. Chunks of masonry falling from the roof of Guarino Guarini's Shroud Chapel cause the chapel to be closed for repairs.

1993   24 February. The Shroud, still in its traditional casket, is removed from its normal home in the Shroud Chapel and installed in a new, temporary display case behind Turin Cathedral's high altar.

1997   11 April. A major fire, probably the work of an arsonist, breaks out in the Shroud Chapel, now in the last stages of its refurbishment. Although the Shroud is not in its normal location it is in danger even in its temporary showcase, which has to be smashed by firemen in order to remove it to safety. 14 April. At the residence of Turin's Cardinal Saldarini the Shroud's condition is checked by a hastily assembled team of experts. It is found to have been unharmed by the fire, and is taken to a secret location. 25 June. New, high-definition photographs of the Shroud are taken by Turin professional photographer Giancarlo Durante.

1998   15 April. The Shroud is brought from its secret location to a hastily built new sacristy adjoining Turin Cathedral. Textile conservator Dr Mechthild Flury-Lemberg and nun Sister Maria Clara Antonini together remove the cloth's nineteenth-century blue satin surround and sew it onto a new white cloth. 17 April. The Shroud is installed in a new conservation case. 18 April. Commencement of a public showing of the Shroud, commemorating the centenary of Secondo Pia's discovery of its photographic properties. 24 May. Pope John Paul II visits Turin Cathedral, prays before the Shroud, and delivers a homily on it. 4 June. The conclusion of the public showings of this year.

2000   2–5 March. A symposium is held at Turin's Villa
        Gualino at which the invited delegates are encouraged
        to 'identify and examine in depth ... any questions
        which may become the subject of future research ... in
        view of a possible future campaign of studies and
        research'. 12 August–29 October. Showings of the
        Shroud to celebrate the Jubilee Year of 2000, attended
        by a million and a half visitors. Afterwards the Shroud
        is transferred to a new conservation case built by Italian
        aerospace company Alenia Spazio. 8 November. In the
        'new' sacristy, a large proportion of the as yet never
        fully seen underside of the Shroud is scanned, using a
        specially adapted flat scanner guided by delicate plexi-
        glass rods. 10 November. A proposal document is
        drawn up proposing the 'removal of the Shroud's
        Holland cloth and patches ... fixing the edges of the
        burn holes; application of a new lining'.

2002   21–25 June. Textile conservators Dr Mechthild Flury
        Lemberg and Irene Tomedi commence implementing the
        programme of conserving the Shroud which Pope John
        Paul II, as the Shroud's legal owner, had formally
        approved the previous November. First they remove the
        backing cloth and patches that had been sewn on to
        the Shroud after the fire of 1532, then stretch the wrin-
        kles on the cloth's underside. This work is followed by
        photography, spectrophotometry and full scanning of
        the Shroud both front and back for the first time. Then
        from 16 to 23 July the two conservators sew on a new
        backing cloth, followed by fresh photography of the
        results, also measurement of the Shroud's dimensions in
        its new form. The Shroud is then returned to its case in
        the left transept of Turin Cathedral, its new 'permanent'
        home following the chapel fire. 9 August. Italian
        journalist Orazio Petrosillo leaks news of the conser-
        vation work in the newspaper *Il Messagero*. This creates
        a media furore. 20–21 September. Turin's Cardinal
        Severino Poletto holds a press conference to present the

conservation work done on the Shroud. After this occasion controlled numbers of those attending are allowed to view the Shroud in its conservation container.

2005   2 April. Death in Rome of Pope John Paul II, the Shroud's legal owner since the death of ex-King Umberto of Italy back in 1983. 19 April. Election of former Cardinal Joseph Ratzinger as Pope Benedict XVI, who becomes the Shroud's new formal owner.

2008   22 January. The specialist Italian photographic company HAL9000 make a multi-section very high resolution digital photograph of the Shroud, which has been made available to them in the new sacristy of Turin Cathedral. 24 January. High Definition filming of the Shroud for the BBC by a British TV team led by producer David Rolfe. June 2. From Rome, Pope Benedict XVI announces a fresh exposition of the Shroud to be held in the spring of 2010.

Notes and References

Chapter One: The Shroud on View

1 The official measurement is 442.5cm × 113.7cm. See Ghiberti, 2002, paragraph 20.

2 Back in the early seventeenth century, Italian artist Giovan Battista della Rovere illustrated this beautifully in an aquatint in which, below a depiction of the angel-borne Shroud, he painted Joseph of Arimathea and Nicodemus wrapping Jesus's body in just this manner after it was brought down from the cross. See Beldon Scott, 2003, pl. 4.

3 Personal interview with Hall at his Oxford laboratory, July 1988.

4 Vikan, 1998.

5 See Fossati, 1984.

6 Barta et al, 2009. Noalejo is a small village in the province of Jaén, in the south of Spain.

7 For some good conservation reasons, the cloth used was actually several decades old.

8 Meacham, 2005.

9 ibid., p.157.

10 Interview with the author, 4 July 2009.

Chapter Two: The World's Most Mysterious Image

1 Some of the new information given here comes from Gian Maria Zaccone, 'Photography and the Shroud over the years', in Zaccone (ed.), 2001, pp.63–95.

2 Chevalier, 1902; Thurston, 1903.

3 Giuseppe Enrie, *La Santa Sindone Rivelata Dalla Fotografia*, Torino, 1938, as translated in Walsh, 1965, p.98.

4 Rinaldi, 1983, p.4.

5 The Italian text of Pope Paul VI's televised address was published in *Rivista Diocesana Torinese*, no.12, 1973, pp.465–6.

6 Quoted in *Amateur Photographer*, 8 March 1967.

7 In 1969, by Italian photographer Gian Battista Cordiglia.

8 Dr John Jackson, interviewed in the 1978 British-made television documentary *The Silent Witness*.

9 Schumacher, 1999.

10 Ibid.

11 The letters stand for Shroud of Turin Research Project.

12 The address is www.shroud.com

13 See chapter 20.

14 Durante, 'The Photographs of 1997 and 2000', in Zaccone (ed.), 2001, pp.109–18.

15 *Time* magazine, 20 April 1998.

16 Ghiberti, 2002.

17 This can be accessed at www.haltadefinizione.com

18 *Shroud of Turin*, BBC2, 23 March 2008.

19 David Rolfe of Performance Films kindly made the HD footage available for me.

20 Interview with the author, 4 July 2009.

21 Craig and Bresee, 1994.

22 Picknett and Prince, 1994.

23 Allen, 1995.

24 *Unravelling the Shroud of Turin*, shown on the History Channel in 2005. Although Heckman was associate

producer for the programme, network executives and senior producers apparently chose to ignore his findings, declaring the medieval camera theory 'possible', whereas Heckman had found it 'nearly impossible'.

25 Email from Sean Heckman to Barrie Schwortz, 9 January 2008.

Chapter Three: Home to a Real-Life Body

1 Dr Robert Bucklin, speaking in the TV documentary *The Silent Witness*, Screenpro Films, 1978.

2 From the reconstruction of Delage's lecture, based on his open letter in *Revue Scientifique* 4, vol. 17 (1902), pp.683–7, and official report in *Comptes Rendus Hebdomadaires des Séances de l'Académie des Sciences*, 134 (1902), pp.902–4, in Walsh, 1965, p.86.

3 *Revue Scientifique* 4, vol. 17 (1902), pp.683–7, from Walsh, 1965, pp.90–1.

4 The translation from the French given in Barbet's book has been changed from the word 'outline'.

5 Barbet, 1954, pp.25–6.

6 As note 1.

7 Ibid.

8 From unpublished notes written by Dr David Willis shortly before his death, in the collection of the author.

9 As note 1.

10 As note 8.

11 Alfonso Paleotto, 1598. For discussion, see chapter 19 of this book.

12 In English-language anatomy, it is simply the junction of the hamate, capitate, triquetral and lunate bones.

13 Bucklin, 1961, p. 9.

14 As note 8.

15 Dr Willis was writing when the repair patches were still in position. There are now only the triangular holes which the patches formerly covered.

16 As note 8.
17 And see Wilson, 1998, pl. 22b.
18 This was the case, for instance, with the Poor Clare nuns who mended the Shroud after the fire of 1532.
19 Zugibe is author of *The Crucifixion of Jesus* (see bibliography).
20 See for instance Hoare, 1978, and Kersten and Gruber, 1994.
21 Pierluigi Baima Bollone, 'Interpreting the Image of the Shroud', in Zaccone (ed.), 2001, pp.119–26.
22 Svensson, 2007.
23 See for instance Straiton, 1989, and Knight and Lomas, 1996.

Chapter Four: Window on the Passion

1 Thurston, 1903, p.19.
2 Luke 22:63.
3 Mark 15:19. Other examples are to be found in Matthew 27:30 and John 19:3.
4 Mark 15:17–18.
5 Jerusalem, because of its elevation, could be cold at night. Later in the day Jesus's disciple Peter is specifically described as warming his hands by a fire (John 18:25). And because of a shortage of timber, dry thorn-branches would have been a logical choice either as tinder or an alternative fuel.
6 John 19:1.
7 Matthew 27:26.
8 Philo, *In Flaccum* 10:75, quoted in O'Rahilly, 1985, p.108.
9 2 Corinthians 11:24.
10 John 19:17.
11 Luke 23:26.
12 In the original New Testament Greek, the word used for 'hand' included the area of the wrist.

13   John 20:25.

14   See translation in Josephus (tr. Williamson), 1981, pp.389–90.

15   Haas, 1970.

16   Zias and Sekeles, 1985.

17   John 19:34.

18   Alternatively, 'folded up'.

19   John 20:7.

20   John 20:8–9.

21   John, chapter 11.

22   Matthew 27:35, Mark 15:24, Luke 23:34, John 1:23.

23   Victor Tunkel, 'A Jewish View of the Shroud', lecture to the British Society for the Turin Shroud, London, 12 May 1983.

24   The surviving code is sixteenth-century, but recognized to embody laws and practices stretching back to ancient times.

25   Gansfried, 1927, vol.IV, ch.CXCVII, Laws Relating to Purification (Tahara nos 9 and 10), pp.99–100.

26   Dr John Robinson, interviewed in the TV documentary *The Silent Witness*.

Chapter Five: Under the Microscope

1   Adler, 'The Shroud Fabric and the Body Image: Chemical and Physical Characteristics', in Scannerini and Savarino (eds), 2000, p.65.

2   Caramello (ed.), 1976.

3   Mottern et al, 1979.

4   Miller and Pellicori, 1980.

5   McCrone, 1976.

6   McCrone, 1980, and McCrone and Skirius, 1981.

7   Heller and Adler, 1981.

8   Heller and Adler, 1980.

9   Cahill et al, 1987.

10   From a report to British film producer David Rolfe, January 1977.

11   Danin et al, 1999.
12   I accompanied him on one of these, in 1977.
13   Kohlbeck and Nitowski, 1986, p.23.

Chapter Six: The Cloth's Own Tale

1   Flury-Lemberg, 2001, p.60.
2   Timossi, 1941/1980.
3   Raes, 1976.
4   Vial, 1989.
5   442.5cm × 113.7cm. See Ghiberti, 2002, paragraph 20.
6   Timossi, 1941/1980, summarized in Crispino, 1991, p.22.
7   The selvedge is the weaver-finished edging at the left and right sides of a piece of fabric as it comes off the loom. The weaver usually binds these edges to prevent any fraying or unravelling.
8   John 19:23.
9   Flury-Lemberg, 2001, p.58. Slight editorial changes have been made to improve the translation.
10   Sheffer and Granger-Taylor, 1994, pp.210–11, figs 111–13; p.169, fig.16.
11   Vial, 1989.
12   Technically a 45° diagonal, comprising an uneven her-ringbone in irregular longitudinal bands about 11mm wide.
13   De Jonghe, 1980–8, pp.81–92 (in Flemish).
14   Sheffer and Granger-Taylor, 1994.
15   Email to the author, 30 April 2009.
16   Flury-Lemberg, 2001, p.70, and Guerreschi and Salcito, 2005.
17   Zaccone (ed.), 2001, p.92, figs 14, 15; p.142, fig.7. This seems to have been done for the benefit of the inaugural meeting of the conservation committee which brought out the Shroud for non-invasive study in that year.
18   Flury-Lemberg, 2003, pp.42–7.

19   Soric, A. Rendic-Miocevic et al, *Katalog Pisati
     Etruscanski*, Muzej MTM Zagreb, 1986.
20   *Secrets of the Dead IV: The Shroud of Christ*. This was
     made for Britain's Channel 4, and WNET PBS in the
     USA.

Chapter Seven: What's in a Date?

1   William Meacham, 'Thoughts on the Shroud 14C
    Debate', in Scannerini and Savarino (eds), 2000,
    pp.443–4.
2   Michael Sheridan and Phil Reeves, 'Turin Shroud Shown
    to be a Fake', *The Times*, 14 October 1988, p.1.
3   Published in a communication by British Shroud
    researcher Vera Barclay in *Sindon* (Journal of the Centro
    Internazionale di Sindonologia, Turin), December 1961,
    p.36.
4   The events leading up to the carbon dating are excellently
    chronicled and documented in Gove, 1996. But see also
    Sox, 1988.
5   The letter is reproduced in full in Gove, 1996,
    pp.213–14.
6   Riggi, with what he subsequently claimed to have been
    Cardinal Ballestrero's approval, kept this for his own
    research purposes, hence at this stage Dr Mechthild
    Flury-Lemberg had no
    opportunity to make her subsequent findings from study
    of this seam.
7   Damon et al, 1989.
8   Kersten and Gruber, 1994.
9   Marino and Benford, 2000.
10 The dorsal half was not, being above my head, because
    of the upright display mode.
11 The opposite end of the cloth was not, being well above
    my head because of the upright mode of display on that
    occasion.

12   Flury-Lemberg, 2007.

13   See note 3.

14   Bowman, 1990, p.27.

15   Ibid., p.56.

16   Personal communication to the author.

17   Tyrer, 1989, p.51.

18   Email communication from Professor Stephen Mattingly to
      the author, 8 April 2009: 'Human skin is shed continuously
      including epithelial cells and attached bacteria. So every-
      thing we touch or rub against, we leave behind a trail of
      skin and microbes. Now, these epithelial cells and microbes
      would not grow on these surfaces, but they would serve as
      a source of nutrients for other microbes in the area. So a
      varnish of microbes will build up. We see this on desert
      rocks (desert varnish) and other rock surfaces.'

19   Stephen Mattingly, email communication to Dr John
      Jackson, copied to the author, December 1999.

20   For a fuller account of this, see Garza-Valdes, 1999,
      chapter 9.

21   Professor Harry Gove, letter to the author, as published
      in the British Society for the Turin Shroud's *Shroud
      Newsletter*, no.40, May 1995, pp.20–2.

22   California-born Dr Loy, distinguished for his work on the
      famous Iceman, worked at the University of Queensland's
      Centre for Molecular and Cellular Biology. Sadly, he died
      in 2005.

23   David, 1978.

24   See the archaeological journal *Current Archaeology*,
      August 1986.

25   Johnson et al, 1985, cited in Meacham, 'Thoughts on the
      Shroud 14C Debate', in Scannerini and Savarino (eds),
      2000, p.444.

26   See note 1.

27   Interview with Professor Christopher Ramsey, Oxford,
      Tuesday 26 May.

28   Lloyd A. Currie, vol.109, March–April 2004, pp.203–4.

Chapter Eight: 'The Saviour's Likeness Thus Imprinted'

1   See Chevalier, 1900, for his assemblage of the most
      relevant documents.
2   Thurston, 1903, p.29.
3   Paris, Bibliothèque Nationale, Collection de Champagne,
      vol.154, fol.138.
4   *Le Livre de Chevalerie*. There are two main manuscripts,
      one in Belgium, in the Royal Library, Brussels, ref. no.
      1124–26; the other in the Bibliothèque Nationale, Paris,
      new French acquisitions no.4736. See Kaeuper and
      Kennedy, 1996.
5   The *Livre Charny*. See Taylor, 1977.
6   Friedlander, 2006, p.462.
7   Archives of Aube I 17, Nicolas Camuzat, *Promptuarium*,
      fol. 422 verso.
8   John 21:25.
9   What remains of this is enshrined in the present-day
      Church of the Holy Sepulchre in Jerusalem. Despite the
      so-called 'Garden Tomb', for which there is a popular
      following, there is a good case for the Church of the
      Holy Sepulchre housing the authentic site. For the latest
      and best scholarly study, see Biddle, 1999.
10   For an excellent discussion, see Thiede and d'Ancona,
      2002.
11   Exodus 20:4.
12   The original manuscript is in the Royal Library,
      Copenhagen, as MS 487; the relevant page is fol.123. For
      the definitive modern-day edition, see Lauer, 1924/1956.
      For a useful translation into English, see McNeal, 1936.
13   Green, 1969.
14   St Jerome, *De Viris Illustribus*, chapter 2.
15   Wardrop and Conybear, 1900, vol.V.
16   Tobler, 1879.

17 St Bede, *De locis sanctis Adamni Abbatis Hiliensis*, Book III, *De locis sanctis, ex relatione Arculfi Episcopi Galli*, X and XI, Acta Sanctorum Ordinis Benedicti, IV, p.456ff.

18 Peter Mallius, *Descriptio Basilicae Vaticanae*, XXV, ed. De Rossi, Inscript. Christ, II, 1,218, n.90.

19 See Wilson, 1991.

Chapter Nine: 'Blessed City'

1 Quoted, except for modernization, from Segal, 1970, p.73.

2 Addai is the Syriac equivalent of 'Thaddaeus'. There was a Thaddaeus among Jesus's twelve disciples, but according to some sources the Edessan Addai was one of the seventy 'outer circle' of Jesus's disciples. Whichever was the case, all traditions agree that this Addai was contemporary with Jesus.

3 Eusebius, tr. Williamson, 1965.

4 Matthew 4:24, Mark 3:7–8, Luke 6:17–18.

5 This exists in two manuscripts. One, from the early fifth century, was discovered in 1848 by the British scholar William Cureton among caravan-loads of manuscripts that had recently been rescued from the Nitrian monastery in the desert of lower Egypt (see the posthumously published Cureton, 1864). The second manuscript, from the sixth century, is more complete, and exists in St Petersburg, Russia. It was translated into English by the nineteenth-century scholar George Phillips (see Phillips, 1876).

6 Guscin, 2009, p.27. Guscin's word 'likeness' has here been replaced by 'Image' to clarify the sense intended.

7 Tacitus, *Annals*, VI, 31.

8 Just to add to the confusion, some historians call him Abgar IX. I have here followed J. B. Segal, author of the most authoritative study of Edessa's history.

9 This chronicle was compiled in the sixth century from

earlier records of which the extract quoted here seems to have been by an eyewitness contemporary. See Cowper, 1864–6.

10   As noted by Segal, the word in the original Syriac is *haikla*, which can also mean 'shrine'.

11   Brock, 2004, p.227.

12   British Museum Catalogue of Greek Coins, pl.XIII,14. Ashmolean Museum Roman Provincial Coinage database temporary number 6491.

13   Email to the author, 12 March 2009.

14   Note that the German New Testament scholar Professor Ethelbert Stauffer pointed this out as early as half a century ago, but has been largely ignored (see Stauffer, 1955).

15   This is all too typical of many of the displayed antiquities in the Şanliurfa Museum. From discussions with the museum director, it appears that local people sell their occasional archaeological finds to the museum on a 'no questions asked' basis, hence the lack of any satisfactory information concerning their provenance.

16   Memory of the location was long preserved by a monastery later built on the site, that of Jacob of the Naphshatha, meaning 'Jacob of the Tomb Tower'. Today the Turks know the site as Deyr Yakup. Mark Guscin and I were most generously and painstakingly led to it by friendly Turks in May 2009. Among many other sites in and around Şanliurfa, the ruins of both the monastery and the earlier tomb complex need proper archaeological investigation.

17   Voobus, 1958, p.7.

18   Moffett, 1992, p.80, n.9.

19   It is notable that Abgar's accompaniment by his apparently pro-Christian son Ma'nu readily corresponds to the Abgar V/Ma'nu V chronology of Edessa's disputed first-century evangelization rather than the Abgar VIII/Abgar IX

succession of a late second-century one.

20   For the full text of Egeria's description of her visit to
      Edessa, see Wilson, 1978, revised ed. 1981, pp.115–17.

21   See his comments to Emperor Constantine the Great's
      sister Constantia in Farrar, 1901, p.56.

22   For example, a mid-third-century Christ as Shepherd in
      Rome's Hypogeum of the Aurelians, and a fourth-century
      example in the Catacomb of Commodilla, Rome.

23   St Augustine, *De Trinitate VIII*, 4, 5, in J. P. Migne,
      Patrologia Latina, vol.42, 1844–55.

Chapter Ten: Rediscovered – as 'King of Kings'

1   This was Kavadh I, born 449, ruled 488 to 531.

2   Trombley, 2000, p.78.

3   As the Byzantine commander Areobindus taunted the
      frustrated Kavadh, 'You have now seen in (your own)
      experience that the city . . . is the city of Christ, who has
      blessed it and has stood against your forces so that they
      may not take control of it.' Ibid., p.79.

4   This was Khusraw I, reigned 531 to 579. Another
      anglicized form of his name is Chosroes.

5   Translation from Whitby, 2000, pp.226–7.

6   Guscin, 2009, p.33.

7   Ibid., pp.35–7.

8   And such a consideration was certainly also in the mind
      of the author of the *Story of the Image of Edessa*: 'I
      would think that
      the priest gave orders to put a tile in front of the Image
      so that no decay from the edifice's mould and no damp
      from the gypsum could make the cloth with the image on
      it deteriorate and suffer the damage done by time'
      (tr. Guscin, 2009, p.35).

9   Karaulashvili, 2007, pp.223–4 and figs 1–3.

10   The reference number is N/Sin-50. For discussion by a
      Georgian scholar, see Karaulashvili, 2007, pp.224–5.

11   Ibid.
12   Certain scholars, most notably J. Chrysostomides (1997),
     have suggested the passage in Evagrius is a later
     interpolation. For a more balanced view, see Whitby,
     2000, Appendix II.
13   This seems to have been a duplicate of the one found
     above Edessa's gate. Because there were times when
     Hierapolis came under
     Edessa's jurisdiction, it is very possible that Abgar also
     ordered a copy for its gate.
14   Today, Hierapolis is dusty, nondescript Membij in
     northern Syria, a pale shadow of its former self as a
     major cult centre, first for the pagan goddess Atagartis,
     then as a sister Christian city to Edessa. Another name
     for it was Mabbog.
15   Taylor, 1933. The face can be seen in the lower
     photograph (2) on plate XI, opposite p.103.
16   I am deeply indebted to Cyprus lawyer George Apostolou
     for his encouraging Mark Fehlmann to make a difficult
     re-entry in a location apparently today infested with
     snakes.
17   The restoration of the mosaic was conducted by the artist
     Toriti during the pontificate of Pope Nicholas IV. The
     inscription recording Toriti's work states that he 'had the
     sacred face of Our Saviour replaced intact in the place
     where it first miraculously appeared to the people of
     Rome when this church was consecrated'.
18   See Oakeshott, 1967, p.70 and pl.VIII.
19   See Pacht, 1961, p.407.
20   This happened in 754 when Rome was being threatened
     by Lombards.
21   Ware, 1963, p.233 (1974 edn).
22   Guscin, 2009.
23   Karaulashvili, 2002. See especially the chart on p.100.
24   Vignon, 1939, pp.131–9.

25   This occurs in a Syriac hymn or *sougitha*. Although some
      modern scholars, such as Andrew Palmer (1988), have
      disputed an allusion to the Image this early, their doubts
      are now dispelled by the evidence of the Georgian
      manuscripts referred to earlier.

26   This villa seems to have been built shortly after the flood
      of 525 because the mosaic flooring lies in the area
      through which the Daisan ran before its re-routing.
      Archaeologists Dr Selcuk Sener and Dr Mehmet Önal
      pointed out several features of the mosaics indicative of
      their sixth-century date.

27   For the latest translation of the hymn, see Palmer, 1988;
      also Dupont-Sommer, 1947, and Grabar, 1968.

28   The letter is included in a three-part edition of Mar
      Ishuyah's Pastoral Epistles. Ishuya, or Isho-yahb, was
      head of the Nestorians between the years 620 and 658.

29   For this reference I am deeply indebted to Mar Gewargis
      Sliwa, Archbishop of Baghdad for the Assyrian Church of
      the East, and Mar Ishuyah's present-day successor.

Chapter Eleven: Trial by Fire

1   Today the small Turkish town of Suruç.

2   The relevant chronicle, that of Dionysius of Tell-Mahre,
      is reconstituted in Palmer, 1993, p.195.

3   Dietz, 1998.

4   In the original Latin, 'sacrum quod habetis lintoleum date
      in mea manu' (quoted in Dietz, op. cit.).

5   Adamnan, *De Situ Terrae Sanctae*, chapter VI, translation
      based on that in Beecher, 1928.

6   An alternative suggestion, argued by Dr Mechthild Flury-
      Lemberg and supported by Dr Karlheinz Dietz, is that the
      holes were caused by a highly corrosive liquid. Dietz has
      plausibly suggested this was the mystery ingredient used for
      'Greek fire', which Mu'awiyah would have experienced at
      first hand during an attempt he had earlier made to capture

Constantinople. Analysis of the carbonized particles taken from around the burn-holes in 2002 may ultimately throw some light on the substance responsible for the burns, but the fundamental interpretation of some kind of 'trial by fire' remains unaffected.

7   Manuscript 266 of the Wadi'n-Natrun collection, from Abd al-Masih, 1947, quoted in Guscin, 2009, p.160. The copy, from an earlier manuscript, was made by the monk Gabriel in Cairo in 1255. The translation has been slightly modified by the author.

8   In the original Latin, 'ante annos ferme ternos'.

9   To clarify that the Image of Edessa is being referred to here, 'Image' has been substituted in place of Segal's word 'portrait'.

10 From Chabot (1899–1924), quoted in translation in Segal, 1970, p.214. Michael the Syrian prefaced this passage by quoting as his source Dionysius of Tell-Mahre: 'Dionysius . . . the historian of these things, says, "I have taken these things from the account of Daniel . . . my maternal grandfather."' According to Drijvers (1998, p.21), Daniel was Daniel bar Moses of Tur Abdin.

11 On exactly where the Image was kept within the Baptistry, a later Armenian chronicle cryptically added this concerning Athanasius: 'He built two churches, one dedicated to the Mother of God, the other to Saint Theodore . . . Between these two churches he constructed underground chapels, and purchased the *sudarium* of Christ, which had become the property of Dadjiks [Arabs], for 50,000 *tahegans*. And he deposited it in this place, where the faithful went up and down via a staircase. The day of the feast [of the *sudarium*] he had it exposed to the gaze of helpers.'

12 Translation from Sendler, 1988, p.20.

13 For an excellent in-depth discussion of this coin issue, see Breckenridge, 1959.

14 Ibid., pl.IX, fig.38.

15 This was one of a number of copycat imprints of Christ rivalling that of Edessa. According to the story, around 554 a woman of Camuliana in Cappadocia discovered an image of Jesus on linen cloth in a water cistern. It quickly acquired its own cult status.

16 See Thompson, 1962.

17 The word used was the Greek *higoumenos*, usually translated as 'abbot'.

18 The word used was the Latin term *chartularius*, keeper of the charters and other official documents.

19 John of Damascus, *De Fide Orthodoxa*, chapter 89, translation from Guscin, 2009, p.152.

20 Mansi 13, 1758–98, 192C, translation from Guscin, 2009, p.179.

21 See, for instance, the tenth-century al-Masudi's reference to it as one of the four wonders of the world, in Sprenger, 1841. The tenth-century Persian geographer Ibn-al-Faqih also wrote 'no other monument in stone can surpass in beauty the church of Edessa'.

22 The great German scholar Ernst von Dobschütz referred to it as the Liturgical Tractate, and treated it as a separate document from the *Story of the Image of Edessa*. However, Mark Guscin found it incorporated with the *Story* in several of the Mount Athos manuscripts, and accordingly presented it in this way in his 2009 book *The Image of Edessa*. The text is to be found in this book between pages 61 and 69.

23 Guscin, 2009, p.65.

24 Mango, 1972, p.184, based on the text in *Anthologia Graeca*, 1.106, ed. H. Beckby, Munich, 1957, 1:152–3.

Chapter Twelve: Baghdad Surrender

1 An eight-volume biography was written about his career, which has not survived.

2   See Bowen, 1928, pp.380–2, and Vasiliev, 1968, pp.296–300.

3   Guscin's translation in Guscin, 2009, p.47.

4   Guscin, 2009, pp.47–9.

5   The present-day Turkish town of Samsat, built to accommodate those displaced, is on higher ground in a different location.

6   Translation based on Guscin, 2009, p.49.

7   Ibid., p.51.

8   An Avar attack in 626, followed by a Russian attack in 860.

9   For the source text, see Belting, 1994, p.511.

10 Translation based on Guscin, 2009, pp.55–7.

11 According to Maguire (1997, p.55ff), 'The church of the Virgin of the Pharos was located close to the imperial apartments and the Chrysotriklinos, the throne room of the Daphne palace complex. It was a small building with a central ribbed dome, three apses, a narthex, and a beautiful atrium. According to Photios, the church was of such great beauty that it could leave the spectator petrified with wonder. The revetment of its exterior façade, completely of white marble, was so well joined that it seemed to be of one piece. The interior was lavishly covered with colorful marbles and gold mosaic *tesserae*. Gilding and silver sheathing decorated the church furnishings, and the altar table was encrusted with precious stones and enamels. When entering, one had the feeling that one had entered heaven itself; a divine and venerable second palace for the Mother of God on earth.'

12 Guscin, 2009, p.57.

13 Ibid., pp.57–9.

14 Ibid., p.61.

15 Migne, *Patrologia Graeca*, CIX, 812–13, translation in Guscin, 2009, p.180.

16 Runciman, 1931, p.250

17 For example, Michael III, Basil I and Leo VI.
18 Csocsán de Várallja, 1987, p.3.
19 See Ahrweiler, 1967, and Mazzucchi, 1983.
20 For source, see Dubarle, 1985, p.55, note 56.

Chapter Thirteen: Not for Common Gaze

1 The Alexandria-Michigan-Princeton Expedition, jointly sponsored by the Universities of Alexandria, Michigan and Princeton.
2 See Weitzmann, 1960 and 1976, pp.94-8.
3 Paul of Thebes and Antonios.
4 Basil and Ephrem.
5 The mural recently found during excavations at the monastery of Deir al-Surian, Egypt, may be a still earlier example. See Van Romfay, 2002. However, besides the dating being far from established, the depiction of the Image in this example is too fragmentary to bear meaningful comparison.
6 See Cormack, 1997.
7 Quoted in Gibbon, Pelican edition, footnote p.625.
8 The manuscript reference number is Cod. Syn. Gr. 183.
9 Yenipinar, 1998.
10 The translation given here is a conflation of Gibbon (Pelican edition, 1963, pp.624-5), and Guscin (2009, pp.135-7).
11 Guscin, 2009, p.33.
12 Ibid., p.21.
13 Ibid., p.25.
14 The earliest usage of the word occurs in the works of the Alexandrian Patriarch Eutychius (877–940), in his *Book of the Demonstration of Eutychius*. See Cameron, 1984, p.90.
15 It has the inscription *to agion manden*.
16 Translation by the author, in order to show the specific usage of *sindon*.

17   Guscin, 2009, p.91. The Greek words used are *ton somatikon autou charaktera*.

18   Vindobonensis [Vienna] bybl Caesar. hist. gr. 45 (olim 14). In this the words used are *panta autou ta mele*.

19   Codex Vossianus Latinus Q69, ff 6r–6v. Translation, with accompanying Latin text, in Guscin, 2009, p.207.

20   Ordericus Vitalis IX, 13. Translation from Chibnall, 1969–80. The original text reads: 'Abgarus toparcha Edessae regnavit; cui dominus Iesus sacram epistolam destinavit, et pretiosum linteum, quo faciei suae sudorem extersit, et in quo eiusdem Salvatoris imago mirabiliter depicta refulget; quae Dominici corporis speciem et quantitatem intuentibus exhibet.'

21   Because none of them is likely to have been genuine (see p. 117–18), the point is purely academic.

22   From the eleventh-century Byzantine chronicler Georgios Cedrenos's account of the reign of Michael and Zoe (see von Dobschütz, 1899, p.176), as quoted in Weyl Carr, 1997.

23   See Weyl Carr, op. cit. The scene is depicted in a miniature from the Madrid Chronicle of John Skylitzes, cat. no.338, fol.210v, illustrated in Evans, 2004, p.11.

24   Graf, vol.2, pp.259–60.

25   This Virgin and Child mosaic we know to have been created in the ninth century, after the overthrow of iconoclasm.

26   Grelot, 1680. For an online photograph of this engraving, go to http://diglit.ub.uni-heidelberg.de/diglit/grelot1680/0166?sid=b59df4af3e65d79364840bfcc805ae1a&zoomlevel=4

27   This same otherwise baffling linkage of the Image of Edessa with the Annunciation is repeated in later Byzantine wall-paintings, most notably in the late twelfth-century church of St Nicholas tou Kastnitze, Kastoria, and the late thirteenth-century chapel of St

Euthymios, Thessaloniki, both in Greece.

28 Tbilisi Institute of Manuscripts Ms A-484.

29 Fol.320 verso. For further background, see Karaulashvili, 2007, pp.234–5.

30 For anyone who finds this unconvincing, today a similar 'token' photograph of the Shroud face accompanies the casket in which the Shroud lies normally hidden from human view in Turin (see pl. 2b).

31 Ciggaar, 1995. Translation in Guscin, 2009, p.182. This same informant adds a further intriguing detail to explain why the Image was so resolutely kept away from the prying eyes of ordinary mortals: 'The case that stored the holy object used to be kept open once, but the whole city was struck by continuous earthquakes, and everyone was threatened with death. A heavenly vision revealed that the city would not be freed of such ill until such time as the linen cloth with the Lord's face on it should be locked up and hidden away, far from human eyes. And so it was done. The sacred linen cloth was locked away in a golden case and carefully sealed up, and then the earthquake stopped and the heaven-sent ills ceased. From that time on nobody has dared to open the case or to see what might be inside it, as everyone believes and fears that if anyone tries to open it the whole city will be struck by another earthquake.' This appears to be a memory of the earthquakes which beset not Constantinople but Edessa in 679 and 717.

32 Weitzmann, 1961.

33 Berkovits, 1969, pl.III.

34 In the original Latin, *manutergium*.

35 In the original Latin, *sine . . . pictura*.

36 In the original Latin, *lintheamen et sudarium*.

37 From Belting, 1994, p.527, after Ciggaar, 1976, p.211ff and esp. p.245ff.

38 Riant, 1878, p.211.

39 The Greek text is in Heisenberg, 1907, p.30.
40 Translation from Phillips, 2005.

Chapter Fourteen: The Templar Secret

1   Kessler, 2000, p.36.
2   Ćurčić and St Clair (eds), 1986.
3   Millet, 1947, pl.CLXXVI/2.
4   Bozzo, 1974.
5   Richard Owen, 'Is This the Face of Christ?', *The Times*, 28 May 2004.
6   *Sainte-Pierre et le Vatican, L'Heritage des Papes*, Album Souvenir, Montréal, 2005, pp.60–4.
7   Runciman, 1931, pp.251–2.
8   Riant, 1878, II, p.135ff.
9   The engraving is reproduced in Morand, 1790, the plate facing p.40.
10 Riant, 1878.
11 See 'The Anonymous of Soissons' in Andrea, 2000, pp.223–38.
12 Michelet, 1841–51, vol.II, p.279.
13 From the Inquisition's *Articles against Singular Persons*, translated in Martin, 1928.
14 Deposition of Jean Taillefer, in Michelet, 1841–51, vol.I, p.190.
15 From the patois version of the king's instructions, quoted in Lizerand, 1923, pp.24–8.
16 Michelet, 1841–51, vol.I, p.399.
17 Ibid., vol.II, p.364.
18 Robert of Oteringham, senior of the Franciscan order, from Addison, 1842, p.272.
19 Nicholson, 2004, p.147.
20 Barber, 1982.
21 Lea, 1887, p.283.
22 See Nicholson, 2004, p.119, pl.4c.
23 Ibid., p.211, pl.8d.

24 Hugh de Peraud, Visitor General of the Order, described one 'idol' as having been supported by two feet at the front and two feet at the back. These may have been to enable a similar copy on a plaque to stand on an altar.

25 'The natural size of a man's head', from a deposition by Templar Jean Taillefer (Michelet, 1841–51, vol.I, p.190).

26 Rule 17 of the Templar Order.

27 Finke, 1907, vol.II, p.334.

28 *Chronicles of St Denis*, art.III, quoted in Du Puy, 1713, vol.I, p.25.

29 Paris, Archives Nationales, J 413 A, n.25, non-numbered pages, ninth page.

30 I have yet to study it. According to Dr Frale the document is not in a good state of preservation, and is difficult to read for anyone who is not familiar with the materials relating to the Templar interrogations. Content-wise it is apparently made complicated by the sexual 'confessions' that are mixed with the information concerning the 'long linen cloth', and Dr Frale rightly intends further study with a view to eventual specialist publication.

31 A similar ritualistic kissing was practised by the successive generations of the dukes of Savoy, when they came to own the Shroud.

32 Frale, 2009 (*Il Templari e la Sindone*).

33 The recent book by Christian Rollat (2008) may offer some clues, but has yet to be accessed by the author.

34 Raynouard, 1813, p.73.

35 Frale, 2004 (*Il Papato e il Processo*).

Chapter Fifteen: The Knight's Tale

1 See Frale, 2009, pp.114–15.

2 This information derives from Geoffrey's 'confession' to the papal commissioners at Chinon in the diocese of Tours, 17–20 August 1308. See Frale, 2004.

3   Barber, 1995, p.274.

4   Frale, 2009, p.116.

5   Ibid., p.117.

6   For much of the family background that follows, by far the most authoritative and reliable source is Contamine, 1992.

7   M. Prinet, 'Armorial de France Composé à la Fin du XIIIe Siècle ou au Commencement du XIVe' (1920), cited in Contamine, 1992, p.108.

8   *Chartularium Culisanense*, fol. no.126, copy created in 1858 and notarized by Monsignor Benedetto d'Acquisisto (1790–1867), Archbishop of Sicily. The original *Chartularium* was exhibited at the National Library of Palermo in 1910, but created little interest at the time, and was destroyed in 1943, a casualty of World War Two.

9   Frale, 2009, p.191.

10   When Otho de la Roche married his wife Elizabeth de Ray he automatically acquired the Ray title along with his own.

11   Piana, 2007, p.15.

12   Ibid., p.20.

13   Taylor, 1977.

14   *Dictionnaire de Biographie Française*, ed. M. Prevost and Roman d'Amat, VIII, Paris, Librairie Letouzey et Ane, 1959, p.614.

15   Contamine, 1992, p.108.

16   See Kaeuper and Kennedy, 1996, pp.5–6.

17   Contamine, 1992.

18   For an excellent account of Edward de Beaujeu's Smyrna expedition, see Setton, 1976, pp.188, 191 and 194.

19   Act of Clement VI, dated 7 July 1344, text in Savio, 1957, p.108.

20   Riley-Smith, 1991, pp.140–1.

21   Bibliothèque de l'Arsenal, Paris, manuscript no. 2251

(153 B.F.). For discussion, see Raffard de Brienne, 2001.

22 *Livre Charny* II, 594–845, printed in Taylor, 1977.

23 Bibliothèque Nationale, P.O.683, dossier Charny en Bourgogne, no.15492, item 5.

24 Bibliothèque de l'Arsenal, manuscript no.2251 (153 B.F.). This document specifically relates that Humbert's expedition took place 'the year after the courageous baron de Beaujeu, who was Marshal of France, Geoffrey de Charny, Bouciquaut, who was Marshal of France, Xaintre and the other brave knights had been in Smyrna'. For this information Raffard de Brienne, 2001, p.25.

25 See, for instance, Kaeuper and Kennedy, 1996, pp.40–1: 'how many of the problems associated with the Shroud would be resolved by the supposition that the cloth was of eastern manufacture, that Charny obtained it while on crusade in 1345–46, and that he considered it a splendid icon, an aid to pious devotion, rather than an actual relic from the life of Christ'.

26 Contamine, 1992, p.115.

27 For an excellent account of these episodes, see Kaeuper and Kennedy, 1996, pp.10–11.

28 For an estimation of date, see ibid., p.22.

29 Ibid., p.195.

30 Ibid., pp.44–5.

31 Quoted in Crispino, Sept/Dec 1988, p.30.

32 Contamine, 1992, p.115.

33 Ibid., p.116.

34 Paris, Bibliothèque Nationale, Collection de Champagne, v.154, fol.138. Translation from the Latin by the Revd Herbert Thurston in Thurston, 1903.

35 Letter to the author, 2 June 1992.

36 Matteo Villani, lib.I, cap.55. Translation from Thurston, 1900, p.58.

37 Contamine, 1973, p.225, n.5. Translation from Kaeuper and Kennedy, 1996, p.15.

38 This is an intriguing phrase. It may mean no more than that the body of Christ was present in the elements of the Mass. But it is also just conceivable that the Shroud was present at the ceremony. Because the Order of the Star became so quickly extinct – directly as a result of the heavy French losses to the English – we are poorly informed about many background details of the Order.

39 Contamine, 1973, p.211. Translation from Kaeuper and Kennedy, 1996, pp.15–16.

40 See chapter 8.

41 A conflation into one narrative of two slightly differing accounts by Froissart, as translated in Kaeuper and Kennedy, 1996, p.17.

42 Ibid.

43 'Le plus prudhomme et le plus vaillant de tous les autres' (Kervyn de Lettenhove, V, p.412).

Chapter Sixteen: Charny Bequest

1 Because even ten years later Geoffrey junior (whom we will call Geoffrey II) had not reached his age of majority he can hardly have been more than a toddler at the time of his father's death.

2 Murray, 1987, pp.16ff.

3 Perret, 1960, p.66, n.78.

4 Geoffrey was declared 'escuyer, moindre d'âge'. For original source, see Perret, 1960, p.67, n.3.

5 Setton, 1976.

6 Vatican Archives, Reg. Avign. 258, fo.468 verso, Latin transcript reproduced in Fossati, 1961, pp.196–8.

7 'Figuram sive representationem Sudarii Domini nostri Jhesu Christi'. As noted by Fossati (1983, p.27), 'The expression does not seem to embrace the idea of a reproduction by hand: we ourselves say "the imprint, the image on the Shroud".'

8 Paris, Bibliothèque Nationale, Coll. de Champagne,

v.154, fol.128, Latin transcript reproduced in Fossati, 1961, pp.198–200. Synthesized translation of the letter's main points from Scavone, 1993, p.218.

9 Paris, Bibliothèque Nationale, Coll. de Champagne, v.154, fol.130. Latin manuscript reproduced in Fossati, 1961, pp.200–3.

10 Froissart, tr. Brereton, 1968, pp.351–60.

11 Perret, 1960, p.69, after Chevalier, 1890, pièces justificatives A, B et F.

12 The original has not survived, only a draft copy.

13 This is described in detail, with sources, in Murray, 1987, p.129.

14 Archives of the Department of the Aube, 9 G4, quoted in Chevalier, 1900, as pièce justificative Q.

15 These included, besides those enumerated in the main text, a gold statue of the Virgin and child, silver statues of John the Baptist and the apostle John, a gilt-silver reliquary containing a piece of the True Cross, a crystal reliquary containing a part of a chemise of the Virgin Mary, and a gold reliquary containing a relic of St Laurence.

16 Hamilton, 2001.

17 Also known as the Buessarts chapel.

18 It carries two other names: 'Pré du Seigneur' or 'Meadow of the Lord', and 'Pré du Mauconseil' or 'Meadow of Bad Advice'.

19 Riedmattin, 2008.

20 For details, see Perret, 1960, pp.77–82.

21 Quoted in Chevalier, 1900, pp.8–10.

22 According to Perret (1960, p.86, n.152), the act is inserted in the account of 1453 of Jean Guyot, housekeeper of the chateau of Miribel, in the department of the Côte-d'Or B4440. Inventory summary of the Archives of the Côte-d'Or, Series B, vol.II, p.264.

Chapter Seventeen: The Wise Mule

1 Worth noting, just in case of a possible relevance, is that this Order, from its very foundation by Amadeus VI, Count of Savoy, back in 1362, was dedicated to the Virgin Mary as 'Our Lady of the Annunciation'. The collar carries a gold medallion depicting the Archangel Gabriel announcing Jesus's birth to Mary. As we have seen, Geoffrey I de Charny gave the same dedication to the Lirey church housing the Shroud.

2 This potentially significant church, the foundations of which are said to date back to the sixth century, was taken over by the Turks as a military base when they invaded northern Cyprus in 1974. It has been inaccessible ever since. My wife and I were quite literally stopped at gunpoint when trying to take a look at it in 1994.

3 Chevalier, 1902, p.16.

4 Chevalier, 1900, pièce justificative Z. See also Perret, 1960, p.90.

5 Sixtus IV, *De Sanguine Christi*, 1473, quoted in Chevalier, 1900, pièce justificative C.

6 See Bouquet-Boyer, 1981.

7 See Monnet, 1926, p.72.

8 Fabre, 1885, pp.55 and 90.

9 Quoted in Sanno Solaro, 1901, note on pp.39–40.

10 Chevalier, 1900, pièce justificative DD.

11 Ibid.

12 Bruchet, 1927, p.15, note 3, and p.139, note 4.

13 Rabelais, ed. Jourda, 1962, vol.I, p.110.

14 The excerpts from Louise de Vargin's account that follow – a conflation of various translations, particularly Crispino (March 1982) – are based on the original French text as published in Gervasio, 1974.

15 For background to the Shroud's travels during this period, see Fossati, 2000, pp.83–5.

16 Ferraris, 1960.

17   Fossati, September 1984, p.8.
18   The words are from Psalm 84:9. The original Latin on the banner reads 'Protector Noster/Aspice Deus/et Respice in Faciem/[Christi Tui]'.
19   Fossati, September 1984, p.9.
20   Don Juan gave the copy to Alcoy's Holy Sepulchre convent in 1574.

Chapter Eighteen: Backbone of a Dynasty

1   From the translation by Dorothy Crispino (March 1988, p.14).
2   Ibid., p.15.
3   Ibid.
4   Reproduced in Beldon Scott, 2003, p.30, fig.22.
5   Paleotto, 1598.
6   Translation from the Italian based on Crispino, 1987, pp.16–17.
7   Ibid.
8   Beldon Scott, 2003, p.68.
9   Ibid.
10   For a truly excellent study of such ducal thinking concerning the architectural projects associated with the Shroud, see Beldon Scott, 2003.
11   Jacopo Grimaldi, Biblioteca Nazionale, Florence, Ms II III 173, fol.106, reproduced in Wilson, 1991, pl.23 (top right).
12   Beldon Scott, 2003, p.169. For his sources, Beldon Scott refers to extensive materials in the manuscript collection of Savoy Court ceremonials preserved in the Royal Library in Turin.
13   Perret, 1960, p.114, after *Oeuvres de Saint François de Sales . . . Edition Complète*, vol.XVI, pp.2 and 177–9.
14   Turin Royal Library, 'Ceremoniale della Real Corte Savoia', Storia *Patria* 726/3, 1690–9, 187r–v. Translation from Beldon Scott, 2003, p.113.

15   Sanno Solaro, 1901.
16   An official record of this, with a sample of the former black silk lining, is preserved in Turin. See Fossati, 1993, for full record in Italian.
17   Fuller details of this episode can be found in Barberis, 1987.
18   One of the key intermediaries was Fr Peter Rinaldi, Italian-born parish priest of Corpus Christi, New York.
19   The will is reproduced in Italian in Fossati, 2000, p.270, after an interview given by Maria Gabriella of Savoy to journalist Maurizio Lupo, published in *La Stampa*, 18 April 1998, p.6.
20   According to the Pope's Polish secretary, Cardinal Stanislaw Dziwisz of Krakow, he accredited the Virgin of Fatima with saving his life following the assassination attempt.
21   Reproduced in colour in Fondazione Umberto II, 1998, p.13.

Chapter Nineteen: A Challenge to our Intelligence

1   Quoted in Fossati, 1990, p.13.
2   According to Fr Peter Rinaldi, who acted as an emissary between the exiled Umberto and those responsible for the Shroud, Umberto made his request conditional upon the Shroud remaining in Turin. See Rinaldi, 1983, p.4.
3   Quoted in Gove, 1996, p.201.
4   Ibid., pp.213–15.
5   The text of the full homily, from the official record of Pope John Paul II's visits to Turin and Vercelli, 23–24 May 1998, is on the Vatican website www.vatican.va
6   The official figure was 2,106,531. See Zaccone (ed.), 2001, p.57, note 9.
7   After Pietro Savarino, 'New Conservation Techniques', in Zaccone (ed.), 2001, p.145.

Chapter Twenty: On the Third Day

1 Belting, 1998, p.9.

2 Fondazione Umberto II, 1998, p.19

3 See, for example, Skhirtladze, 1998, pp.72–3.

4 Gerstel, 1999.

5 Karaulashvili, 2002; also her doctoral thesis for the Central European University, kindly made available to me by Dr István Perczel.

6 See Drijvers, 1982. The mosaic is no longer in evidence in Şanliurfa. The museum director simply shrugged his shoulders when he was asked its whereabouts.

7 See Balicka-Witakowska, 2004, p.118.

8 Walter, 1995, p.221.

9 Report by Vesna Ivanovska, Dvenik, Skopje, Macedonia, 11 April 2002.

10 Richard Owen, 'Missing Turin Shroud was in Knights' Keeping', The Times, 7 April 2009.

11 Guscin, 1998.

12 Wilson, 1998, pp.20–6.

13 Danin et al, 1999 and 2008.

14 Marion and Courage, 1997.

Bibliography

Abd al-Masih, Yassa, 'An Unedited Bohairic Letter of Abgar', *Bulletin de l'Institut Français d'Archéologie Orientale du Cairo*, 45, 1947, pp.65–80

Addison, C. G., *The History of the Knights Templars, the Temple Church, and the Temple*, Longman, Brown, Green, and Longmans, 1842

Ahrweiler, Hélène, 'Un Discours Inédit de Constantin VII Porphyrogénète', in *Travaux et Mémoires* (Centre de Recherche d'Histoire et Civilisation Byzantine), 2, 1967, pp.393–404

Allen, Nicholas P. L., 'Verification of the Nature and Causes of the Photo-negative Image on the Shroud of Lirey-Chambéry-Turin', *De Arte*, 51, UNISA, Pretoria, 1995, pp.21–35

Andrea, Alfred J., *Contemporary Sources for the Fourth Crusade*, Brill, Leiden, Boston & Köln, 2000

Balicka-Witakowska, Ewa, 'The Holy Face of Edessa on the Frame of the Volto Santo of Genoa: the Literary and Pictorial Sources', in Jan Olof Rosenqvist (ed.), *Interaction and Isolation in Late Byzantine Culture*, Papers read at a Colloquium held at the Swedish Research Institute in Istanbul, 1–5 December 1999, Swedish Research Institute in Istanbul, *Transactions* 13, 2004

Barber, Malcolm, 'The Templars and the Turin Shroud', in *The Catholic Historical Review*, April 1982

——*The New Knighthood: A History of the Order of the Temple*, Cambridge, Cambridge University Press, 1995

Barberis, Adolfo, 'The Secret Chamber: An Episode in Shroud History', in *Shroud Spectrum International*, 22, March 1987, pp.15–19 (translation by Dorothy Crispino from the original article in Italian in *Sindon*, no.1, October 1959)

Barbet, Dr Pierre, *The Passion of Our Lord Jesus Christ*, tr. the Earl of Wicklow, Dublin, Clonmore & Reynolds, 1954

Barta, César, José M. Orenga and Daniel Duque, 'The Noalejo Shroud Copies', in *Shroud Newsletter*, 69, British Society for the Turin Shroud, June 2009, pp.3–16

Beecher, P. A., *The Holy Shroud: Reply to the Rev. Herbert Thurston S.J.*, Dublin, M. H. Gill, 1928

Beldon Scott, John, *Architecture for the Shroud: Relic and Ritual in Turin*, Chicago and London, University of Chicago Press, 2003

Belting, Hans, *The Image and its Public-Likeness and Presence: A History of the Image before the Era of Art*, Chicago and London, University of Chicago Press, 1994

——'In Search of Christ's Body: Image or Imprint?', in Kessler and Wolf (eds), 1998, pp.1–11

Berkovits, Ilona, *Illuminated Manuscripts in Hungary, XI–XVI Centuries*, tr. Zsuzsanna Horn, Shannon, Ireland, Irish University Press, 1969

Bertelli, Carlo, 'Storia e Vicende dell'Imagine Edessena', in *Paragone*, no.217, Arte, 1968, pp.3–33

Biddle, Martin, *The Tomb of Christ*, Stroud (Glocs), Sutton, 1999

Bouquet-Boyer, Marie-Thérèse, *Itinerari Musicali della Sindone: Documenti per la Storia Musicale di una Reliquia*, Torino, Centro Studi Piemontesi, Fondo Carlo Felice Bona, 1981

Bowen, Harold C., *The Life and Times of Ali ibn Isa, 'the Good Vizier'*, Cambridge, Cambridge University Press, 1928

Bowersock, G. W., 'Notes on the New Edessene Mosaic of Prometheus', in *Hyperboreus*, 7 (2001, published 2002), fasc.1–2, pp.411–16

——*Mosaics as History: The Near East from Late Antiquity to*

Islam, Cambridge, Massachusetts, The Belknap Press of Harvard University Press, 2006

Bowman, Sheridan, *Radiocarbon Dating: Interpreting the Past*, London, British Museum Publications, 1990

Bozzo, Colette Dufour, *Il 'Santo Volto' di Genova*, Rome, Istituto Nazionale d'Archeologia e Storia dell'Arte, 1974

Breckenridge, James D., *The Numismatic Iconography of Justinian II*, New York, The American Numismatic Society, 1959

Brock, Sebastian, 'Iconoclasm and the Monophysites', in A. Bryer and J. Herrin (eds), *Iconoclasm*, Papers of the 1975 Symposium, Birmingham University, Centre for Byzantine Studies, 1977

——'Eusebius and Syriac Christianity', in Harold W. Attridge and Gohei Hata (eds), *Eusebius, Christianity and Judaism*, Wayne State University Press, 1992

——'Transformations of the Edessa Portrait of Christ', in *Journal of Assyrian Academic Studies*, vol.18, no.1, 2004, pp.46–56

Bruchet, M., *Marguerite d'Autriche, Duchesse de Savoie*, Lille, 1927

Bucklin, Dr Robert, 'The Medical Aspects of the Crucifixion of Christ', in *Sindon*, December 1961, pp.5–11

Cahill, T. A., R. N. Schwab, B. H. Kusko, R. A. Eldred, G. Möller, D. Dutschke and D. L. Wick, 'The Vinland Map Revisited: New Compositional Evidence on its Inks and Parchment', in *Analytical Chemistry*, 59, 15 March 1987, pp.829–33

Cameron, Averil, 'The Sceptic and the Shroud: An Inaugural Lecture in the Departments of Classics and History delivered at King's College, London on 19 April 1980', London, 1980

——'The History of the Image of Edessa: The Telling of a Story', in Cyril Mango and Omeljan Pritsak (eds), *Okeanos, Essays Presented to Ihor Sevcenko on his Sixtieth Birthday*, Cambridge, Massachusetts, 1984, pp.80–94

Caramello, Pietro (ed.), *La S. Sindone, Ricerche e Studi della Commissione di Esperti Nominata dall'Arcivescovo di*

436THE SHROUD

Torino, Card. Michele Pellegrino, nel 1969, Supplement, in *Rivista Diocesana Torinese*, Turin, January 1976

Cassannelli, Antonio, *The Holy Shroud, A Comparison between the Gospel Narrative of the Five Stages of the Passion (Flagellation, Crowning with Thorns, Way of the Cross, Crucifixion and Burial), and the Shroud as Evidence*, tr. Brian Williams, Leominster, Gracewing & Rome, Novum Millennium Romae, 2002

Chabot, J. B. (ed.), *Chronique de Michel le Syrien Patriarche Jacobite d'Antioche (1166–1199)*, Paris, 1899–1924

Chevalier, Ulysse, *Le Saint Suaire de Turin, est-il l'Original ou une Copie? Etude Critique*, Chambéry, Ménard, 1890

——*Etude Critique sur l'Origine de Saint Suaire de Lirey-Chambéry-Turin*, Paris, A. Picard, 1900

——*Le Saint Suaire de Turin: Histoire d'une Relique*, Paris, A. Picard, 1902 Chibnall, Marjorie (ed.), *Ordercus Vitalis Historia Ecclesiastica*, 6 vols, Oxford, 1969–1980

Chrysostomides, J., 'An Investigation Concerning the Authenticity of the *Letter of the Three Patriarchs*', in J. A. Munitiz et al, *The Letter of the Three Patriarchs to Emperor Theophilos and Related Texts*, Camberley, Porphyrogenitus, 1997, pp.xvii–xxxviii

Ciggaar, Krijnie N., 'Une Description de Constantinople Traduit par un Pélerin Anglais', in *Revue des Etudes Byzantines*, 34, 1976

——'Une Description de Constantinople dans le Tarragonensis 55', in *Revue des Etudes Byzantines*, 53, 1995, pp.117–40

Contamine, Philippe, 'L'Oriflamme de Saint-Denis aux XIVe et XVe siècles,' in *Annales de l'Est*, 7, 1973, pp.179–244

——'Geoffroy de Charny (Début du XIVe siècle–1356), "Le Plus Prudhomme et le Plus Vaillant de Tous les Autres"', in *Histoire et Société: Mélanges Offerts à Georges Duby/Textes Réunis par les Médiévistes de l'Université de Provence*, Aix-en-Provence, Université de Provence, 1992, pp.107–21

Cormack, Robin, *Painting the Soul: Icons, Death Masks and Shrouds*, London, Reaktion, 1997

Cowper, B. H., 'The Chronicle of Edessa – Selections from the

Syriac', in *Journal of Sacred Literature*, vol.V, 1864–6, pp.28–45

Craig, Emily, and Randall R. Bresee, 'Image Formation and the Shroud of Turin', in *Journal of Imaging Science and Technology*, vol.34, no.1, January–February 1994

Crispino, Dorothy, 'Why Did Geoffrey de Charny Change his Mind?', in *Shroud Spectrum International*, 1, 1982, pp.28–34. (This now defunct publication can be accessed online at www.shroud.com)

——'The Report of the Poor Clare Nuns', in *Shroud Spectrum International*, 2, March 1982, pp.19–27

——'Perceptions of an Antecessor', in *Shroud Spectrum International*, 23, June 1987, pp.16–21

——'The Letter of Agostino Cusano', in *Shroud Spectrum International*, 26, March 1988

——'To Know the Truth – A Sixteenth-Century Document with Excursus', in *Shroud Spectrum International*, 28/9, Sept/Dec 1988, pp.24–40

——'A Chronological Survey of Observations on the Shroud Textile', in *Shroud Spectrum International*, 38/9, June 1991, pp.21–7

Csocsán de Várallja, Eugene, 'The Turin Shroud and Hungary', in *Ungarn-Jahrbuch, Zeitschrift für die Kunde Ungarns und verwandte Gebiete*, Munich, Dr Rudolf Trofenik, vol.15, 1987, pp.1–49

Ćurčić, Slobodan, and Archer St Clair (eds), *Byzantium at Princeton: Byzantine Art and Archeology at Princeton University, Catalogue of an Exhibition at Firestone Library, Princeton University, August 1 through October 26, 1986*, Princeton, NJ, Princeton University Press, 1986

Cureton, W., *Ancient Syriac Documents Relative to the Earliest Establishment of Christianity*, Edinburgh, 1864

Currie, Lloyd A., 'The Remarkable Metrological History of Radiocarbon Dating [II]', in *Journal of Research of the National Institute of Standards and Technology*, 109, pp.185–217, 2004

Damon, P. E., et al, 'Radiocarbon Dating of the Shroud of

Turin', in *Nature*, vol.337, no.6208, 16 February 1989, pp.611–15

Danin, Avinoam, Alan D. Whanger, Uri Baruch and Mary Whanger, *Flora of the Shroud of Turin*, St Louis, Missouri, Missouri Botanical Garden Press, 1999

——and H. Guerra, *L'Uomo della Sindone. Un Botanico Ebreo Identifica Immagini di Piante della Terra Santa sulla Sacra Sindone*, Rome, Edizioni ART, 2008

David, Rosalie, *Mysteries of the Mummies: The Story of the Manchester University Investigation*, London, Cassell, 1978

Dietz, Dr Karlheinz, 'The Caliph and the Shroud', in *Sindone e Scienza, Atti III Congresso Internazionale di Studi sulla Sindone*, Turin, 5–7 June 1998

Dimitrova, Elizabeta, *The Monastery of Matejce*, Skopje, Kalamus, 2001

Drijvers, H. J. W., 'A Tomb for the Life of a King: A Recently Discovered Edessene Mosaic with a Portrait of King Abgar the Great', in *Le Museon*, 95, 1982, pp.167–89

——'The Image of Edessa in the Syriac Tradition', in Kessler and Wolf (eds), 1998

Du Puy, Pierre, *Histoire de l'Ordre Militaire des Templiers ou Chevaliers du Temple de Jérusalem depuis son Etablissement jusqu'à sa Décadence et sa Suppression*, Brussels, 1713

Dubarle, A. M., *Histoire Ancienne du Linceul de Turin jusqu'au XIIIe Siècle*, Paris, OEIL, 1985

Dupont-Sommer, A., 'Une Hymne Syriaque sur la Cathédrale d'Edesse', in *Cahiers Archéologiques*, II, 1947, pp.29–39

Eusebius, *The History of the Church from Christ to Constantine*, tr. G. A. Williamson, Harmondsworth, Penguin, 1965

Evans, Helen C. (ed.), *Byzantium, Faith and Power (1261–1557)*, New York, The Metropolitan Museum of Art, 2004

Evans, Helen C. and William D. Wixom (eds), *The Glory of Byzantium; Art and Culture of the Middle Byzantine Era. AD 843–1261*, New York, The Metropolitan Museum of Art, 1997

Fabre, A., *Trésor de la Sainte Chapelle des Ducs de Savoie au Château de Chambéry d'après des Inventaires Inédits des XVe et XVIe Siècles. Etude Historique et Archéologique*, Lyon, N. Scheuring, 1885

Farrar, Frederic W., *The Life of Christ as Represented in Art*, London, A. & C. Black, 1901

Ferraris, Giuseppe, *La S. Sindone Salvata a Vercelli*, Convegno Regionale del Centro Internazionale di Sindonologia, Vercelli, 1960

Finaldi, Gabriele (ed.), *The Image of Christ; The Catalogue of the Exhibition Seeing Salvation*, London, National Gallery Company Ltd, 2000

Finke, Heinrich, *Papsttum und Untergang des Templeordens*, Münster, 2 vols, 1907

Flury-Lemberg, Mechthild, 'The Linen Cloth of the Turin Shroud: Some Observations of its Technical Aspects', in *Sindon*, new series, no.16, December 2001, pp.55–76

——*Sindone 2002, L'Intervento Conservativo, Preservation, Konservierung*, Turin, Editrice ODPF, 2003

——'The Invisible Mending of the Shroud in Theory and Reality', in *Shroud Newsletter*, 65, British Society for the Turin Shroud, June 2007, pp.10–27

Fondazione Umberto II, *La Sindone nei Secoli nella Collezione di Umberto II*, Gribaudo, 1998

Fossati, Don Luigi, SDB, *La Santa Sindone, Nuova Luce su Antichi Documenti*, Borla Editore, Torino, 1961

——'The Lirey Controversy', in *Shroud Spectrum International*, 8, September 1983, pp.24–34

——'Copies of the Holy Shroud: Part I', in *Shroud Spectrum International*, 12, September 1984, pp.7–23

——'Copies of the Holy Shroud: Parts II & III', in *Shroud Spectrum International*, 13, December 1984, pp.23–39

——'The Shroud: From Object of Devotion to Object of Discussion', in *Shroud Spectrum International*, 37, December 1990, pp.9–18

——'L'Ostensione del 1868', in *Collegamento Pro Sindone*, July/August 1993

——*La Sacra Sindone, Soria Documentata di una Secolare*

Venerazione, Turin, Elledici, 2000

Frale, Barbara, *Il Papato e il Processo ai Templari: l'Inedita Assoluzione de Chinon alla luce della Diplomatica Pontificia*, Bologna, il Mulino, 2004

——*I Templari*, Bologna, il Mulino, 2004, translated into English and published as *The Templars: The Secret History Revealed*, Dunboyne (Ireland), Maverick, 2009

——*I Templari e La Sindone di Cristo*, Bologna, il Mulino, 2009

Friedlander, Alan, 'On the Provenance of the Holy Shroud of Lirey/Turin: A Minor Suggestion', in *The Journal of Ecclesiastical History*, Cambridge University Press, 2006, 57:3, pp.457–77

Froissart, *Chronicles*, tr. Geoffrey Brereton, Harmondsworth, Penguin, 1968

Gansfried, Solomon, *Code of Jewish Law (Kitzur Shulchan Aruch)*, tr. Hyman E. Goldin, New York, Hebrew Publishing Company, 1927

Garza-Valdes, Leoncio, *The DNA of God?*, New York, Doubleday, 1999

Gerstel, Sharon E. J., *Beholding the Sacred Mysteries: Programs of the Byzantine Sanctuary*, College Art Association in association with University of Washington Press, Seattle and London, 1999

Gervasio, Riccardo, 'Validità ed Attualità di Due Antiche Descrizioni della Santa Sindone', in *Sindon*, 19, Centro Internazionale di Sindonogia, Turin, April 1974

Ghiberti, Giuseppe, *Sindone le Immagini 2002 Shroud Images*, Editrice ODPF, Torino, 2002

Gibbon, Edward, *The Decline and Fall of the Roman Empire*, abridged D. M. Low, Harmondsworth, Pelican Books, 1963

Gove, Harry E., *Relic, Icon or Hoax? Carbon Dating the Turin Shroud*, Bristol and Philadelphia, Institute of Physics Publishing, 1996

Grabar, A., 'La Sainte Face de Laon et le Mandylion dans l'Art Orthodoxe', in *Seminarium Kondakovianum*, Prague, 1935, pp.5–37

——'Le Témoignage d'une Hymne Syriaque sur l'Architecture de la Cathédrale d'Edesse au VIe Siècle et sur la

Symbolique de l'Édifice Chrétien', in *L'Art de la Fin de l'Antiquité et du Moyen Age*, Collège de France Fondation Schlumberger pour des Etudes Byzantines, 1968

Graf, Georg, *Geschichte der Christlichen Arabischen Literatur*, 5 vols, Vatican City, 1944–53

Green, Maurus, 'Enshrouded in Silence', in *Ampleforth Journal*, LXXIV, 1969, pp.319–45

Grelot, Guillaume-Joseph, *Relation Nouvelle d'un Voyage de Constantinople*, Paris, 1680

Guerreschi, Aldo, and Michele Salcito, 'Further Studies on the Scorches and the Watermarks', paper presented at the Dallas Congress on the Shroud of Turin, 2005 (available online at www.shroud.com)

Guscin, Mark, *The Oviedo Cloth*, Cambridge, Lutterworth Press, 1998

——*The Image of Edessa*, Brill, Leiden & Boston, 2009

Haas, N., 'Anthropological Observations on the Skeletal Remains from Giv'at ha-Mivtar', in *Israel Exploration Journal*, vol.20, nos 1–2, 1970, pp.38–9

Hallam, Elizabeth (ed.), *Chronicles of the Age of Chivalry*, London, Weidenfeld & Nicolson, 1987

Hamilton, Bernard, 'Our Lady of Saidnaiya: an Orthodox Shrine Revered by Muslims and Knights Templar at the Time of the Crusades', in *The Holy Land, Holy Lands and Christian History: Studies in Church History*, 26, Ecclesiastical History Society, 2001, pp.207–15

Heisenberg, A., *Nicholas Mesarites – Die Palasrevolution des Johannes Comnenos*, Würzburg, 1907

Heller, J. H., and A. D. Adler, 'Blood on the Shroud of Turin', in *Applied Optics*, vol.19, no.16, pp.2742–4, 14 August 1980

——'A Chemical Investigation of the Shroud of Turin', in *Canadian Society of Forensic Science Journal*, 14, no.3, 1981, pp.81–103

Hoare, Rodney, *The Testimony of the Shroud*, London, Quartet, 1978

Johnson, R. A., J. J. Stipp, M. A. Tamers, G. Bonani, M. Suter and W. Wolfli, 'Archaeological Sherd Dating: Comparison of Thermo-luminescence Dates with Radiocarbon Dates by

Beta Counting and Accelerator Techniques', paper presented at the International Radiocarbon Conference, Trondheim, Norway, 1985

Jonghe, Daniel de, 'Particularités de Tissage d'une Toile Attribuée à Matin de Vos', in *Bulletin de l'Institut Royal du Patrimoine Artistique, Brussels*, vol.XVIII, 1980–8, pp.81–92

Josephus, *The Jewish War*, tr. G. A. Williamson, revised E. Mary Smallwood, Harmondsworth, Penguin, 1981

Jumper, E. J., A. D. Adler, J. P. Jackson, S. F. Pellicori, J. H. Heller and J. R. Druzik, 'A Comprehensive Examination of the Various Stains and Images of the Shroud of Turin', in J. B. Lambert (ed.), *Archaeological Chemistry III, Advances in Chemistry Series*, vol.205, American Chemical Society, 1984, pp.446–76

Kaeuper, Richard W., and Elspeth Kennedy, *The Book of Chivalry of Geoffroi de Charny, Text, Context and Translation*, University of Pennsylvania Press, Philadelphia, 1996

Karaulashvili, Irma, 'The Date of the Epistula Abgari', in *Apocrypha*, 13, 2002, pp.85–111

——'The Abgar Legend Illustrated: The Interrelationship of the Narrative Cycles and Iconography in the Byzantine, Georgian and Latin Traditions', in Colum Hourihane (ed.), *Interactions: Artistic Interchange between the Eastern and Western Worlds in the Medieval Period*, Princeton: Index of Christian Art, Department of Art and Archaeology, Princeton University, University Park, Pa., in association with Penn State University Press, 2007

Kersten, Holger, and Elmar Gruber, *The Jesus Conspiracy: The Turin Shroud and the Truth about the Resurrection*, Shaftesbury, Element, 1994

Kervyn de Lettenhove, Baron (ed.), *Oeuvres Complètes de Froissart*, 26 vols, Brussels, 1867–77

Kessler, Herbert, *Spiritual Seeing: Picturing God's Invisibility in Medieval Art*, University of Pennsylvania Press, 2000

——and Gerhard Wolf (eds), *The Holy Face and the Paradox of Representation: Papers from a Colloquium held at the*

Bibliotheca Hertziana, Rome and the Villa Spelman, Florence, 1996, Bologna, Nuova Alfa Editoriale, 1998

Kitzinger, Ernst, *Studies in Late Antique & Byzantine Art, vol.II: Studies in Medieval Western Art and the Art of Norman Sicily*, Pindar Press, 2003

Knight, Christopher, and Robert Lomas, *The Hiram Key, Pharaohs, Freemasons and the Discovery of the Secret Scrolls of Jesus*, London, Century, 1996

Kohlbeck, Joseph A., and Eugenia L. Nitowski, 'New Evidence May Explain Image on Shroud of Turin', in *Biblical Archaeology Review*, July–August 1986

Lauer, Philippe, *La Conquête de Constantinople*, Paris, Classiques Français du Moyen Age, 40, 1924, reprinted 1956

Lavoie, Gilbert R., *Unlocking the Secrets of the Shroud*, Allen, Texas, Thomas More, 1998

Lea, H. C., *History of the Inquisition in the Middle Ages*, vol.III, New York, 1887

Lidov, Alexei, 'The Mandylion over the Gate: A Mental Pilgrimage to the Holy City of Edessa', in *Routes of Faith in the Medieval Mediterranean, History, Monuments, People, Pilgrimage Perspectives*, Proceedings of an International Symposium, Thessalonike, 7–10 November 2007, Thessalonike, 2008

Lizerand, G., *Dossier de l'Affaire des Templiers*, Paris, 1923

McCrone, W. C., 'Chemical Analytical Study of the Vinland Map', report to Yale University Library, Yale University, New Haven, Connecticut, 1974

——'Authenticity of Medieval Document Tested by Small-Particle Analysis', in *Analytical Chemistry*, 48, 8, 1976, pp.676–9

——'Light-Microscopical Study of the Turin Shroud', III, *The Microscope*, 29, 1981, pp.19–38

——and C. Skirius, 'Light-Microscopical Study of the Turin Shroud', I & II, *The Microscope*, 28, 1980, pp.1–13

McNeal, Edgar H., *The Conquest of Constantinople, Translated from the Old French*, New York, Columbia Records of Civilisation 23, 1936

Maguire, Henry, *Byzantine Court Culture from 829 to 1204*, Dumbarton Oakes Research Library, Harvard University Press, 1997

Mango, Cyril (ed.), *The Art of the Byzantine Empire, 312–1453; Sources and Documents*, Englewood Cliffs, NJ, Prentice-Hall, 1972

Marino, J. G., and M. S. Benford, 'Evidence for the Skewing of the C-14 Dating of the Shroud of Turin Due to Repairs', Worldwide Congress 'Sindone 2000' in Orvieto, Italy, 28 August 2000

Marion, André, and Anne Laure Courage, *Nouvelles Découvertes sur le Suaire de Turin*, Paris, Albin Michel, 1997

Martin, E. J., *The Trial of the Templars*, London, 1928

Mazzucchi, C. M., 'La Testimonanzia la più Antica dell'Esistenza di una Sindone a Constantinopoli', in *Aevum*, 57, 1983, pp.227–31

Meacham, William, *The Rape of the Turin Shroud: How Christianity's Most Precious Relic Was Wrongly Condemned and Violated*, Lulu.com, 2005

Meek, H. A., *Guarino Guarini and his Architecture*, Yale University Press, 1988

Michelet, M., *Procès des Templiers*, 2 vols, Paris, 1841–51

Miller, V. D., and S. F. Pellicori, 'Ultraviolet Fluorescence Investigation of the Shroud of Turin', in *X-ray Spectrometry*, vol.9, no.2, 1980, pp.40–7

Millet, Gabriel, *Broderis Religieuses de Style Byzantin*, Paris, 1947

Mingana, A., *The Chronicle of Arbela*, tr. Msiha-Zkha, Press of the Dominican Fathers at Mosul, 1907

Moffett, Samuel Hugh, *A History of Christianity in Asia, vol.I: Beginnings to 1500*, San Francisco, Harper San Francisco, 1992

Monnet, G., *Bayard et la Maison de Savoie*, Paris, Bossard, 1926

Morand, S.-J., *Histoire de la Sainte-Chapelle Royale du Palais*, Paris, 1790

Mottern, R. W., R. J. London and R. A. Morris, 'Radiographic

Examination of the Shroud of Turin – a Preliminary Report', in *Materials Evaluation*, vol.38, no.12, 1979, pp.39–44

Murray, Stephen, *Building Troyes Cathedral: The Late Gothic Campaigns*, Indiana University Press, 1987

Nicholson, Helen, *The Knights Templar, A New History*, Stroud (Glocs), Sutton, 2004

Nicolle, David, *Poitiers 1356: The Capture of a King*, Oxford, Osprey, 2004

O'Rahilly, Alfred, *The Crucified*, ed. J. Anthony Gaughan, Mount Merrion, Co. Dublin, Kingdom Books, 1985

Oakeshott, Walter, *The Mosaics of Rome*, London, Thames and Hudson, 1967

Pacht, Otto, 'The Avignon Diptych and its Eastern Ancestry', in *De Artibus Opuscula, Essays in Honour of Erwin Panofsky*, New York, 1961, pp.402–17

Paleotto, Alfonso, *Esplicatione del Sacro Lenzuolo ove fu involto il Signore, & ell Piaghe, in esso impresse col suo pretioso sangue confrontate con la Scrittura Profeti e Padri. Con la notitia di molte Piaghe occulte, & numero de' Chiodi . . .*, Bologna, 1598

Palmer, Andrew, 'The Inauguration Anthem of Hagia Sophia in Edessa, a New Edition and Translation with Historical and Architectural Notes and a Comparison with Contemporary Constantinopolitan Kontakion', in *Byzantine and Modern Greek Studies*, 12, 1988, pp.117–68

——*The Seventh Century in the West-Syrian Chronicles*, introduced, translated and annotated by Andrew Palmer, Liverpool, Liverpool University Press, 1993

Perret, M., 'Essai sur l'Histoire du S. Suaire du XIVe au XVIe Siècle', in *Mémoires de l'Académie des Sciences, Belles Lettres et Arts de Savoie*, IV, 1960, pp.49–121

Phillips, George, *The Doctrine of Addai*, London, Trubner & Co., 1876

Phillips, Jonathan, '1204: The Crusader Sack of Constantinople', in *MHQ*, autumn 2005

Piana, Alessandro, 'The Missing Years of the Shroud', in *Shroud*

Newsletter, 66, British Society for the Turin Shroud, December 2007, pp.9–31

Picknett, Lynn, and Clive Prince, *Turin Shroud: In Whose Image? The Shocking Truth Unveiled*, London, Bloomsbury, 1994

Procopius, *History of the Wars*, tr. H. B. Dewing, 1916

Rabelais, *Oeuvres Complètes*, ed. P. Jourda, Paris, 1962

Raes, Gilbert, 'Appendix B-Rapport d'Analise', in *La S. Sindone* supplement to *Rivista Diocesana Torinese*, January 1976

Raffard de Brienne, Daniel, 'Document: Voyage de Geoffroy de Charny en Orient', in *Revue Internationale du Linceul de Turin*, 14, 2001, pp.23–5

Raynouard, F., *Monuments Historiques Relatifs à la Condemnation des Chevaliers du Temple*, Paris, 1813

Riant, Comte de, *Exuviae Sacrae Constantinopolitanae*, Geneva, 1878

Riedmattin, Pierre de, 'Le Linceul est Passé à Saint Hippolyte sur le Doubs, mais pas Besançon', in *Montre Nous Ton Visage*, 2008

Riggi di Numana, Giovanni, *Rapporto Sindone 1978–87*, Milan, 3M Edizioni, 1988

Riley-Smith, Jonathan, *The Atlas of the Crusades*, New York, 1991

Rinaldi, Peter M., 'Humbert II of Savoy 1904–1983', in *Shroud Spectrum International*, 7, June 1983, pp.2–5

Rogers, R. N., 'Studies on the Radiocarbon Sample from the Shroud of Turin', in *Thermochimica Acta*, 425, 2005, pp.189–194

Rollat, Christian, *L'Affaire Roussillon dans la Tragédie Templière*, Christian Rollat, 2008

Ross, Steven K., *Roman Edessa: Politics and Culture on the Eastern Fringes of the Roman Empire, 114–242 CE*, London and New York, Routledge, 2001

Runciman, Steven, 'Some Remarks on the Image of Edessa', *Cambridge Historical Journal*, 3, 1931, pp.238–52

Sanno Solaro, G. M., *La S. Sindone che si Venera a Torino Illustrata e Difesa*, Turin, 1901

Savio, P., *Ricerche Storiche sulla Santa Sindone*, Turin, SEI, 1957

Scannerini, Silvano, and Piero Savarino (eds), *The Turin Shroud, Past, Present and Future: International Scientific Symposium, Torino, 2–5 March 2000*, Torino, Effata Editrice, 2000

Scavone, Daniel, 'The Turin Shroud from 1200 to 1400', in W. J. Cherf (ed.), *Alpha to Omega: Studies in Honor of George John Szemler on his Sixty-Fifth Birthday*, Chicago, Ares, 1993

Schumacher, Peter M., 'Photogrammetric Responses from the Shroud of Turin', paper delivered at the Shroud of Turin International Research Conference, Richmond, Virginia, Friday, 18 June 1999

Segal, J. B., *Edessa, The Blessed City*, Oxford, Clarendon, 1970

Sendler, Egon, *The Icon Image of the Invisible: Elements of Theology, Aesthetics and Technique*, tr. Fr Steven Bigham, Oakwood Publications, Redondo Beach, California, 1988

Setton, Kenneth M., *The Papacy and the Levant, 1204–1571*, Philadelphia, American Philosophical Society, 1976

Sheffer, Avigall, and Hero Granger-Taylor, 'Textiles', in *'Masada IV', The Yigael Yadin Excavations 1963–1965, Final Reports*, Jerusalem, 1994

Skhirtladze, Zaza, 'Canonizing the Apocrypha: The Abgar Cycle in the Alaverdi and Gelati Gospels', in Kessler and Wolf (eds), 1998

Sox, David, *The Shroud Unmasked: Uncovering the Greatest Forgery of All Time*, Basingstoke, Lamp Press, 1988

Sprenger, Aloys (tr.), *El-Mas'udi's Historical Encyclopaedia Entitled 'Meadows of Gold and Mines of Gems'*, London, John Murray and Parbury, Allen & Co., 1841

Stauffer, Ethelbert, *Christ and the Caesars*, London, SCM Press, 1955

Straiton, M., 'The Man in the Shroud: a 13th-century Crucifixion Action-replay', in *Catholic Medical Quarterly*, August 1989, pp.135–43

Svensson, Niels, *Det Sande Ansigt: Jesus og Ligklaedt i Torino*, Gyldendal, 2007

Taylor, Joan du Plat, 'A Water Cistern with Byzantine Paintings,

Salamis, Cyprus', in *Antiquaries Journal*, vol.XIII, April 1933, pp.97–108

Taylor, Michael Anthony, *A Critical Edition of Geoffroy de Charny's Livre Charny and the Demandes pour la Joute, les Tournois et la Guerre*, PhD thesis, University of North Carolina at Chapel Hill, 1977. (Available from University Microfilms International.)

Thiede, Carsten, and Matthew d'Ancona, *The Quest for the True Cross*, New York, Palgrave, 2002

Thompson, R. W., 'An Eighth-Century Melkite Colophon from Edessa', in *Journal of Theological Studies*, NS, vol.XIII, pt.2, October 1962, pp.249–58

Thurston, Revd Herbert, *The Holy Year of Jubilee, an Account of the History and Ceremonial of the Roman Jubilee*, London, Sands & Co., 1900

——'The Holy Shroud and the Verdict of History', in *The Month*, CI, 1903, pp.17–29

Timossi, Virginio, 'Analisi del Tessuto della S. Sindone', in *La Santa Sindone nella Ricercha Moderne*, Acts of the Congress of Studies, Turin, 1939; and Rome/Turin, 1950; Turin, Berruti, 1941; Marietti reprint, 1980, pp.105–11

Tobler, T., *Itineraria Hierosolymitana et Descriptiones Terrae Sanctae*, Geneva, 1879

Trombley, Frank R., and John W. Watt (eds & tr.), *The Chronicle of Pseudo-Joshua the Stylite*, Translated Texts for Historians, vol.32, Liverpool University Press, Liverpool, 2000

Tyrer, John, 'Is it Really a Fake?', in *Textile Horizons*, March 1989, pp.51–2

Van Romfay, Innemée and Lucas, 'Deir al-Surian (Egypt), New Discoveries of 2001–2002', in *Hugoye: Journal of Syriac Studies*, vol.5, no.2, July 2002

Vasiliev, A. A., *Byzance et les Arabes*, vol.II, Brussels, Fondation Byzantine, 1968

Vercelli, Pietro, 'The Cloth of the Holy Shroud: a Technical Product Analysis of the Cloth and its Reproduction with Similar Characteristics', in Scannerini and Savarino (eds), 2000, pp.169–75

Vial, Gabriel, 'Le Linceul de Turin – Etude Technique', in *Bulletin du CIETA*, 67, Lyon, 1989, pp.11–35

Vignon, Paul, *Le Saint Suaire de Turin devant la Science, l'Archéologie, l'Histoire, l'Iconographie, la Logique*, Paris, Masson, 1939

Vikan, Gary, 'Debunking the Shroud: Made by Human Hands', in *Biblical Archaeology Review*, July/August 1998, pp.27–9

Von Dobschütz, Ernst, *Christusbilder, Untersuchungen zur Christlichen Legende*, Leipzig, 1899

Voobus, Arthur, *History of Asceticism in the Syrian Orient I, The Origin of Asceticism: Early Monasticism in Persia*, Corpus Scriptorum Christianorum Orientalium, vol.184, Subsidia, tomus 14, Louvain, 1958

Walsh, John, *The Shroud*, New York, Echo Books, 1965

Walter, C., 'The Abgar Cycle at Mateic', in B. Borkopp, B. Schelleward and L. Theis (eds), *Studien zur Byzantinischen Kunstgeschichte. Festschrift für Horst Hallensleben zum 65. Geburtstag*, Amsterdam, A. M. Hakkert, 1995, pp.221–32

Wardrop, M. and J. O., and F. C. Conybear, 'The Life of St Nino', in *Studia Biblica et Ecclesiastica*, Oxford, 1900

Ware, Timothy, *The Orthodox Church*, Harmondsworth, Penguin, 1963

Weitzmann, Kurt, 'The Mandylion and Constantine Porphyrogennetos', in *Cahiers Archéologiques*, XI, 1960, pp.164–84

——'The Origins of the Threnos', in *De Artibus Opuscula*, XL, *Essays in Honor of Erwin Panofsky*, New York, 1961, pp. 476–90 & plates 161–6

——*The Monastery of St Catherine at Mount Sinai, The Icons I, From the Sixth to the Tenth Century*, Princeton, NJ, Princeton University Press, 1976

Weyl Carr, Annemarie, 'Court Culture and Cult Icons in Middle Byzantine Constantinople', in Maguire, 1997.

Whitby, Michael (tr.), *The Ecclesiastical History of Evagrius Scholasticus*, Liverpool, Liverpool University Press, 2000

Wilkinson, John, *Egeria's Travels to the Holy Land*, Warminster, Aris & Phillips, revised edition 1981

Wilson, Ian, *The Turin Shroud*, London, Gollancz, 1978

——*Holy Faces, Secret Places: The Quest for Jesus' True Likeness*, London, Doubleday, 1991

——*The Blood and the Shroud*, London, Weidenfeld & Nicolson, 1998

Yenipinar, H., *Paintings of the Dark Church*, Istanbul, 1998

Zaccone, Gian Maria (ed.), *The Two Faces of the Shroud: Pilgrims and Scientists Searching for a Face*, Turin, Editrice ODPF, 2001

Zias, Joseph, and Eliezer Sekeles, 'The Crucified Man from Giv'at ha-Mivtar: A Reappraisal', in *Biblical Archaeologist*, September 1985

Zugibe, Fred, *The Crucifixion of Jesus: A Forensic Inquiry*, New York, M. Evans & Co., 2005

Picture
Acknowledgements

Colour Plates

1a Scene in the Turin Cathedral sacristy, 21 April, 1998. From *Il Telo, Rivista di Sindonologia*, September/December 2000, p. 28.

1b Cardinal Poletto inspecting the Shroud in its conservation container. Courtesy of the Archdiocese of Turin.

2a The Shroud as it appears today. Courtesy of the Archdiocese of Turin.

2b The Shroud's present-day repository. Celina Slodowy, Missionaries of the Holy Face, Melbourne, Australia.

3a Artist's copy of the Shroud, 1516. Church of St Gommaire, Lierre, Belgium. Author's photo archive.

3b Close-up of set of triple burn-holes. From high-definition filming of the Shroud by Performance Films for the BBC, 24 January 2008. Courtesy of the Archdiocese of Turin and Performance Films.

4a Secondo Pia. Author's photo archive.

4b The Shroud face, natural appearance. As 3b, courtesy of the Archdiocese of Turin and Performance Films.

4c Dr Emily Craig replication. Courtesy of Dr Emily Craig, Kentucky State Medical Examiner's department.

4d Dr Nicholas Allen's replication (detail). Courtesy of Dr Nicholas

Allen, Faculty of Art & Design, Port Elizabeth Technikon, Republic of South Africa.

4e Dr Luigi Garlaschelli's replication. Courtesy of Dr Luigi Garlaschelli, University of Pavia, Italy.

5   The Shroud face in negative. Giuseppe Enrie. Author's photo archive.

6   Shroud front-of-body imprint in negative. Barrie Schwortz.

7   Shroud back-of-body imprint in negative. Barrie Schwortz.

8a The Shroud image viewed via the VP8 Image Analyzer. Barrie Schwortz.

8b The filming of the Shroud in high definition, 24 January 2008. From 'Shroud of Turin', a Performance Films production for the BBC, © 2008, courtesy of Performance Films.

9a Detail of crown of thorns bloodstains. As 3b, courtesy of the Archdiocese of Turin and Performance Films.

9b Dr Niels Svensson. As 8b, courtesy of Performance Films.

9c Marks from the whipping. As 3b, courtesy of the Archdiocese of Turin and Performance Films.

9d Reconstruction of Roman flagrum, from P. Vignon, 1939, p. 56, fig. 27.

9e Roman Republican coin of Titus Didius. Collection of the Museum of Cultural History, the University of Oslo, Sweden. Courtesy of the VRoma Project (www.vroma.org).

10a Bloodstains from wrist-nailing. As 3b, courtesy of the Archdiocese of Turin and Performance Films.

10b (Inset) bones of the wrist. Courtesy of Dr Robert Bruce-Chwatt.

10c Bloodstains as from nailing in the feet. As 3b, courtesy of the Archdiocese of Turin and Performance Films.

11a Bloodstain as from bladed weapon in the chest. As 3b, courtesy of the Archdiocese of Turin and Performance Films.

11b Roman lancea, Landesmuseum, Zurich. Author.

11c The lance-wound seen in negative. Giuseppe Enrie. Author's photo archive.

11d Identical wound on 'Dying Gaul' statue. Capitoline Museum, Rome. Leonardo Ferri.

11e Back-of-body bloodstains. As 3b, courtesy of the Archdiocese of Turin and Performance Films.

12a Close-up of area of the nose. As 3b, courtesy of the Archdiocese of Turin and Performance Films.

12b As 4b.

12c Area of bloodstain greatly magnified. ©1978, Mark Evans.

13a Dr Mechthild Flury-Lemberg working on the Shroud. Courtesy Archdiocese of Turin.

13b The underside of the Shroud, revealed 2002. Courtesy of the Archdiocese of Turin.

13c The section of the Shroud removed for carbon dating, April 1988. Giovanni Riggi. Author's photo archive.

14a Şanliurfa today, photographed from the Citadel. Author.

14b King Abgar receiving his 'wonderful vision'. Detail from codex Syn gr. 183 (Vlad. 382], fol. 192 verso. The State Historical Museum, Moscow.

15a Coin of Abgar VIII with Christian cross on tiara. British Museum Edessa, catalogue number 13, as featured on the website of the Ashmolean Museum, Oxford, Roman Provincial Coinage Online, temporary number 6491, accessible at http://rpc.ashmus.ox.ac.uk/coins/6491.

15b Sculpted lion, Şanliurfa Museum. Author.

15c Şanliurfa's fish-pools. Judith Wilson.

16a Ruins at Der Yakup. Author.

16b Linguist Mark Guscin. Judith Wilson.

16c Sixth-century mosaic, Haleplibahce, Şanliurfa. Author.

17a Beardless Christ, fifth-century sarcophagus, Istanbul Museum. Author.

17b Sixth-century rediscovery of the Image of Edessa. Icon of Fjodor Zubov, Cathedral of Christ the Saviour, Moscow, as reproduced in Briusova, V. G.: *Fedor (Fjodor) Zubov*, Izobraziteljnoe Iskusstvo, Moscow 1985.

17c Sixth-century face of Christ, silver vase from Homs, Syria. Louvre Museum, Paris.

17d Sixth-century Christ Pantocrator icon, Monastery of St

Catherine, Sinai. Professor Kurt Weitzmann and the Michigan-Princeton-Alexandria Expeditions to Mount Sinai.

18a Sixth-century face of Christ, wall-painting, Spring of Nicodemus, Salamis, Cyprus. Copy by Monica Bardswell, circa 1936, displayed in the Limassol Castle Mediaeval Museum, Limassol, Cyprus. George Apostolou.

18b Ruins of the Georgian monastery of Ishan, north-eastern Turkey. Judith Wilson.

18c Mosaic face of Christ, nineteenth-century restoration after sixth-century original, apse of the basilica of St John Lateran, Rome. Author's photo archive.

19a Sixth-century face of Christ, mosaic fragment, Şanliurfa Museum. Reproduced from *Şanliurfa sanat tarih ve turizm dergisi*, January 2009, p. 48.

19b Sixth-century face of Christ, wall-painting, Spring of Nicodemus, Cyprus, Marc Fehlmann, Eastern Mediterranean University, northern Cyprus.

19c Sixth-century face of Christ, Ss. Sergius & Bacchus icon (detail). Museum of Western & Oriental Art, Kiev, Ukraine.

20a Gold coin of the Byzantine emperor Justinian II. Obverse. Numismatic Museum, Athens, © Hellenic Ministry of Culture and Tourism.

20b Wall-painting of Christ Pantocrator from the catacomb of St Ponziano, Rome. Author's photo archive.

20c Detail of Shroud face natural appearance, black and white. Giuseppe Enrie. Author's photo archive.

20d Detail of the above.

21a Hagia Sophia, Istanbul. Judith Wilson.

21b Gold solidus of Byzantine emperor Constantine VII Porphyrogennetos. Obverse. Specimen featured on the Forum Ancient Coins website: www.forumancientcoins.com

21c Icon of king Abgar holding Image of Edessa (detail), Monastery of St Catherine, Sinai. Nikos & Kostas Kontos, after Svensson 2007, p. 107.

21d Gold solidus of Byzantine emperor Constantine VII

Porphyrogennetos. Reverse. Specimen featured on the Forum
Ancient Coins website: www.forumancientcoins.com

22a Typical rock church location, Göreme, Cappadocia. Judith
Wilson.

22b: Image of Edessa, wall-painting, Sakli Church, Göreme. Judith
Wilson.

22c Image of Edessa, manuscript illumination. Greek Patriarchal
Library,   Alexandria, Egypt, codex gr. 35, fol. 142 v.
Author's photo archive.

22d Image of Edessa, wall-painting. Karanlik, or 'Dark' church,
Göreme. Mark Guscin.

23a Sanctuary barrier, Sakli church, Göreme. Composite photo by
the author from sections photographed by Judith Wilson.

23b Icon of the Annunciation. Twelfth-century. Monastery of St
Catherine, Sinai. Bruce White. By kind permission of Saint
Catherine's Monastery, Sinai, Egypt.

23c Detail of the above.

24a Engraving of the interior of Hagia Sophia, Istanbul (detail),
from Guillaume Joseph Grelot, *Relation Nouvelle d'un Voyage
de Constantinople*, Paris, 1680. Courtesy of the University
Library, Heidelberg.

24b Equivalent area of the Hagia Sophia interior today. Author.

24c The Image of Edessa in its casket. Alaverdi Tetraevangelion,
National Centre of Manuscripts, Tbilisi, Georgia, ms A-484,
fol. 320v. Courtesy of the National Centre of Manuscripts,
Tbilisi.

25a Entombment of Christ and Visit of the Holy Women,
illumination from Pray manuscript, National Szechenyi Library,
Budapest. Professor Jérôme Lejeune.

25b Image of Edessa, wall-painting. Church of Hagia Sophia,
Trabzon. Author.

25c The Church of Hagia Sophia, Trabzon. Author.

26a Image of Edessa, wall-painting. Monastery of Sopocani, Serbia.
Courtesy of Slavisa Djurdje Jevtic.

26b Equivalent face area of Turin Shroud. Detail of pl. 2a, with burns from 1532 fire digitally removed by the author.

26c   Hand-over of the Image of Edessa, wall-painting in the church of Matejce, Macedonia. Courtesy of Darko Nikolovski, Kalamus, Skopje.

27a   Epitaphios of Milutin II Uros, Museum of the Serbian Orthodox Church, Belgrade. Author's photo archive.

27b   Carrying of the epitaphios. Wall-painting, Kosovo, courtesy of Slavisa Djurdje Jevtic.

27c   Epitaphios from Thessaloniki, Greece. Byzantine Museum, Athens. Author's photo archive.

28a   Panel painting. Church of St Mary, Templecombe, Somerset, England. Author's photo archive.

28b   Barbara Frale, photo, Arcade Publishing. Seal of Templar Master Frederick Wildergrave. Bayerisches Hauptstaatsarchiv, Steingaden Urk, 126/II. Photo: Bayerisches Hauptstaatsarchiv.

28c   Templar leaders being burned at the stake. From British Library ms. Royal 20 C VII, f. 48. The British Library.

29a   Pilgrim's souvenir badge. Musée du Moyen Age, Paris. Dr Niels Svensson.

29b   Geoffrey I de Charny ambushed at Calais. Manuscript illumination, Jean Froissart *Chroniques*, Bibliothèque Municipale Toulouse, ms.511, fol. 117 r. Accessible via the Internet at the excellent website www.enluminures.culture.fr, selecting first 'Toulouse' then 'ms.511'.

29c   Shroud, facial section. Wall-painting. Church of Saint-Léger, Terres-de-Chaux. Thierry Boillot. For further wall-paintings in the same church, see Thierry Boillot's website Racines Comtoises at www.racinescomtoises.net

30a   Showing of the Shroud, Chambéry, *c.* 1520, stained-glass window, Chapel of Pérolles, Fribourg, Switzerland. From Marcel Strub, *Les Monuments d'art et d'histoire du Canton de Fribourg*, Bâle: ed. Birkhäuser, 1959.

30b   Showing of the Shroud, Turin, 1582. Engraving. Civic Library of Turin.

31a   Showing of the Shroud in Turin, 1684. Oil painting by Pieter

Bolckmann, 1686. Castello di Racconigi, Turin. Courtesy Soprintendenza per i Beni Ambientali e Architettonici del Piemonte, Turin.

31b Altar of the Shroud designed by Bertola. Author's photo archive.

31c Showing of the Shroud on the steps of Turin Cathedral, 1933. Giuseppe Enrie. From P. Vignon, 1939, p. 11.

32a Pope John Paul II and King Umberto II, 14 May, 1982. From Fondazione Umberto II, 1998, p. 13.

32b Fire engulfs the Shroud Chapel, 11 April 1998. Courtesy of the Archdiocese of Turin.

32c Fireman Trematore smashes the Shroud display case. Courtesy of the Archdiocese of Turin.

32d John Paul II in prayer before the Shroud, 24 May 1998. From *Il Telo, Rivista di Sindonologia*. May/June 1998, p. 13.

Figures reproduced in the text

1 Shroud photo courtesy of the Archdiocese of Turin; ground plan: John Beldon Scott, School of Art & Art History, University of Iowa; reconstruction: author, after an aquatint by G. B. della Rovere.

2 Artist's copy of the Shroud on cloth, Church of Notre-Dame de Chambéry. Author's photo archive.

3 The Nicholas Allen theory. Drawing by the author after a photo by Professor Allen.

4 Up and down motion of crucifixion victim. Reconstruction by the author.

5 Table of pollen findings. List adapted by the author from one published posthumously on Dr Max Frei's behalf by Professor Heinrich Pfeiffer, Gregorian University, Rome.

6 Crystallographic 'signature' of Shroud limestone compared to that of Jerusalem. As reproduced in *Biblical Archaeology Review*, Jul/Aug 1986, p. 23.

7 How the Shroud was originally woven on a much wider piece of cloth. Courtesy of Dr Mechthild Flury-Lemberg.

8 Technical drawing of the seam on the first-century fabric found at Masada. Courtesy of Dr Mechthild Flury-Lemberg.

9 The Shroud's herringbone twill weave, and plain weave, compared. Drawing: author, based on diagrams by the late John Tyrer.

10 The Shroud's folding arrangement at the time of the 1532 fire. Drawing based on reconstruction by Aldo Guerreschi, Turin, courtesy of Aldo Guerreschi.

11 The Shroud's folding arrangement at the time of creation of the triple burn-hole damage. Drawing based on reconstruction by the author.

12 The Shroud's folding arrangement at the time of creation of the 'second incident' water stains. Drawing based on reconstruction by Aldo Guerreschi, Turin, courtesy of Aldo Guerreschi.

13 Plan of the pieces taken during the carbon-dating sampling. The author, using photos kindly provided by the Archdiocese of Turin, the Arizona, Oxford and Zurich radiocarbon-dating laboratories, Dr Leoncio Garza-Valdes of San Antonio Texas and Professor Gilbert Raes, Institute of Textile Technology, Ghent.

14 Detail of showing of the Shroud, showing the corner from which the carbon-dating sample was taken. From a print on silk in the collection of Sherborne Castle, Dorset, England. H. Tilzey, courtesy Simon Wingfield Digby, Sherborne Castle.

15 The Veronica cloth. From a late-fifteenth-century woodcut in Hartmann Schedel's *Liber Chronicarum* (more popularly known as the *Nuremberg Chronicle*), Nuremberg, Anton Koberger, 1493.

16 Missionary Journey of Addai to Edessa. Map originated by the author.

17 Text of Jesus's letter to Abgar, inscription found at Kirk Magara. From J. B. Segal, 1970, pl. 31b, after H. von Oppenheim and F. Hiller von Gaetringen, *Sitzungsber. D. Königl. Preuss. Akad. D. Wissensch. zu Berlin*, phil.-hist. KI., 1914, p.824.

18 Fourth-century mosaic of Jesus's face, from Roman villa in Dorset. British Museum. Author's photo archive.

19 Persian 'king of kings' in war finery. Drawing in the author's photo archive, after the original in the Cabinet des Médailles, the Bibliothèque Nationale, Paris.

20 Map of sixth-century Edessa. Map originated by the author, based on location data in J. B. Segal, 1970, plan 1.

21 Typical example of early depiction of the Keramion. Bob Barrow, the Mani, Greece.

22 (a) Sagalasos image. Drawing by the author after reconstruction in P. Mueller et al, 'Photo-realistic and detailed 3D modelling: the Antonine nymphaeum at Sagalassos, Turkey)', www.vision.ee.ethz.ch/~pmueller/documents/caa04_pmueller.pdf; (b) mythological heads from Hatra, from R. Ghirshman, *Iran*, Penguin Books, 1954, fig. 81; (c) reconstruction of hiding place: author.

23 (a) Ankhiskhati icon with cover removed, from a reconstruction by Ahalva Amiranashvili as reproduced in Karaulashvili, 2007, p. 224; (b) 'Acheropita' icon of the Lateran, Rome with cover removed, after C. Cecchelli, entry 'Acheropita' in *Enciclopedia Italiana*, Rome, 1949, vol. I, p. 312.

24 Roundel depicting the Image of Edessa, from a wall-painting in the Church of the Holy Cross, Telovani, Georgia. After drawings reproduced in Z. Skhirtladze, 1998, fig. 2.

25 Doubled-in-four mode of presentation of the Shroud as the Image of Edessa. Reconstruction: author.

26 The so-called Vignon markings. Author, after P. Vignon, 1939, pp. 133–5.

27 The Great Palace at Constantinople, with map of Constantinople. After a drawing by C. Vogt in the Mosaics Museum, Istanbul.

28 Entombment scene from pulpit by Nicholas of Verdun, Chorherrenstift Klosterneuberg Stiftsmuseum. Photo AV Medienstelle der Arzdiese, Vienna.

The author wishes to acknowledge the generous help of Don Giuseppe Ghiberti of the Archdiocese of Turin, and David Rolfe of Performance

Films, for making available from Performance Films high-definition filming of the Shroud from which footage was created the details from the Shroud featured in this book. Also needing special mention are Professor Bruno Barberis, Director of the International Centre of Sindonology, Turin; Gian Maria Zaccone, Scientific Director of the Museum of the Shroud, Turin; Barrie Schwortz, creator of the website www.shroud.com; Professors Ann R. Raia and Barbara F. McManus, both of the College of New Rochelle, New Rochelle, New York; Professor Håkon Ingvaldsen, Department of Archaeology, University of Oslo; Professor Annemarie Weyl Carr, Southern Methodist University, Dallas, Texas; Dr Andrew Peacock, British Institute of Archaeology, Ankara; Turin-based publisher Kim Williams; Dr Niels Svensson of Maribo, Denmark; Anna Voellner of the University of Heidelberg; Dr Tamar Gegia of the National Centre of Manuscripts, Tbilisi; barrister George Apostolou of Nicosia, Cyprus; John Chapman, authority on the Mani region of Greece; Pierre de Riedmattin, President of Montre Nous Ton Visage, and others, for their most helpful guidance and assistance with the task of obtaining photos and permissions for various illustrations featured in this book. Every effort has been made to obtain such permissions at the time of this book going to press, but in the case of any omissions the author and the publishers will be pleased to make the appropriate acknowledgements in any future edition.

Index

The Genesis Enigma

Andrew Parker

'An astounding work which seeks to prove that the ancient
Hebrew writers of the Book of Genesis knew all about
evolution – 3,000 years before Darwin'
DAILY MAIL

*A biblical enigma exists that is on the one hand so cryptic it
has remained camouflaged for millennia, and on the other so
obvious one cannot miss it . . .*

IT TAKES JUST one page in the Bible to describe the creation of the
universe, the Earth, the sky, the seas and all life on our planet.
For thousands of years, Judeo-Christian belief has accepted this
progression as truth. But the *Genesis* account has no *right* to be
correct. The author or authors could not have known these
things happened in this order, and in the detail science has come
to recognize.

In a fascinating and controversial scientific detective story,
Andrew Parker reveals how the latest discoveries of science
corresponds in unerring detail with the creation account in
Genesis.

His astonishing conclusions will revolutionize how many of us
view the debate between science and religion, and he asks the
question: in a scientific world is there still a place for God?

9780552775281

The Greatest Show on Earth
Richard Dawkins

'This is a magnificent book of wonderstanding: Richard
Dawkins combines an artist's wonder at the virtuosity of
nature with a scientist's understanding of how it comes to be'
MATT RIDLEY, author of *Nature via Nurture*

CHARLES DARWIN, WHOSE 1859 masterpiece *On the Origin of
Species* shook society to its core, would surely have raised an
incredulous eyebrow at the controversy over evolution still
raging 150 years later.

The Greatest Show on Earth is a stunning counter-attack on
creationists, followers of 'Intelligent Design' and all those who
still question evolution as scientific fact. In this brilliant *tour de
force* Richard Dawkins pulls together the incontrovertible
evidence that underpins it: from living examples of natural
selection to clues in the fossil record; from plate tectonics to
molecular genetics.

The Greatest Show on Earth comes at a critical time as
systematic opposition to the fact of evolution flourishes as never
before in many schools worldwide. Dawkins wields a devastating
argument against this ignorance whilst sharing with us his
palpable love of science and the natural world. Written with
elegance, wit and passion, it is hard-hitting, absorbing and
totally convincing.

'A voice of reason in irrational times, Richard Dawkins
is both theorist and explainer of one of the greatest
discoveries of the human mind'
THE TIMES

9780552775243

The Story Of God

A personal journey into the world of science and religion
Robert Winston

FROM THE TINIEST microchip to the information superhighway, the modern world is dominated by and dependent upon science. Yet whether we realize it or not we live in an age where faith is still an important influence in our lives. The majority of Americans profess a belief in a Christian God and Islam acts as a unifying, energizing force for many of the world's most dispossessed people. In the UK congregations may be shrinking, but popular belief in the supernatural – ghosts and spirits, fortune-telling, faith healing – is stronger than ever.

In *The Story of God* Robert Winston examines a relationship between science and religion across time, beginning with the primitive worship of early ancestors and concluding with a vivid portrait of faith in the modern world.

Grand in scope, adventurous in tone – and written from the perspective of a respected scientist who is also committed to Judaism – this groundbreaking work traces a line across continents, cultures and eras.

'Intelligent and readable'
TIMES LITERARY SUPPLEMENT

'Lively and accessible. An enthusiastic starting point for the study of a fascinating area of human philosophy and psychology'
WATERSTONE'S BOOKS QUARTERLY

9780553817430

Genghis Khan

John Man

GENGHIS KHAN IS one of history's immortals: a leader of genius and the founder of the world's greatest land empire – twice the size of Rome's. His mysterious death in 1227 placed all at risk, so it was kept a secret until his heirs had secured his conquests. Secrecy has surrounded him ever since. His undiscovered grave, with its imagined treasures, remains the subject of intrigue and speculation.

Today, Genghis is by turns scourge, hero and demi-god. To Muslims, Russians and Europeans, he is a mass-murderer. Yet in his homeland, Mongols revere him as the nation's father; Chinese honour him as dynastic founder; and in both countries, worshippers seek his blessing.

This book is more than just a gripping account of Genghis' rise and conquests. John Man uses first-hand experiences to reveal the khan's enduring influence. He is the first writer to explore the hidden valley where Genghis may have died, and one of the few westerners to climb the sacred mountain where he was probably buried.

The result is an enthralling account of the man himself and of the passions that surround him today. For in legend, ritual and controversy, Genghis lives on . . .

'A fine introduction to the subject, as well as a rattling good read'
INDEPENDENT

'A fine, well-written and well-researched book'
MAIL ON SUNDAY

'Fascinating . . . history doesn't come much
more enthralling than this'
YORKSHIRE EVENING POST

9780553814989